Water Resources Development and Management
Series Editors: Asit K. Biswas and Cecilia Tortajada

AF181246

Editorial Board

Dogan Altinbilek (Ankara, Turkey)
Chennat Gopalakrishnan (Honolulu, USA)
Jan Lundqvist (Linköping, Sweden)
Alexandra Pres (Berlin, Germany)
Anthony Turton (Pretoria, South Africa)
Olli Varis (Helsinki, Finland)

Anthony R. Turton
Hanlie J. Hattingh
Gillian A. Maree
Dirk J. Roux
Marius Claassen
Wilma F. Strydom
Editors

Governance as a Trialogue: Government-Society-Science in Transition

With 45 Figures

Springer

CSIR
our future through science

Dr. Anthony R Turton
Gibb-SERA Chair in IWRM
CSIR
PO Box 395
Pretoria
0001 South Africa
Email: aturton@csir.co.za

Ms. Hanlie J. Hattingh
Council for Scientific and Industrial Research (CSIR)
PO Box 395
0001 South Africa
Email: hhattingh@csir.co.za

Ms. Gillian A. Maree
Council for Scientific and Industrial Research (CSIR)
PO Box 395
Pretoria
0001 South Africa
Email: gmaree@csir.co.za

Dr. Dirk J. Roux
Council for Scientific and Industrial Research (CSIR)
PO Box 395
Pretoria
0001 South Africa
Email: droux@csir.co.za

Dr. Marius Claassen
Council for Scientific and Industrial Research (CSIR)
PO Box 395
Pretoria
0001 South Africa
Email: mclaasse@csir.co.za

Ms. Wilma F. Strydom
Council for Scientific and Industrial Research (CSIR)
PO Box 395
Pretoria
0001 South Africa
Email: wstrydom@scir.co.za

ISBN 978-3-642-07966-5 e-ISBN 978-3-540-46266-8

This work is subject to copyright. All rights are reserved, whether the whole or part of the material is concerned, specifically the rights of translation, reprinting, reuse of illustrations, recitation, broadcasting, reproduction on microfilm or in any other way, and storage in data banks. Duplication of this publication or parts thereof is permitted only under the provisions of the German Copyright Law of September 9, 1965, in its current version, and permission for use must always be obtained from Springer-Verlag. Violations are liable to prosecution under the German Copyright Law.

Springer is a part of Springer Science+Business Media
springer.com
© Springer-Verlag Berlin Heidelberg 2007
Softcover reprint of the hardcover 1st edition 2007

The use of general descriptive names, registered names, trademarks, etc. in this publication does not imply, even in the absence of a specific statement, that such names are exempt from the relevant protective laws and regulations and therefore free for general use.

Cover design: E. Kirchner, Heidelberg
Cover photo: Ms. Ernita van Wyk (CSIR)

Preface

The last two decades have been marked by a dramatic increase in global attention to the concept of governance, especially in relation to the effective and sustainable management of natural resources. During this period, issues of water governance have received particular attention, for example in relation to the provision of reliable water supplies as a catalyst for poverty eradication. Within the context of the Millennium Development Goals, and against a backdrop of an increasing frequency of water crises (ranging from widespread flooding to severe water scarcity), it is essential that each country (and particularly those countries that face development challenges) should be in a position to ensure that access to water is available to those who need it most. This issue has been highlighted by the Global Water Partnership, which stated that the water crisis facing the world is in reality a crisis of governance.

The South African political environment has changed dramatically in recent years, and the central concepts of social equity and the right to a healthy environment are now entrenched in the country's Constitution. These concepts are supported by several new laws, in particular the National Environmental Management Act and the National Water Act, which, in turn, are based on the principles of sustainable development. However, despite the highly desirable attributes of these landmark pieces of legislation, South African authorities are still struggling to implement the requirements of these Acts almost a decade after their promulgation.

Investigations by the South African CSIR into possible reasons for non-implementation of legislation and government tools revealed the pivotal role of governance issues. Importantly, the investigations also indicated that an incomplete understanding of the importance of governance was a central reason for the lack of successful implementation. This research confirmed the findings reported in the international literature that the concept of "governance" had not really been properly defined and fully explored. Countries and regions differed in their understanding and interpretation of "governance", whilst equally wide differences were recorded in countries that had different levels of socio-economic and political development.

In an effort to unpack the so-called "*black box*" of governance, a group of specialists from different countries was invited to review governance issues related to their areas of technical specialization, covering different levels of development and maturity of democracy. Each specialist was also challenged to interrogate the new 'Trialogue' hypothesis on governance during the International Symposium on Ecosystem Governance that was held in South Africa from 10-14 October 2005.

The Trialogue hypothesis was based on research conducted by the CSIR and is explained in Chapter 1.

Although the CSIR's research has focussed on integrated water resource management (IWRM), this cannot, and should not, be considered in isolation. Several other important factors influence IWRM and help to shape ecosystem health. Consequently, the symposium took a holistic view of ecosystem governance, with the main objectives of the symposium to:

- Unpack "governance", through the exchange of ideas and experience between specialists from different disciplines and countries; and
- Develop a research agenda on ecosystem governance.

The solicited manuscripts presented at the symposium addressed specific topics and were written by selected authors; each manuscript was reviewed by at least two external referees before being selected for publication. Selected manuscripts were chosen for publication in a special edition of the journal "*Water Policy*" entitled: "Case studies of Government-Society-Science as a Trialogue: Towards a Governance Research Agenda", to be published by *World Water Council*. The other manuscripts presented at the symposium were selected for publication in this book.

The manuscripts published in this book focuses on interrogating the Trialogue Model and hypotheses from different perspectives. The final part of book revisits the Trialogue model and hypotheses and summarises a new research agenda for ecosystem governance.

Special thanks are due to the group of key southern African stakeholders – Department of Water Affairs and Forestry (South Africa), Global Water Partnership Southern Africa, UNESCO and the Water Research Commission (South Africa) – who partnered the CSIR in arranging and supporting the International Symposium on Ecosystem Governance. Without their enthusiastic backing and commitment, the subsequent publications and the research agenda that emerged from the symposium would not have been possible. Thanks also to Prof Asit Biswas and Springer-*Verlag* who have graciously allowed us to publish this book.

The Editors

Acknowledgements

The editors would like to thank the following people and organizations for making both the International Symposium on Ecosystem Governance and this publication possible:

Authors and Participants

Without the willingness, enthusiasm, patience and dedication of all the authors and participants, the symposium and publications would not have been possible. We trust you will enjoy the outcomes of our combined efforts.

Funding partners

The Council for Scientific and Industrial Research (CSIR) would like to thank their funding partners for supporting this initiative:

- Department of Water Affairs and Forestry, South Africa
- Global Water Partnership Southern Africa
- UNESCO
- Water Research Commission
- Universities Partnership for Transboundary Waters (UPTW)
- Distell Ltd., South Africa.

Publishers

These publications are largely due to the support received from:

- Prof. Asit Biswas, Editor in Chief of the International Journal of Water Resources Development and President of the Third World Centre for Water Management; and
- Dr Jerome Delli Priscoli, Editor in Chief of Water Policy and Member of the World Water Council

External Reviewers

Thank you for the efficient, timeous and constructive manner in which the peer reviews were completed. Your participation has ensured a quality product. The peer reviewers were:

Name	Affiliation	Country
Dr Alfonso Rivera	Geological Survey of Canada, Natural Resources	Canada
Ms Anna Matros	Global Water Partnership	Namibia
Dr Anthony Turton	CSIR/AWIRU	South Africa
Dr Bill Rowlston	Department of Water Affairs and Forestry	South Africa
Dr Bob Scholes	CSIR	South Africa
Prof Charles Breen	University of KwaZulu Natal	South Africa
Dr David le Maitre	CSIR	South Africa
Dr Dirk Roux	CSIR	South Africa
Mr Duard Barnard	Duard Barnard and Associates	South Africa
Dr Emmanuel Manzungu	University of Zimbabwe	Zimbabwe
Prof Eugene Cloete	University of Pretoria	South Africa
Prof Geoffrey D. Gooch	Jean Monnet Chair of European Political Integration	Sweden
Mr Harry Biggs	SANParks	South Africa
Mr Irene Saayman	CSIR	South Africa
Mr Jan Koster	Development Bank of South Africa	South Africa
Ms Jane Burt	Rhodes University	South Africa
Dr Jeffta Sakupwanya	Global Water Partnership	Zimbabwe
Dr Jerome Delli Priscoli	World Water Council and US Army Core of Engineers	USA
Dr Johan de Beer	CSIR	South Africa
Dr Ken Conca	University of Maryland	USA
Dr Kevin Murray	Insight Modelling	South Africa
Dr Kevin Pietersen	Water Research Commission	South Africa
Prof Larry Swatuk	Harry Oppemheimer Okavango Research Centre	Botswana
Ms Linda Godfrey	CSIR	South Africa
Dr Luis E Marin	Mexican National University: Geophysics Institute	Mexico
Ms Marina Joubert	Southern Science	South Africa
Prof Michael Campana	University of New Mexico	USA
Dr Nevil Quinn	University of the West of England	South Africa
Mr Nyambe Nyambe	University of KwaZulu Natal	South Africa
Dr Owen McIntyre	University College Cork	Ireland
Prof Peter Ashton	CSIR	South Africa
Mr Piet Heyns	Ministry of Agriculture, Water and Rural Development	Namibia
Mr Quinton Espey	Department of Water Affairs and Forestry	South Africa
Ms Raya Stephan	UNESCO	France
Prof Richard Cowling	Nelson Mandela Metropolitan University	South Africa
Dr Rob Fincham	University of KwaZulu Natal	South Africa
Prof Rob O'Donoghue	Rhodes University	South Africa
Sandile Zeka	University of KwaZulu Natal	South Africa
Ms Sandra Fowkes	Ukuvuka Trust	South Africa
Dr Steve Mitchell	Water Research Commission	South Africa
Mr Willie Enright	Department of Water Affairs and Forestry	South Africa

CSIR Symposium Organising Team and Editorial Committee

The symposium organising team was responsible for the planning, organisation, logistics and administration, right from conceptualising the symposium to the drafting the Proceedings. The editorial committee identified the authors and topics, screened the draft manuscripts, commented on the second draft manuscripts, selected manuscripts for publication, assisted with the publication's layout and co-authored the prefaces and conclusions. Your support, enthusiasm and commitment made this event a success.

The Symposium Organising Team was: Hanlie Hattingh, Marelize Moolman, Peter MacMillan, Gillian Maree, Wilma Strydom, Joy Leaner, Thabo Sekonyela, Marius Claassen, Dirk Roux, and Anthony Turton.

The Editorial Committee was: Hanlie Hattingh, Gillian Maree, Wilma Strydom, Joy Leaner, Marius Claassen, Dirk Roux, Peter Ashton and Anthony Turton.

Mike Coke (language editing), **Amanda van der Mescht** (format editing) and **Gillian Maree** (co-ordinating the peer review and editing processes) are specifically acknowledged for their contribution and effort, within tight time frames and under stressful circumstances.

Editors' Biographies

Dr. **Anthony Turton** occupies the Gibb-SERA Chair in IWRM at the Council for Scientific and Industrial Research (CSIR) in South Africa. He also heads the African Water Issues Research Unit (AWIRU) at the University of Pretoria and is the current President of the Universities Partnership for Transboundary Waters.

Ms **Hanlie Hattingh** is integrated water resource management and environmental governance specialist at the Council for Scientific and Industrial Research (CSIR) in South Africa. She is the project leader for the governance initiative within the CSIR and is a member of IAIAsa, with experience in implementation of legislation. She registered for her PhD on the topic during 2005.

Ms **Gillian Maree's** research experience covers the fields of conservation ecology, spatial planning, ecosystem governance and knowledge management. She currently works as a researcher at the CSIR within a Research Group focusing on Water Resources Governance Systems.

Dr **Dirk Roux's** scientific experience covers the fields of aquatic toxicology, conservation planning, design of monitoring programmes, environmental reporting, adaptive resource management, policy development, and institutional learning and knowledge sharing processes. He works as a principle researcher and project leader for the CSIR.

Dr **Marius Claassen** is experienced in water resource management, risk assessment and decision support and has been involved in projects in the region, and internationally. He managed the Water Resource Management group at the CSIR and was recently appointed as the Strategic Planning Manager for the Business Unit. He completed his PhD thesis during 2005.

Ms **Wilma Strydom** is from the CSIR and research focuses on communicating science for the River Health Programme and other environmental projects. She has a keen interest in the impact of science communication on water resource management and river conservation.

Contents

Introduction

Part 1: An Overview of Governance

Conclusion

Foreword

Inaugural Speeches at the International Symposium on Ecosystem Governance, 10-13 October 2005, Kwa Maritane Bush Lodge, Pilanesberg, South Africa

Phera S Ramoeli

Senior Programme Manager: Water
SADC Secretariat: Directorate of Infrastructure & Services: Water Division

Chairperson, Renowned Experts of the Water Fraternity, Distinguished Delegates, Ladies and Gentlemen: It is not often that one gets invited to such an esteemed gathering of experts and to share in the discussion of what I consider a central issue in the process of Integrated Water Resources Management. I feel very honoured to have been given this opportunity to make a few remarks on this occasion and to share with you my own ideas and experience with regard to ecosystems governance. I therefore wish to thank the organisers of this symposium for inviting me and putting at our disposal these impressive facilities.

Because I come from a government type of institution, I will address this topic with a slight emphasis on government's role and the need for an interface with the other two Trialogue components. While I will not go into great detail in my exposé at this stage, as there will be specific experts who have prepared more detailed presentations, I would however still like to highlight some of SADC's experiences in this area in some kind of detail.

You may all be aware that the institution I represent, the Southern African Development Community (SADC), is an inter-governmental grouping of fourteen sovereign southern African states, namely Angola, Botswana, the Democratic Republic of the Congo (DRC), Lesotho, Madagascar, Malawi, Mauritius, Mozambique, Namibia, South Africa, Swaziland, the United Republic of Tanzania, Zambia and Zimbabwe.

The main basis for this grouping and the overall objective as defined in the SADC Declaration and Treaty of August 1992, as amended, are to:

- Promote sustainable and equitable economic growth and socio-economic development that will ensure poverty alleviation with the ultimate goal of its eradication, enhance the life of its citizens and support the socially-disadvantaged through regional integration;

- Promote common political values, systems and other shared values through institutions that are democratic, legitimate and effective; and
- Consolidate, defend and maintain democracy, peace, security and stability.

These are premised on the five main principles that govern the relationships between and among SADC member states and their peoples, namely:

1. Sovereign equality of member states;
2. Solidarity, peace and security;
3. Human rights, democracy and rule of law,
4. Equity, balance and mutual benefit; and
5. Peaceful settlement of disputes.

Governance issues relating to SADC are therefore codified in this main instrument and its subsidiary instruments, the protocols. To effect these governance issues, SADC has established requisite institutions and structures and allowed for the inclusivity of all stakeholders in the management and implementation of the SADC programme of action. Some of these structures include:

- The Summit of Heads of State and Government;
- The Council of Ministers;
- The Integrated Committee of Ministers;
- The Standing Committee of Officials;
- The SADC Secretariat; and
- The SADC National Committees.

While participation in the above structures is, by and large, intergovernmental, there are specific fora and facilities through which other groups such as NGOs, the private sector, civil society and workers and employers associations can participate. The strongest area of participation is through the SADC national committees, which is where programmes and projects should be initiated and implementation facilitated. It is worth noting at this stage that the SADC Treaty does make specific mention of the involvement and participation of stakeholders in the process of regional integration in Article 23. It stipulates that “SADC shall seek to involve fully, the people of the region and key stakeholders in the process of regional integration.”

Chairperson, Ladies and Gentlemen: I am giving you this information in order to set a background and context for water governance in SADC and to demonstrate that this has been codified into law and is upheld by all

SADC member states through their signature and ratification. You may be aware that, prior to March 2001, when the SADC Summit took a decision to restructure its institutions, SADC was a decentralised organisation, with the coordination of various programmes and projects being allocated to specific member states for coordination. This situation changed after the 2001 decision was taken and has seen the centralisation of all sectors and their clustering into four specific directorates including:

- Trade, Industry, Finance and Investment;
- Food, Agriculture and Natural Resources;
- Infrastructure and Services; and
- Human Resources Development and Special Programmes.

I have to indicate at this stage that, although the coordination and facilitation of the implementation of programmes and projects was centralised, programme implementation is still decentralised and has been made possible through the establishment of the SADC national committees who are seen as the drivers of the SADC development agenda.

Chairperson, Ladies and Gentlemen: Let me now focus on water governance, the subject of this symposium, and highlight some of the salient features of this issue within the SADC organisation. As indicated above, the treaty that established SADC (in Article 22) provides for the negotiation and conclusion of sector specific protocols, such as in the case of water. These protocols "spell out the objectives and scope of and institutional mechanism for cooperation and integration" in the various areas of cooperation. The Protocol on Shared Watercourse Systems was the first such protocol to be developed and adopted by the Summit in August 1995 in Johannesburg, South Africa. It was subsequently ratified by the required number of member states and came into force in September 1998. This protocol was revised in 2000 to take into account the new developments in international water law, for example the adoption by the United Nations General Assembly of the UN Convention on the Law of the Non Navigational Use of Shared Watercourses. The Revised Protocol on Shared Watercourses came into force in September 2003 and repealed the old Watercourse Systems Protocol. There are specific governance issues that the revised protocol provides for to which the member states, party to the protocol, have agreed.

The overall objective of the protocol is to "foster closer cooperation for judicious, sustainable and coordinated management, protection and utilisation of shared watercourses." This will be achieved through:

- Promotion and facilitation of the establishment of shared watercourse agreement and requisites institutions for the management of the shared watercourses;
- Advancement of the sustainable, equitable and reasonable utilisation of the shared watercourses;
- Promotion of research and technology development, information exchange, capacity building, and the application of appropriate technologies in shared watercourse management; and
- Promotion of coordinated and integrated environmentally sound development management of shared watercourses.

The above approaches recognise the importance of the management of water resources at the lowest appropriate level and the use of the river basin as the unit of management. Technology and science are very important in ensuring the achievement of this kind of management and the protocol already promotes the interface between science and the management of shared watercourses.

The SADC region has recently adopted a regional water policy at the highest level which sets out the principles for the management and utilisation of, as well as participation in, the water resources of shared watercourses. The policy principles for water resources management for the SADC region as presented in the policy with direct relevance to governance are:

- Water is an instrument for peace, cooperation and regional integration;
- Effective public consultation and involvement of users;
- Focus on integrated, people-centred planning;
- Further development of SADC water resources through the joint planning and construction of strategic water infrastructure, in order to rectify historical imbalances and promote water supply for irrigation and poor communities;
- Efficient use of water through demand management, conservation and re-use, and the efficient use of water for agriculture;
- Recognition of the environment as a legitimate user of water, as well as a resource base;

- The protection of the environment through appropriate user-charges and the enforcement of the “the polluter pays” principle, taking into account equity and social justice; and
- Integration of water supply, sanitation and health and hygiene education programmes.

The policy emphasises the principle of proper stakeholder participation in the whole spectrum of water resources development and management that takes on board views of all stakeholders, especially disadvantaged groups such as women, youth, rural communities etc.

In recognition of the imperative of putting these principles into practice, the water sector, focusing specifically on programme development, became very consultative and participatory after it had been set up as an independent and dedicated sector. The sector developed a comprehensive action plan based on inputs from stakeholders; these became manifest in the 1998 Regional Strategic Action Plan on Integrated Water Resources Management and Development (RSAP-IWRMD). This plan was implemented over a period of five years (1999-2004) and focused very much on creating an enabling environment for IWRM in the region. This entailed the establishment of requisite institutions at basin level and capacity building programmes that targeted all stakeholders including specific projects on NGOs and CBOs, women, policy makers and the private sector to enhance their participation in the water sector. The implementation strategy for the RSAP was based on decentralised implementation, using implementing agencies and project management units.

The plan has now been reviewed to take stock of the successes and failures in its implementation and, having identified those, has been refocused to address the challenges that were identified. Within the new plan, which is clustered into four specific thematic areas, there are specific initiatives dealing with:

- Regional Water Resources Planning and Management;
- Infrastructure Development Support;
- Water Governance; and
- Capacity Building.

As can be seen here, the region still sees issues of water governance as being of high priority, that need focused attention. We still see a lot challenges ahead, especially as regards the implementation and rolling out of the various programmes that are aimed at enhancing water governance. There is still a need to strengthen the newly established river basin institutions and to operationalise and strengthen the SADC national committees, especially the clusters that deal with water. This symposium will assist in shaping our approach to this very important issue and develop research that will further enhance our interaction in these areas. We indeed see merit in and subscribe to the Trialogue model that is being promoted in the seminar and in the global water sector arena.

I wish everyone a good few days of deliberations and I encourage us to have active participation and frank debates, because a lot depends on the outcomes of this symposium. I wish to once again thank the organisers for what promises to be an exciting three days of discussions and the participants for their efforts in taking time from their busy schedules to be present at this symposium.

I thank you all very much for your kind attention!

Khungeka Njobe

Executive Director: CSIR: Natural Resources and The Environment

Research and Learning for Ecosystem Governance

Natural Resources and the Environment Distinguished speakers and guests, good evening and welcome to the International Symposium on Ecosystem Governance. An extended word of welcome also goes to our event partners and funders; the Department of Water Affairs and Forestry (DWAF), the Global Water Partnership, the Water Research Commission, the University of Zimbabwe and the Universities Partnership for Transboundary Water.

I am honoured to stand here today and officially welcome all of you to this Symposium on Ecosystem Governance. A few days ago, on 5 October 2005, I attended the CSIR's Excellence Awards, which also marked the organisation's 60th Anniversary. I was struck by something that was said by the master of ceremonies, and CSIR Group Executive in Institutional Planning and Operations, Mr Vishnu Pillay. He noted that most countries or nations are known for specific attributes or qualities. For example, he said, one would easily associate Japan with quality, Italy with luxury and so on. South Africa, being a young democracy, however, did not have one such official attribute. He therefore, suggested that the country, in order to have a competitive edge, should position itself as a "learning country", and consequently that the CSIR should position itself as a "learning organisation". It makes perfect sense, therefore, for the Natural Resources and the Environment (NRE) Unit of the CSIR to host an event such as this one, where we are afforded the opportunity to learn from each other. I hope that we will have much to learn from each other, over and beyond the next three days.

The process leading to this event has also been a learning curve. The research reports, compiled through proactive collaboration and engagement in working sessions, have given us substantial information and knowledge in the areas of ecosystem governance here in South Africa and elsewhere in the world. We have learned that societal, human and organisational behaviour, as well as their interactions and governance structures are becoming increasingly important in terms of the management of ecosystems that support human activities. This pattern follows a world-wide trend in the decentralisation of natural resource management and decision-making responsibilities, and subsequently, the inclusion of society at large in natural resource decision-making.

South African water policies and legislation (notably the National Water Act, No 36 of 1998) echo these developments and make provision for new governance structures (e.g. Catchment Management Agencies and Water User Associations) and processes to achieve fundamental change in society, towards greater equity and sustainability in both society and ecosystems. The new role of natural resource governance systems has particular significance in South Africa because of historical disparities in the distribution of costs and benefits associated with access to and use of natural resources.

As would be expected, a young democracy such as ours would encounter what others may refer to as "teething problems". And the country has certainly experienced some teething problems and challenges as far as ecosystem governance is concerned. So, what are some of these key challenges?

Firstly, due to fundamental changes in the South African political environment, South Africa has been going through a period of rigorous revision and development of government tools. Government offices, for a number of reasons, are experiencing difficulties in effectively implementing various government tools that have been developed since 1994. Previously, limited research was done to investigate why policies were not operational. We have concluded that it is not about making policies operational, but rather about translating new policies into new operational practices and then implementing the new practices. The issue of funding ties in directly to the discussion above. How much of its total Gross Domestic Product is South Africa allocating to Research and Development (R&D) initiatives? Well, according to figures taken from the R&D strategy document of 2001, the South African total (public and private sector) expenditure on R&D amounted to approximately 0,7% (government 0,29%) of GDP, whereas the average Organisation for Economic Co-operation and Development (OECD) country expenditure is 2,15% of GDP. The report notes that Finland, whose economy is the same size as that of South Africa, spends 3,5% of GDP on R&D. Perhaps we ought to adopt these global trends.

In addition, a number of strategic and tactical obstacles to the implementation of good governance have also been identified through defining the context of water resource management in South Africa and by examining the concept of government versus governance. Included amongst some of the strategic obstacles are, as stated earlier, issues around our young democracy, for example, the level of its maturity, the struggle between adopting either a Westernised or African Democracy, adopting participative government strategies, and decentralising government structures. Other issues include the loss of institutional memory and norms and capacity to act, complexity and uncertainty, the ingenuity gap, the knowing-doing gap and the knowledge divide, resistance to re-examination of core assumptions as well as change management.

Amongst the tactical obstacles we can distinguish amongst such issues are: the reluctance to make decisions, the lack of access to appropriate information, the appropriateness of logistics and capacity, inadequate communication, as well as misinterpretation of the word and intent of legislation.

South Africa is a very progressive nation and, as a result, South African water and environmental laws are also very progressive. The CSIR, with its refocused vision of adhering to its founding purpose, which is grounded in science and technology, has a responsibility to deploy its science capacity to assist the South African government with delivery on such issues. Through fora such as this one it is hoped that strategies that will ensure our country protects its limited natural resources, such as water, will be drafted. As you know, South Africa is a water scarce country. Water availability has the potential to limit the country's future economic growth and development potential. We are therefore aware that we need to be very smart about the way we manage our resources.

While South Africa is acknowledged as a leader in the development of law, specifically as it relates to human rights and the environment, this symposium is not only about South Africa and its challenges with regard to ecosystem governance. The concept of governance, specifically water governance, is receiving significant attention globally. Amongst other things, this relates to the objective of providing water as a catalyst for poverty eradication. In the context of the Millennium Development Goals, and in the face of the ever-increasing world water crisis (from flooding to severe water scarcity), it is essential that each country (especially countries that face development challenges) be in the position to ensure that access to water is available to those who need it most. The Global Water Partnership has stated that the world is not experiencing a water crisis, but rather a crisis of governance.

Ladies and Gentlemen, to conclude my welcome tonight, I would just like to add that I have full confidence that this symposium will achieve its objectives of:

- Unpacking the issue of governance, through specialists exchanging ideas,
- Developing a research agenda on ecosystem governance, and

- Compiling a book on governance based on the manuscripts submitted for the symposium.

I thank you, and wish you an enjoyable learning experience over the next two days.

Jerome Delli Priscoli, PhD

Editor in Chief Water Policy
Senior advisor Institute for Water Resources USACE

Five Challenges for Water Governance

Introduction

Southern Africa is the perfect place to take up the world dialogue on water and governance, and I am honoured to participate. No other country is attempting to change its water rights system completely and to define special reserved rights on the scale attempted here. The links between water governance, development, ecology and social justice are dramatically revealed to the world as it observes what is happening here in Southern Africa. All of us have a stake in these events. As we begin, I hope the following five challenges and ten recommendations will help our discussion.

Five Challenges

First we should ask: what is water governance? What are we governing: the water resources, delivery of service or something else? To what end are we governing: to preserve, to develop and change, to right injustices or in some way equalise them or for another reason altogether?

We know that good social/political governance can increase water availability and that good water availability might not mean good social/political governance. So then, does water governance really mean 'water management' by another name? Does water management determine a social/political system? Do social/political systems determine water management? Can we have good governance with bad water management? Or can we have bad governance with good water management?

I believe we can assert that fundamental water reform substantially involves social and political change, no matter what rhetoric is used and also that major social/political change does drive water management changes. Indeed, governance and water management are linked; this is one reason why the words "civil" and "engineering", in the discipline civil engineering, are so closely linked in our western history.

Thus, thinking of governance and water leads one to ask: are the developed rich because of investment in water, or do the developed have good water-investment because they are rich? Since political systems and water governance and management are linked, is the soft path, e.g. behaviour change and management, more likely to result in authoritarian or democratic political cultures? Is the hard path, e.g. using technology and structural interventions, more likely to lead to authoritarian or democratic political culture? These questions are not simple or obvious to answer. Thus I would suggest that the framing paradigm we are using should also include the missing dimensions of political culture, nature and ecology.

Second, we should acknowledge that water is central to our civic cultures. In the 1930s Karl Wittfogel, a student of Max Weber, drew systematic attention to this in his “Theory of Oriental Despotism”. He attempted to explain the political authoritarianism he found in many Asiatic systems by the drive to centralisation caused by continued imperatives and rationalisation in the water sector. In short, water sector reform drove his view of political centralisation. Recent writers have used this thesis to describe the development of water in the US west. Subsequent scholars have found the Wittfogel thesis to be deficient, as many Sinhalese irrigation systems indicating the opposite were not examined. However, the general notion that water is a central driver in political culture stands.

More modern experiences have continued to reveal this linkage between political culture and water sector management. For example, one of the oldest continuous western democratic institutions is the Valencia Water Tribunal. It operates on a rotating basis of water owners and operators resolving conflicts among other owners and operators. Variations of this system, most likely from Arab North Africa via medieval Spain, were actually transported to the early southwestern part of North America in the form of the aeseqia water rights systems. The Dutch water boards have operated since the Middle Ages, and are widely acknowledged as having provided a model for modern Dutch democracy.

Today, an examination of the political and cultural drivers of water sector change also supports this critical linkage. For example, water reforms have been most advanced where macro-economic reform, open markets, less corruption and more participation have prospered. Macro-economic crises, such as in Mexico or India, are examples. Political restructuring, such as here in South Africa, has generated water sector change. Liberalisation policies, such as in Chile, Brazil, China and other countries, have also resulted in water sector change. Meeting EU standards and the subsequent requirement for water sector reform has become important to several countries petitioning to enter the EU, such as Spain, Poland, Hungary and others. International lenders and donors have somewhat belatedly begun to understand that water sector lending and ESA support is not a purely technical exercise.

We do not have to stick with the political to see these points. Throughout thousands of years of history and across the time and space of many different and distant cultures, poets have repeatedly written allegorically that “Water is … humanity’s carrier of its collective memory.” Indeed, today, scientists in their own way seem to support these allegorical writings as they speak of “…the same water recycling through us over time and space…and of molecules carrying information....” CJ Jung noted that water was the symbol of the unconscious.

Third, I believe our discussions should reconnect water and development. Doing that will elicit considerable controversy. Reconnecting water to development is a necessary condition if we are to avoid water-related conflicts among the world’s most vulnerable people, and if we are to capture water’s immense potential to build community, to mitigate conflict and to act in second-track diplomacy.

Sustainable development is offered as a meeting ground among those who warn against repeating the mistakes of the past and those who remind us that many in the world need the benefits that water can offer. It is a meeting-point on a continuum between the extremes of behavioural change, ecological preservation and structural intervention. Each place on the scale implies values, assumptions and often distinct views on governance actions. Thus we need to remember, especially here in southern Africa, that infrastructure interventions do matter.

The ethical parameters of the world’s water access situation, as set up by geographic reality, are well mirrored here. Most of those without access to water live in places where the rain falls in only 20%, or less, of the year. These same people are those who are bearing the brunt of water-related disasters, of floods and droughts. These are disasters which the UN continues to identify as accounting for the vast majority of disaster-related deaths and damages suffered throughout the world. And these are damages that are not reimbursed and disasters that often follow on each other’s heels in the same areas. Also, the numbers of people involved are staggering; they are in the hundreds of millions, or billions. As many in the environmental and NGO communities have noted, relocating people is all too often degrading to human dignity, corrupt and unfair. So, what will we do with those hundreds of millions, or billions, living in places without water?

Indeed, infrastructure really matters, and will matter even more as we seek to flatten the peaks and valleys of the hydrographs that occur in the areas populated by these hundreds of millions and billions. The answer to how it matters will be provided by the water governance responses to these phenomena, as witnessed here in Southern Africa.

There are strong correlations between public capital investment and movements in private sector productivity. The ratio of non-structural/behavioural measures to structural measures matters. If it is too high, extreme events can crack social systems, as leaders have no tools to respond with, other then draconian, authoritarian measures. If it is too low, the ecological costs are too high. However, we need to be wary of the governance myth that "the soft path will be more democratic". Political leaders, when faced with crises of drought or flood, will ask their water managers or engineers to do something. This usually means using some residual storage or other means to meet the crisis, or buying time while minimising the social shock. But if no such measures exist, leaders easily fall into authoritarian measures of forcing behavioural change, or of manipulating people or, worse, distributing benefits in starkly authoritarian ways to the benefit of the tribe, community or political advantage.

Recent World Bank data have begun to discern clear patterns between the ability to flatten the peaks and valleys of hydrographs and the capacity to break out of poverty and hopelessness. Data on rainfall and growth in Zimbabwe between 1978 and 1993, in Ethiopia between 1982 and 2000 and in Kenya between 1956 and the 1990s, all show how fluctuations in rainfall closely parallel the fluctuations in GDP in each country. Their data also suggest that between 25% and 30% of the variability of the GDP in Mozambique and Kenya and, by implication, other poor African countries as well, is accounted for by the inability to manage the peaks and valleys of the hydrograph variability. Indeed, one Finance Minister in India noted, "Every one of my budgets was largely a gamble on rain."

Actually, these relationships have been known in the richer north and, indeed, have been the basis for much investment. For example, U.S. data show that while flood damages have increased over time, flood damages as a percentage of GDP, nationally and regionally, have continued to decline. Multiple dollar returns per dollar of investment have been measured over both the short and long term. Similar data are also available on post-war Japan. Disaster damage as a percentage of GDP is a measure of the viability of socio-economic activity and of the capacity not only to absorb but also to stay on a prosperous trajectory. Indeed, World Watch notes that disaster damages in developed countries vastly exceed those in poor countries. However, the damages suffered by the poor countries, as a percentage of *their* GDPs, is three to four times as great as in the rich west.

This experience in the north, coupled with evidence emerging in the poorer south, suggests once again that infrastructure which allows the integration of flood defense measures, water storage and electricity generation, ecological flows and other uses, is key to breaking the poverty cycle. The evidence is an argument for IWRM or what was once called multi-objective water operations and planning. It can also suggest that there is a minimum platform of water resources infrastructure and institutions necessary to achieve some type of platform for sustained economic growth and poverty eradication.

Fourth, I think that water governance needs to initiate a new dialogue between rich and poor. Prescriptions of the rich north reflect the values of their producers. These are based on assumptions stemming from experience that is very different in time and place from that of those to whom they prescribe. Today, the rich north is far more concerned with less infrastructure investment and more management. The opposite is true in the poor countries that the North seeks to help. However, prescriptions often come across as "Do as I know I must do". To the poor, this begins to sound like a new kind of imperialism – ecological imperialism. The West starts becoming dismayed as it has to explain that the environment is not just something you take into consideration *after* you develop.

Indeed, the experience-bases are very different at this time. In the rich North, the active self-helping citizen, so central to democratic political culture, has become complacent, often passive, risk-averse, status quo-orientated, and often demands rights but forgets obligations. To be sure, he/she has often been lured into this mindset through the paternalistic promises of engineers that they will take care of everything. Thus, the latter's prescriptions tend to revolve around mitigating the rate of loss against limited fixed resources, minimisation of, or no human impact, reducing the use of resources, increasing efficiencies and reallocating resources to fit new demographics. Also, they often sound like this: "Do as I say – not as I have done…."

Those in the developing world to whom the rich prescribe are engaged in promoting the existence of an active self-helping citizen as the foundation for democratic civic culture; such a citizen will be someone who will choose risk and obligations in a situation where there is a lack of means. He/she will usually be more interested in fundamentally changing the status quo. Thus, water is seen as a means to change and growth, and not only as an end in itself, to be preserved. Supply and availability are critical and human interventions are not separable luxuries but instead are necessary.

Essentially, an often unconscious transfer of value assumptions becomes the centre of the dialogue. For example, the US investment in water supply and infrastructure in the New Deal by the PWA consisted of 2600 water projects at around $312 million (in 1930s terms!). FERA, CWA and WPA accounted for another $112 million for municipal water (again in 1930s terms!). Between 1972 and 1990, more than $650 billion in federal grants for sewage treatment, and more than $20 billion from the states, was awarded. Also, these numbers still do not come close to the actual expenditure for water in the US over the last half of the twentieth century. Still, the WEF estimates that we need $23 billion/year for 20 years to meet EPA standards. By comparison, over 100 countries in the world, those without adequate sanitation, have an annual budget of less then $23 billion! Thus it is easy to see how unchecked assumptions can become driven prescriptions.

Water governance needs to recognise such underlying dynamics of the rich-poor dialogue. The rich should be talking about the unforeseen costs recognised over time and how to build that experience into the design of water investment, rather than simply portraying their investment in water infrastructure as a false vector for the poor to seek today. They should be helping to form what the Kuznets curve identifies as 'adaptive investment'. This is investment which reduces and mitigates, if not eliminates, the ecological costs. Much of the dialogue risks pushing the poor into a normative governance arena that is vaguely defined, if at all, with no demonstrable performance measures.

Fifth, water governance needs to help clarify the new rhetoric and language surrounding the world water debate. This new rhetoric and language have both positive and negative aspects. For example:

Sustainable development, as an avenue of dialogue has increased the space for dialogue among interests not prone to talking or cooperating. It has thus managed to help the governance dialogue. However, as an analytical tool, it looks like a contradictory term. 'Sustainable', de facto, seems to indicate no change or minimum change. 'Development', however, means change. Thus sustainable development becomes equal to "...no change – change...."

On the other hand, adaptive management brings governance to focus on design and performance criteria. Ecologists are challenged to develop indicators. Decision-makers are challenged to use feedback in decisions, and thus it pushes us to a conscious choice of ends. This is critical to any governance, and especially to water governance.

It is hard to argue with the words "integrated" and "holistic". They have little relevance to governance unless they are applied to specific levels such as the river basin, the watershed or the city, etc. Also, when this occurs, they do not mean the same in each area. Often, the use of these words disguises political agendas and changes in power-relationships. Since political–technical relationship improvement is central to good water governance, it does little good to disguise such issues. They are immediately recognised by the political sphere, anyway. It does little good to use these terms and then to appear surprised when the political sphere does not implement them. Water governance needs to encourage a more explicit dialogue on the implied power and political implications of suggestions. Frequently, this revolves around the discrepancies derived from starting with a geographic, scientific or human jurisdictional point of reference.

Calls for financing and full cost recovery, now repeated around the world, seem admirable and rational. However, most of the rich countries, who are the instigators of these requests, have not themselves implemented them carefully, especially during periods of intense capital water infrastructure investment. It seems that water, with its capital-intensive needs and long-term investments, will not attract substantial money from private capital markets, unless, of course, such investments are guaranteed by the public sector. Subsidies are a way of life for water and they address both political and technical questions of water governance. Indeed, the secret to the TVA and the Columbia systems in the US is the cross-subsidies from revenue-producing aspects of multi-purpose uses, such as hydropower, to other public benefit uses, such as flood-control, ecological flows and restoration. The question for water governance is really about making the flow of money transparent and accountable, so that it is clear what is subsidising what. Indeed, in the US during the New Deal, FDR said that rural electricity would entail cost-recovery, but not at market prices, since the private companies had been unable to bring electricity to those areas. The result was that rural electrification in the US went from around 40% to 95% in 10 years. Water governance needs to balance such concerns.

Decentralisation, or subsidiarity, has become an important objective of the water governance debate. However, it too needs to be better defined. Generally it seems to mean taking responsibilities from the central government and placing them in the hands of localities and regions. However, again the issue is about what decisions need to be taken at what levels, which is also the issue for water governance as a whole.

Since water management and governance everywhere are moving towards more risk-based decision-making, the ever present "precautionary principle" is becoming more important. But what does it really mean? Not to decide is a decision which has consequences as well, since nature itself is change. What are the likely impacts or consequences if no action is taken? Too often, the ethics of governance are portrayed as revolving around the costs of doing *something*, rather than weighing them against the likely cost of doing *nothing*.

What are the ethics of not taking decisions, in the face of needing to decide? How much must we know in order to decide; 100%, 90%, 80%...? If the precautionary principle, essentially, holds 100% certainty, as some absolute or asymptotic optimal, then the principle would defeat the very basis for risk-based water governance. But, more important for governance, if we essentially believe that we will never know all the complex interactions, can exercising the precautionary principle itself be ethical? Or is the precautionary principle actually a synonym for no action, minimum action, stasis or no-risk? These are important questions if we are to strive for better water governance.

Preservation and nature, of course, have long histories in water governance, but they too carry some profound importance for today's water governance debate, especially for the water dialogue between the North and South. At the turn of the century in the US, two strands of environmentalism emerged. One was championed by Gifford Pinchot, friend of Teddy Roosevelt and founder of the US Forest Service, and focused on the *wise use*, or utilitarian approach, to water and environment. The second was championed by John Muir, founder of the Sierra Club, and focused on the *preservation* of nature and creation. Indeed, John Muir often used religious illustrations to present his opposition to wise-use approaches to the public, such as with the famous Hetch-Hetchy Reservoir fight in California in the early 1900s. The Pinchot utilitarian approach was dominant for most of the 20th century. Then, in the 1970s, the Muir's preservationist approach came to be dominant. Its primary reason for dominance lay in its adoption of a call for the regulation versus the planning of water resources in the US.

Each approach is built on different assumptions of nature. The preservationist notions have come to portray human actions as reducing nature or using up limited resources, often at alarming rates. The utilitarian approaches have essentially assumed that nature is changing and that human efforts need to use the resources in ways that do not disrupt the viability of the resources. Water is very amenable to this second approach, since it can be used in many ways that actually create benefits and values, both in the water and beyond the water.

So, what is this preservation or creation – a static and limited resource, or a grand process of creation? Are we engaged in something like the Second Law of Thermodynamics, where we are in a game of minimising the rate of inevitable loss of limited resources? Much of the response of the public in the rich North seems influenced by this notion. Or are we part of a changing nature, where we are co-creating with nature and adding value and benefit? The first approach puts water governance into a situation of redistributing within a limited pie. The second approach puts water governance into the arenas of using the resources to increase the pie. The first really portrays an inevitability of conflict while the second looks optimistically to new forms of evolution of nature and humans. My own studies of water and civilisation point towards the second. However, what we prescribe concerning water governance depends greatly on the assumptions we make on this issue, and it does little good to hide these assumptions.

Conclusions and Recommendations

Finally, I conclude with the following 10 recommendations for our water governance dialogue:

1. We should strive to build a new ideological and ethical consensus on water, one that focuses on the common ground of engineering means and environmental ends. It should be one that moves beyond the equilibrium, status quo and preservationist notions of ecology to co-designing with nature and to choosing desired future ecologies.
2. We need to reformat the messages and conflicting perceptions that we, especially in the rich North, are sending. These include, for example:
   - Keeping water resources conditions "natural";
   - Labelling structures as 'bad';
   - Controlling populations, but reducing poverty;
   - Seeing privatisation as the solution; and
   - Seeing external funding as the solution for reducing the revenue gap.
3. We need to distinguish between water management and water services and we need to create political will to act.
4. We need to acknowledge the role of structures and infrastructures in water governance. Yes, we need to build infrastructure, but, no, not exactly as we did it. We need to include lessons learned about costs, and to design mitigation and ecology concerns and costs into infrastructure.
5. We need to reconnect water as the vital tool for economic and social development, and to go beyond seeing it only as a human right or an ecological good. We need to investigate the minimum platform of water resources infrastructure and institutions to achieve water security. This includes investigating governance structures, potential infrastructure costs, financing options, consequences of inaction, transboundary opportunities and actions. We need to promote the importance of achieving essential water security for sustained economic growth and poverty eradication. To this end, we should promote the notions of multipurpose uses and planning for water and use aid as a leverage for this.
6. Water governance should engage our public water agencies more directly in our water aid, both in its design and review. We also need to focus more on public agencies reform and capacity-building in the water sectors. To this end we should establish direct public-to-public and agencies-to-agencies relationships between technical experts and decision-makers.
7. We in the North need to understand our own history and context of water before we prescribe it to others.

8. We should understand water as humanity's learning-ground for building communities and forming society. In this regard, we should connect democracy and institutional capacity-building to water aid programmes. This means providing meaning to the "civil" in "civil society", and to building the democratic civic culture, as well as facilitating partnerships and dialogue among professionals and civil society.
9. Water people are not going to solve the water crises. The water crises are not being prioritised within countries. We need to look to finance ministers and the macro-economic and social development picture in order to solve the problems associated with water.
10. We must understand that water decisions are also ethical decisions. Our water debates mirror debates of social ethics. For example:
    - Water as a common good;
    - Water and human dignity;
    - Water as a facilitator of wellbeing;
    - Rights and responsibilities to access of water;
    - Water and social justice; and
    - Wealth-generation roles of water

Indeed, water, as a symbol of reconciliation, healing and regeneration, is found in most of the main-stream faith traditions known to humans.

Over 2000 years ago, in China, the philospher Lao Tze noted:

...The sage's transformation of the World arises from solving the problem of water. If water is united, the human heart will be corrected. If water is pure and clean, the heart of the people will readily be unified and desirous of cleanliness. Even when the citizenry's heart is changed, their conduct will not be depraved. So the sage's government…consists of talking to people and persuading them, family by family. The pivot (of work) is water.

Perhaps the challenge of water governance is best captured in the following chart, which reflects another piece of ancient wisdom from China. It shows the words for 'river' and 'dike' added together. Taken together, they do not mean 'water management', as we might expect. Rather, they mean 'political order'. I think this sums up the agenda for water governance.

References

Priscoli JD and Hassan F (1996) Water and civilization. UNESCO, Paris

Priscoli JD and Llamas MR (2001) International perspectives on ethical dimensions in the water industry. In: Davis CK and McGinn RE (eds) Navigation rough waters: ethical issues in the water industry. AWWA, Denver, pp 41-65.

Priscoli JD (1998) Water and civilization: using history to reframe water policy debates and to build a new ecological realism. Water Policy, vol 1, pp 623-636.

World Bank (2006) Water for growth and development: a framework for analysis. Baseline document for WWF4, Mexico City

List of Contributors

Ashton, Peter J., Research Fellow and Universities Partnership for Transboundary Waters, CSIR: Natural Resources and the Environment, PO Box 395, Pretoria 0001, South Africa, email: pashton@csir.co.za

Breen, Charles, Professor, University of KwaZulu-Natal, Centre for Environment, Agriculture and Development, P/Bag X01, Scottsville 3209, Pietermaritzburg, South Africa

Campana, Michael E., Professor, Institute for Water and Watersheds, Oregon State University, Corvallis, OR 97331-2208 USA and Universities Partnership for Transboundary Waters, email: aquadoc@unm.edu

Claassen, Marias., Manager: Water Resources, CSIR: Natural Resources and the Environment, PO Box 395, Pretoria 0001, South Africa, email: mclaassen@csir.co.za

Conca, Ken, Professor, Department of Government and Politics, University of Maryland, USA, email: kconca@gvpt.umd.edu

Doolan, Jane M., Executive Director: Environmental Water Reserve and River Health, Department of Sustainability and Environment, PO Box 500, East Melbourne 3002, Department of Sustainability and Environment Victoria, Australia, email: Jane.Doolan@dse.vic.gov.au

Eloff, Estie, Researcher: Aquatic Ecosystems, CSIR: Natural Resources and the Environment, PO Box 395, Pretoria 0001, South Africa

Falkenmark, Malin, Professor, Stockholm International Water Institute, Drottninggatan 33, SE-111 51 Stockholm, Sweden, email: malin.falkenmark@siwi.org

Fincham, Robert, University of KwaZulu-Natal, Centre for Environment, Agriculture and Development, P/Bag X01, Scottsville 3209, Pietermaritzburg, South Africa

Fowkes, Sandra, Metaplan (Pty)Ltd, PO Box 30306, Tokai 7966, South Africa, email: sandra@metaplan.co.za

Godfrey, Linda, Research Leader: Integrated Waste Management and Industrial Ecology, CSIR: Natural Resources and the Environment, PO Box 395, Pretoria 0001, South Africa, email: lgodfrey@csir.co.za

Gooch, Geoffrey D., Professor of Political Science, Jean Monnet Chair of European Political Integration, Director of the Masters Program in International and European Relations, Linköping School of Management, Linköping University 58183, Linköping, Sweden, email: geogo@eki.liu.se

Hall, Alan W., Network Coordinator, Global Water Partnership, Drottninggatan 33, SE-111 51 Stockholm, Sweden, email: Alan.Hall@gwpforum.org

Hattingh, Hanlie, Senior Scientist: Water Resources Governance Systems, CSIR: Natural Resources and the Environment, PO Box 395, Pretoria 0001, South Africa, email: hhattingh@csir.co.za

Hill, Liesl, Researcher: Aquatic Ecosystems, CSIR: Natural Resources and the Environment, PO Box 395, Pretoria 0001, South Africa, email: lhill@csir.co.za

Klise, Geoffrey T., Water Resources Program, University of New Mexico, Albuquerque, NM 87131-0001 USA and Universities Partnership for Transboundary Waters

Maree, Gillian A., Researcher: Water Resources Governance Systems, CSIR: Natural Resources and the Environment, PO Box 395, Pretoria 0001, South Africa, email: gmaree@csir.co.za

Murray, Kevin, Insight Modelling Services, PO Box 38953, Garsfontein East, 0060, South Africa, email: kmurray@mweb.co.za

Neir, Alyssa M., Water Resources Program, University of New Mexico, Albuquerque, NM 87131-0001 USA and Universities Partnership for Transboundary Waters

Nyambe, Nyambe, University of KwaZulu-Natal, Centre for Environment, Agriculture and Development, P/Bag X01, Scottsville 3209, Pietermaritzburg, South Africa, email: NyambeN@ukzn.ac.za

Roux, Dirk J., Principal Scientist CSIR: Water Ecosystems, Natural Resources and the Environment, PO Box 395, Pretoria 0001, South Africa, email: droux@csir.co.za

Schreiner, Barbara, Senior Executive Manager: Policy and Regulation, Department of Water Affairs and Forestry, South Africa, P/Bag X313, Pretoria 0001, South Africa

Simalabwi, Alex, Regional Project Manager: IWRM/WE Plans GWPSA, 141 Creswell Street, Wead Park, Silverton, Pretoria, South Africa, email: alex@gwpsa.org.za

Stephan, Raya Marina, Lawyer, UNESCO-International Hydrological Program, 7 place de Fontenoy, 75352 Paris 07 SP France, e-mail: r.stephan@unesco.org

Strydom, Wilma F., Researcher: Aquatic Ecosystems, CSIR: Natural Resources and the Environment, PO Box 395, Pretoria 0001, South Africa, email: wstrydom@csir.co.za

Turton, Anthony R., Research Leader: Water Resource Governance Systems and Universities Partnership for Transboundary Waters, CSIR: Natural Resources and the Environment, PO Box 395, Pretoria 0001, South Africa, email: aturton@csir.co.za

Van Wyk, Ernita, Senior Researcher: Water Resources Governance Systems, CSIR: Natural Resources and the Environment, PO Box 395, Pretoria 0001, South Africa, email: evwyk@csir.co.za

Introduction

Towards a Model for Ecosystem Governance: An Integrated Water Resource Management Example

Anthony R Turton, J Hattingh, Marius Claassen, Dirk J Roux and Peter J Ashton

Abstract

The concept of governance, and especially good governance, is pivotal to the achievement of Integrated Water Resource Management (IWRM). The concepts of *governance* used in much of the current literature on IWRM indicate that it is often used in a contradictory way and sometimes used interchangeably with the word *government*. This chapter proposes a Trialogue Model of governance that is structured around three groups of actors – government, society and science – and discusses the dynamic interactions between these groups. The interfaces between these three groups of actors, or actor-clusters, and the dynamics of their interactions, provide the basis for a critical assessment of governance as a concept. The chapter isolates four specific elements of scale that are relevant to governance: economic, political, administrative and international; as well as three structural aspects: mechanisms, processes and institutions. In addition, the chapter identifies four processes: articulating interests, exercising legal rights, discharging legal obligations and mediating disputes, and analyses the central role of norms and values in good governance. Finally, an analytical distinction is made between governance as a *process* and governance as a *product*, and a new definition of ecosystem governance is offered. Evidence is presented to demonstrate the highly dynamic nature of governance processes, with clear differences that distinguish mature democracies and fledgling democracies.

Keywords: ecosystem management, governance, fledgling democracy, mature democracy, Integrated Water Resource Management (IWRM), Trialogue

Introduction

In recent years, growing awareness that the world's fresh water supplies are vulnerable to human activities, has been matched by the realisation that water resources need to be managed in an integrated and systematic way, to ensure these resources can continue to meet the current and future needs of society (Falkenmark 1989; Biswas 1993; Gleick 1999). This consciousness has been accompanied by increasing recognition of the mounting difficulty and expense involved in providing sufficient supplies of wholesome water, to meet the rapidly growing needs of communities and countries that are fuelled by rapid population growth, and increased rates of urbanisation and industrialisation (Falkenmark 1991). These problems are particularly acute in countries with low levels of economic development, and also in countries located in the drier regions of the world, where water supplies are relatively scarce (Pallett 1997; Gleick 1999).

Internationally, water resource managers have responded to these challenges by adopting Integrated Water Resource Management (IWRM) approaches, since this offers the greatest potential to ensure that water resources are managed effectively and efficiently over the long-term (World Water Forum 2000). Importantly, acceptance of the principles of IWRM is driven by the recognition of two key issues: first, that all the components of the water cycle need to be managed as a single unit, rather than as separate components; and, secondly, that all stakeholders should be more closely involved in decision-making processes, to ensure that management outcomes have greater acceptance and legitimacy. The growing demands for water in water-scarce regions, coupled to the increased prevalence of deteriorating water quality in many areas, has exerted additional pressure on water resource management authorities to adopt this more holistic approach to water resource management. Indeed, the consensus of opinion at the Second World Water Forum indicated that the current crisis in water is not about having *too little* of the resource to satisfy our needs, but rather a crisis of how we *manage* the available water resources (World Water Forum 2000). More recent studies have again stressed the importance of effective local stakeholder participation in decision-making processes that affect their lives and livelihoods (Pegram et al. 2005).

Much of the earlier impetus for these changes can be traced back to the International Conference on Water and the Environment (ICWE) held in Dublin in 1992, which was a prelude to the United Nations Conference on Environment and Development (UNCED) held later that same year in Rio de Janeiro. The ICWE produced a set of key management concepts, known as the Dublin Principles (ICWE 1992a; 1992b) that were then espoused and endorsed at the UNCED meeting (UNCED 1992). The Dublin Principles form the core upon which the principles of IWRM have subsequently been built. At a technical level, the changes in management philosophy have prompted a relatively gradual shift away from more traditional engineering approaches, where water supply infrastructure was developed to provide sufficient water to meet people's needs, to a more integrated planning approach that incorporated both conventional and non-conventional options for the reconciliation of supply and demand, including water conservation and demand management measures (Pallett 1997; Rosegrant 1997; Vermillion & Merrey 1998; Turton 2002; Pegram et al. 2005).

Now, a decade after IWRM approaches first began to be adopted by water resource managers, it is clear that the full suite of the potential benefits of IWRM have yet to be realised. There is an emerging consensus that this failure could be due to inadequate attention being paid to ensuring that appropriate governance systems are in place; and this appears to be due to very varied understandings as to what constitutes good governance (WWC 2000).

This chapter seeks to provide a conceptual foundation for a Trialogue Model of governance that can be applied in various socio-cultural and political-economic settings to foster consensus among water resource management and policy professionals. This Trialogue Model will be interrogated in subsequent chapters of this book in more detail, in order to determine whether it can stand up to the demands of real world applications. Effective IWRM is used as an example of the intended outcome of what we call good governance.

Drivers of Change in Integrated Water Resource Management

International acceptance of the principles of IWRM has enabled water resource managers to develop and apply a suite of interlocking management approaches and options that can be configured to suit specific circumstances. However, it is also important to recognise that acceptance of IWRM principles requires water resource managers to ensure that appropriate institutional structures and stakeholder participation processes are able to accompany and complement the more traditional engineering aspects of water resource management.

As a result, a process of institutional decentralisation and democratisation has accompanied this shift in management approaches to embrace the principles of IWRM – what the Dublin Principles refer to as subsidiarity (ICWE 1992a, 1992b) – to facilitate and strengthen local stakeholder participation in decision-making for water resources management (Pegram et al. 2005). These institutional changes are most easily visible in the emergence of Catchment Councils, Water User Associations and Catchment Management Agencies in different parts of the world (Turton and Meissner 2000; Turton 2002). Taken together, the technical, economic and institutional components of the IWRM philosophy, require water resource managers to have a more thorough understanding of the functioning of the complex inter-linkages between ecosystems, water resource management options, and human activities that impact on the water resource (Pegram et al. 2005). Conceptually, the degree to which water resource management is centralised, and the degree to which management focus is directed towards supply-side or demand-side options, can be considered as important drivers of change in the water sector. These drivers can also be represented as axes on a matrix, with the implications of the interactions between these drivers being presented schematically in Figure 1.1.

Figure 1.1 shows the general trend of change in the breadth of focus of water resource management over time. This is shown as the progressive broadening of the scope of management envelopes from (A) to (C) over time. The upper left-hand quadrant represents the position occupied by most countries during their earlier phases of development. Here, the locus of management tends to be highly centralised and bureaucratic, with its primary focus on supply-side options that can provide water with a high assurance of supply (e.g. envelope "A" in Figure 1.1). This has been called the "hydraulic mission" phase of society (Waterbury 1979; Reisner 1993), where water resource infrastructure provides the foundation for all future economic development (Turton et al. 2004). The hydraulic mission should also be seen as a necessary, albeit initial and insufficient condition, for the sustainable development of any modern economy. Since the primary emphasis is on the construction and operation of engineering structures, the institutional needs also reflect this importance and the management cadre consists predominantly of engineers and hydrologists. Management in this phase can also be seen as focussing on first-order resources, where the resource consists solely of water.

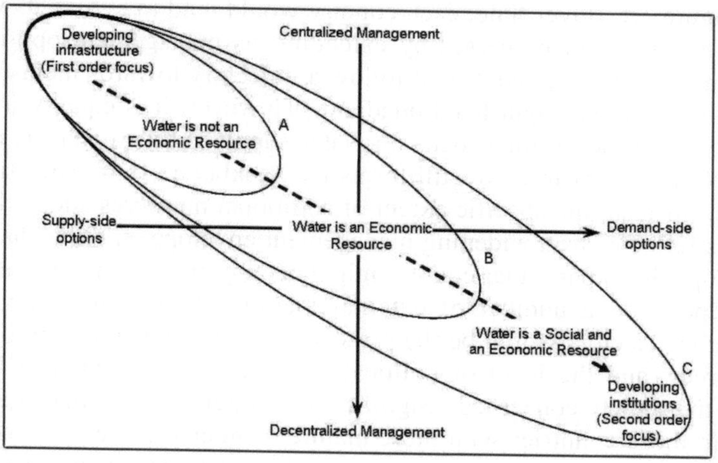

Fig. 1.1. *Conceptual model illustrating the general trend of change as water resource management approaches expand from their original centralised focus to include increasingly more decentralised approaches (A), (B) and (C), and the primary focus widens from strictly supply-side options to include more demand-side options such as water conservation policies (Modified from Turton 2002)*

The lower right-hand quadrant of Figure 1.1 shows the changes that occur as the locus of management becomes progressively more decentralised over time (e.g. envelopes "B" and "C"). Here, the primary focus of management places increasingly greater emphasis on the effectiveness of decentralised institutional structures, (subsidiarity on the vertical axis) and efficiencies of water utilisation patterns (on the horizontal axis). More attention is also paid to water conservation measures, such as those associated with standard water demand management strategies, including new policy applications such as inter-sectoral water allocation, and efficiency measures such as intra-sectoral allocative incentives. In addition, activities in the lower right-hand quadrant emphasize institution-building and the efficiency and effectiveness of operational procedures. This requires a much softer approach and can be seen as a second-order resource, where the principal resource being mobilised is social capital or social adaptive capacity (Ohlsson 1999; Ohlsson & Turton 1999; Turton 1999). In this phase, it is logical that the management cadre is much more heterogeneous, consisting of policy specialists, social scientists, economists, lawyers, engineers, hydrologists, ecologists and geohydrologists, to name but a few specialist disciplines (Turton 1999).

In this analysis it is important to note that countries would generally start their water resource management systems in the upper left-hand quadrant of Figure 1.1. Over time, each country would tend to evolve its water resource development processes by extending its management options to include additional components that follow a trajectory towards inclusion of activities in the lower right-hand quadrant. However, it is equally important to note that the original focus on water supply infrastructure is never abandoned; this infrastructure still forms the backbone of all water supply measures. Instead, the specific extent of additional measures adopted by a country is reflected as a widening management envelope, and the shape of the envelope that a particular country might occupy at a given point in time would depend on a number of external factors. Particularly important amongst these factors would be the physical nature and availability of water resources, and the level of political "maturity" (for want of a better word) of the society concerned. Significantly, water resource management systems in those countries with more mature democracies, tend to include a wider range of positions located towards the lower right-hand quadrant, while those of countries with less mature, or emerging, democracies, tend to occupy narrower positions, located closer to the upper left-hand quadrant. This distinction also emphasizes the differences between the needs of developing countries, or those with fledgling democracies – where the level of infrastructural development may be so low that the assurance of supply is inadequate for sustainable development to take place – and those of more developed countries with more mature democracies, where the transition from an agricultural economy to an industrialised economy has already taken place.

An additional distinction can be made on the basis of the types and variety of skilled personnel available in the different countries. Those countries with mature democracies, tend to have a strong and well-established intellectual base of multidisciplinary specialists in management and other positions, who are able to address the complexities posed by IWRM. However, the same is seldom true within developing countries, or those that are characterised as having fledgling democracies. Here, the management cadre is usually far smaller, and consists mainly of engineers and hydrologists, engaged in primary supply-side management activities. Explicit recognition of this aspect has provided an important impetus for change in these countries, but the necessary multi-disciplinary intellectual base is often slow to develop.

The Role and Importance of Governance

At the global level there is growing consensus among policy-makers and water resource managers about the important catalytic role played by governance issues in achieving effective water resource management. For example, during the Second World Water Forum at The Hague, the Global Water Partnership (GWP) noted that the crisis in the water sector is most often a crisis of *governance* (GWP 2000). This sentiment was reflected in the ensuing Ministerial Declaration that committed the signatory governments to ensuring "good governance", including effective public involvement in decision-making, so that stakeholder interests would be included in the management of water resources (WWC 2000).

In recent years, the term *governance* has been used to describe a wide array of situations that incorporate concerted or directed actions and behaviours, structural elements, institutional settings, legal or statutory instruments and idealised participative or collaborative processes. Many of these descriptions of governance have also been linked to specific considerations, where governance is seen either as a process, a structure, a system of values, or a specific product or outcome. While each of these uses may be entirely appropriate and legitimate within the specific circumstances under discussion, the sheer diversity of these uses has created considerable confusion about the underlying meaning and purpose of governance as a process and, in particular, good governance as a product. In the context of this chapter, therefore, it is important to understand and contextualise the use and interpretation of the term governance correctly in relation to water resource management. This will provide greater clarity for subsequent discussions in the different chapters of this book, whilst also ensuring greater alignment and consistency in the use of this terminology.

What is governance?

Essentially, governance describes the relationships between people, the ways that they interact with each other in the context of their environment, and the systems of principles, rules and norms that are set up to guide these interactions. For example, in a typical commercial enterprise, the term corporate governance is used to describe the suite of internal and external relationships, roles, responsibilities and accountabilities that guide interactions between external stakeholders, staff and corporate office bearers. Similar applications have been used to describe the functioning of small and large organisations, communities and groups of professionals. All of these applications share the recurring themes of appropriate behaviour, structured discourses and consensus on key roles and responsibilities.

We have shown (Figure 1.1) how the transition from a first-order focus to one that gradually includes a greater degree of second-order focus in water resource management within a given country, requires added emphasis on wider stakeholder participation in policy issues and decision-making, as well as the deployment of a more diverse set of skills. In effect, this equates to a greater emphasis on issues of governance and, importantly, the transition also places greater demands on the *intellectual base* of the country. Unfortunately, in many developing countries the need for new intellectual skills can seldom be met by the tertiary education system, and may even require the importation of skilled professionals from other countries. Effectively, the higher demands placed on *governance* and the *intellectual base* of a country is the core of the crisis identified by the GWP (2000) and others (e.g. Cosgrove and Rijsberman 2000), as being *the* main problem in contemporary integrated water resource management.

Good governance, as a global issue, can be regarded as the cornerstone of what is often referred to as effective ecosystem and integrated water resource management. Unfortunately, this pivotal role of good governance has often been obscured by the wide diversity of published opinions on, and definitions of, governance approaches and systems in different settings. In order to develop some scientific rigour in this quest, it is necessary to start off with an adequate statement of the problem. What is actually meant by the concept of governance? What are the key elements of the concept and how do they relate to each other?

Different Definitions of Governance

For the purposes of this chapter, there are three broad types of governance, namely corporate governance, co-operative or network governance, and the notion of adaptive governance.

Corporate governance generally refers to a system for the promotion of corporate honesty, fairness, transparency and accountability to shareholders. The corporate governance structure specifies the relations and the distribution of rights and responsibilities among four groups of participants – the board of directors, managers, workers, and shareholders (Williamson 1988).

Network governance refers to the means for achieving direction, control, and coordination of individuals and organisations that have varying levels of autonomy to advance the interests or objectives to which they jointly contribute. It involves the following: configuring governmental and non-governmental organisations; statutes; organisational, financial and programmatic structures; administrative rules and routines; resource levels, and; institutionalised rules and norms. It also involves formal organisational structures, personal relationships, and judgement by those individuals working in the complex space of administering public programmes. It is inherently political and involves bargaining, negotiation, and compromise (Imperial 2004).

Adaptive governance is a process of creating adaptability and transformability in social-ecological systems (SESs). Adaptability refers to the capacity to absorb disturbance and reorganise while undergoing change so as to retain essentially the same core function, structure, identity and feedbacks. Transformability refers to the capacity to create a fundamentally new system (e.g. new ways of making a living) when ecological, economic, or social (including political) conditions make the existing system untenable (Walker et al. 2004). Adaptive governance relates strongly to adaptive management (Walters 1986), which has widely been promoted as a necessary basis for sustainable development, but has frequently failed because the existing governance structures have not allowed it to function effectively (Walters 1997).

Water Governance More Specifically Defined

A critical review of ten years of the Stockholm Water Symposia has identified the problem within the global water sector as being located outside of the actual provision of water (SIWI 2000). Stated differently, this suggested that the problem was not about the engineering aspects of water resource management, but that something else was. This hinted at the notion of governance, but did not use the word as such.

The GWP, citing the United Nations Development Program (UNDP) as the source, published a book a few years later, in which governance as a general concept was defined as:

"the exercise of economic, political and administrative authority to manage a country's affairs at all levels…comprises the mechanisms, processes and institutions through which citizens and groups articulate their interests, exercise their legal rights, meet their obligations and mediate their differences" (Rogers and Hall 2003:7).

The GWP has also noted that water governance, as a specific "species" of what is generically called governance, can be defined as;

"*the range of political, social, economic and administrative systems that are in place to develop and manage water resources, and the delivery of water services, at different levels in society*" (Rogers and Hall 2003:7).

When applying this to the notion of IWRM, governance is clearly a fundamental part of the overall process, specifically when emphasis is placed on the "integrated" part of the concept. This is evident from the definition of IWRM, seen to be:

"*a process which promotes the co-ordinated development and management of water, land and related resources in order to maximize the resultant economic and social welfare in an equitable manner without compromising the sustainability of vital ecosystems*" (Jønch-Clausen 2004:14).

Gilman et al. (2004) define water resource management as any effort to plan or control the human use of freshwater ecosystems, or the provision of services by those ecosystems; or any deliberate human activity that temporarily or permanently alters the biological or hydrological function of a freshwater ecosystem. Integrated River Basin Management (IRBM) is widely advocated as the most appropriate tool for managing water resources. IRBM is defined as the process of coordinating conservation, management and development of water, land and related resources across sectors within a given river basin, in order to maximise the economic and social benefits derived from water resources in an equitable manner while preserving and, where necessary, restoring freshwater ecosystems (Jones et al. 2003).

With respect to governance in general, Hattingh et al. (2004) initially defined it as "the participation of civil society in *testing, monitoring, auditing and providing feedback* on government tools." In essence, interest groups, corporations and NGOs at international, national and local levels test the fiduciary trust under which those in authority exercise that responsibility. Seen in this light, *governance* is a *process* that supports the *legitimacy* of government, by holding elected officials accountable for the aggregation of interests articulated by interest groups in society at large (Hattingh et al. 2004). This specifically refers to the auditing and monitoring functions within the governance process, and the critical link between the needs of civil society and government.

These definitions seem to deal with different *elements of scale,* but significantly, do not mention all of the key *structural aspects* associated with governance in general, the distinct *process aspects*, or the *norms and value* that order and structure relationships. Consequently, this leaves water resource professionals in somewhat of a dilemma, because they have little clear guidance as to how the various *structural aspects* of governance should best be arranged, in order to meet the complex series of demands arising from the four distinct *process aspects*, or how the *norms and values* are to be arranged in a hierarchy that can best inform and support the decision-making process.

Redefining the Problem

Reflexivity can be regarded as the condition that arises when a society becomes concerned by the unintended consequences of their historic developmental trajectories, and actively seeks to do something about altering the outcome in a more environmentally-sustainable way (Giddens 1990). Within the context of IWRM, this is being projected, and sometimes even forced onto the aid-dependent developing countries. The resulting tension lies at the very heart of the governance *problematique* under review, specifically as being experienced by fledgling democracies that have emerged from periods of political instability and low economic growth. It is therefore argued that this is one of the reasons why IWRM strategies driven by countries with mature democracies are seldom truly effective when they are "transplanted", unchanged, into countries with fledgling democracies.

Governance, in fledgling democracies, will need to differ from that in developed democracies, because the norms and values within fledgling democracies have not yet been fully legally codified or bureaucratically institutionalised. Consequently, some degree of ingenuity will be needed to overcome administrative bottlenecks or bureaucratic inertia. Furthermore, little is said in any of the different definitions about the specific norms and values that serve to inform the key management processes, which are:

- Coordination between water and land in the broadest sense;
- The maximisation of economic and social welfare;
- The notion of equity or feelings of fairness;
- Ecosystem thresholds, which are vital if environmental quality objectives are to be set, as broad parameters for sustainable development; and
- Informed decision-making.

Central to supporting these key management processes are at least four critical elements:

- What norms and values form the foundation of the core interests of specific stakeholders?
- What norms and values are used as the basis of the legal system?
- What norms and values are used to base decisions on?
- What norms and values are applicable in the mediation of disputes?

The core problem to be addressed can thus be stated as follows:

- In terms of *governance as process*, what are the key structural aspects, process aspects, norms and values that comprise governance as applicable to what we know as IWRM?

- In terms of *governance as product*, how are these structural aspects and process aspects configured when the output of the governance process can be called "good" or "effective"?
- What *norms and values* provide the qualitative aspects of governance as product?
- How are *elements of scale* best taken care of if the desired outcome of governance as product is said to be "good" or effective?
- Do these *structural aspects* and *process aspects* evolve in all countries in the same way, or are there a variety of evolutionary paths?

By synthesising key elements of these various definitions, it is now possible to generate a new, more encompassing definition of governance. To this end, *governance* can now be redefined as being: *the process of informed decision-making that enables trade-offs between competing users of a given resource so as to balance protection with beneficial use in such a way as to mitigate conflict, enhance equity, ensure sustainability and hold officials accountable.* Seen in this way, *governance as process* involves a number of distinct elements, including decision-making about potentially contestable outcomes, while *governance as product* can be seen as the quality of those outcomes, specifically with respect to the legitimacy of trade-offs and the level to which these are contested or accepted by society at large.

A New Perspective: The Trialogue[1] Model of Governance

In an attempt to introduce more scientific rigour into the development of the processes that normally occur in the black box of governance, it is necessary to present a series of hypotheses. These have been used to develop an understanding of the key relationships between variables within the black box of governance, and between that black box and the broader socio-economic environment in which it operates. These will be revisited in the concluding chapter of this book to assess their validity.

[1] Professor Malin Falkenmark is acknowledged as the first person to use the term Trialogue to describe the interaction between these thee actor-clusters. This occurred during a conversation with one of the authors in Stockholm during the 2004 Stockholm Water Symposium.

These hypotheses are:

1. Governance is a *process*, with effective ecosystem (or IWRM) management being the *product* of "good" governance.
2. There are *three clusters of actors* in the context of governance as process that can generically be defined as government, society and science.
3. Government, society and science represent three different *communities of practice* that are complementary only to the degree that they interface with one another.
4. Governance as process involves decision-making, but given the inherent complexity of the real world (specifically with respect to natural resource management), decisions tend to be made against a background of different sets of *norms and values*, which vary within different stages of economic development and political evolution.
5. In the context of *governance as product*, good governance occurs when the interfaces between the three clusters of actors – what can be called a Trialogue – are effective, as this allows for appropriate feedback loops and exchange of information with which to inform the decision-making process. This is also reflected in sound ecosystem governance.
6. Governance is dynamic in nature and is enmeshed in the social, economic, biophysical and political landscape in which it occurs.
7. Good governance is more likely to occur where there is a prevailing political culture of democracy.
8. Given the dynamic nature of governance, the evolution of what can be described as good governance occurs over time, with potentially different trajectories or pathways being possible in what can generically be described as "mature" democracies and "fledgling" democracies.
9. Governance is not the same as government, nor is it solely the purview of government authorities.

From the nine hypotheses noted above, we consider the three actor-clusters to be extremely important in terms of the governance model that is proposed. The notion of a cluster is used because it best captures the observation that in the real world none of these groups of actors is monolithic in either form or function.

Government Actor-Cluster

The first actor-cluster in the Trialogue is what we can generically call government. In reality, this cluster consists of three distinct elements, none of which are monolithic in their own right, but which are sufficiently coherent to be treated separately:

- Rule-making that resides in the legislative branch of government;
- Rule-application that occurs in the executive branch of government; and
- Rule-adjudication that is performed by the judicial branch of government.

The government actor-cluster is similar in most modern states, with the three arms of government being reflected in different forms, but always in existence. Technically, in political science literature this is known as the *trias politicas*, reflecting the three core elements of the sovereign authority of a country. The relationship between these three arms of government is normally defined in the constitution of the country concerned (Hattingh et al. 2005a; 2005b).

The government actor-cluster has the capacity to direct and enable the science actor-cluster to develop solutions to increasingly complex problems. In a democratic system, government should ideally base policy formulation on the articulated needs of society. It is therefore the role of the government actor-cluster to provide an enabling environment and institutions that maximise socio-economic development through appropriate scientific and social inputs (Hattingh et al. 2005a; 2005b).

In this regard there are two major schools of thought (Gleditsch 2003; Brauch 2005:13). The Malthusian school views social collapse as being inevitable, as the result of population growth, which is seen inevitably to outstrip the capacity of society to adapt (Hardin 1968; Clark 1977; Homer-Dixon 1995; 1996). This argument is sometimes reflected in the explanation of the brain drain phenomenon from the fledgling democracy countries to the mature democracy countries. The Cornucopian school essentially sees human ingenuity as being boundless, arguing that any problem that arises can be solved. Both of these schools are reflected in the sustainable development discourse (Wolf 1999, Wolf et al. 2003).

Society Actor-Cluster

The second actor-cluster, which can be generically called society, is constituted of three major sub-elements (Hattingh et al. 2005a):

- Civil society that represents people, who collectively have interests that they seek to articulate in one form or another. In this context, civil society includes non-governmental organisations, industry, community based organisations and individuals, all of whom can potentially be organised in some form;
- The economy that employs these people has interests of its own, many of which are articulated through both formal and informal channels; and

- The environment in which society and the economy is embedded.

The society actor-cluster includes the environment because, in developing countries, there is a more intimate link between livelihood-flows from direct access to natural resources, than in the developed countries (Allan 2000:314). This is not exclusive to the developing countries however, because the global environmental movement, which successfully placed ecological concerns onto the international agenda, came from reflexive elements inside developed democracies. This has been successful to the extent that triple bottom line accounting is now an internationally accepted form of corporate governance applicable to large multi-national corporations. For this reason, the sustainable development discourse is firmly embedded in this cluster, consisting of a dynamic interaction between society, the economy and the environment, each of which have now become separate indicators of sustainability.

Science Actor-Cluster

The third actor-cluster can generically be called science, which is defined as "the organised, systematic enterprise that gathers knowledge about the world and condenses the knowledge into testable laws and principles" (Wilson 1998). Reductionism – the breaking apart of nature into its natural constituents – has long been the primary and essential activity of science. However, scientists do not only engage in dissection and analysis. Also crucial are the activities of synthesis and integration, which need tempering by philosophical reflection on significance and value. This is particularly evident in the emerging field of what is being called Sustainability Science.

For the purposes of this chapter, science can be thought of as consisting of three distinct sub-components:

- *Fundamental or basic research,* also termed *Classic, Mode 1 or Type A research.* Gibbons et al. (1994) refer to scientific investigations that primarily serve the advancement of understanding, rather than the solving of specific problems. Knowledge in this domain is highly systematised and organised along disciplinary lines. Quality control is dominated by intensive codification and peer review. Some refer to the outcome of this kind of research as generally new knowledge, research papers, post graduate degrees and science and technology (S&T) platforms (OECD 1994).
- *Applied, Mode 2* or *Type B research* is trans-disciplinary, heterogeneous and directed at solving practical problems experienced by society.

Knowledge-creation is driven by its perceived usefulness and is highly contextual, but not necessarily any less original. Successful application requires a development and design phase, in which the knowledge is specifically packaged to address the needs of potential adopters. Some refer to the outcome of this kind of research as generally new products, designs, patents, solutions, methodologies, software packages, research papers and post graduate degrees (OECD 1994).

- *Type C research* relates to technology transfer and specialist services (knowledge application, leading to new enterprises, product lines, consultancy reports and feasibility studies). The outputs from Type C activities are generally pilot project results, training materials, software packages, investigative reports, feasibility study reports, specialist interpretation and advice, and policy studies.

These sub-components apply to the three larger science actor-clusters, namely: natural sciences, life sciences and social sciences, and all of the underpinning science fields or disciplines within these clusters such as mathematics and philosophy. The science actor-cluster has the explicit responsibility of addressing the needs of society and informing government of technical solutions, thus implementing the results of the three sub-components and integrating across the different scientific fields. It is the structure and function of the science actor-cluster that forms the core logic in Homer-Dixon's argument about the role of ingenuity in sustaining society in a socially and politically stable manner (Homer-Dixon 1994; 1995; 1996; 1999; 2000).

Nested Trialogues and Governance

Closer examination of these three actor-clusters shows that while they are all structured in some way with one another, in a configuration that can best be called a Trialogue, each cluster is a form of triangle in its own right (Hattingh et al 2005a), as illustrated in Figure 1.2.

This raises an important conceptual element within the model of governance that is being developed here. If these three actor-clusters are all individually important within a given society, then they are all locked into some form of relationship with one another at more than one level and in more than one place (Hattingh et al. 2005a). That relationship is based on communication and feedback loops, so the effectiveness of the relationship is clearly a function of the interface between each of these three actor-clusters. In this regard, three important interfaces can be identified:

- The Government-Society interface determines the needs and requirements of society, the legitimacy of the political process, and the permeability of government to new ideas from civil society and the corporate world. The interface also represents the degree to which the needs of society are satisfied by government;
- The Government-science interface determines the extent to which science and technology form the basis of the political economy, and the extent to which scientific knowledge informs the decision-making processes that are a core function and output of the government actor-cluster. Government facilitates and enables the scientific process through policy initiatives, resource allocations and overall strategic direction. This interface is critical as it has major implications for social stability and economic growth, making it a key issue for effective governance in the developing countries with fledgling democracies; and
- The Science-Society interface can be thought of as science in the service of society, consisting of a number of elements, including the way that scientific knowledge is diffused into society. In a developed country with a mature democracy this is visible as the technology-base of the economy, eventually manifesting as comparative advantage in the global economy. In developing countries with fledgling democracies, this is reflected in the effectiveness with which the science and technology-base is harmonised with the overall needs of society, and becomes a key determining factor in the success of the emerging economy as it overcomes historic and structural comparative disadvantages.

Fig. 1.2. *The three main actor-clusters constitute a form of triangle in their own right (revised from Hattingh et al. 2005a)*

These three actor-clusters interact dynamically with one another, each being connected to the others via a two-way interface. It is reasonable to conclude that the quality of these interfaces is a key determining factor, to the extent that any serious study of governance is likely to benefit by treating these as independent variables in their own right. This is a core assertion of the Trialogue Model of Governance, which is presented schematically in Figure 1.3.

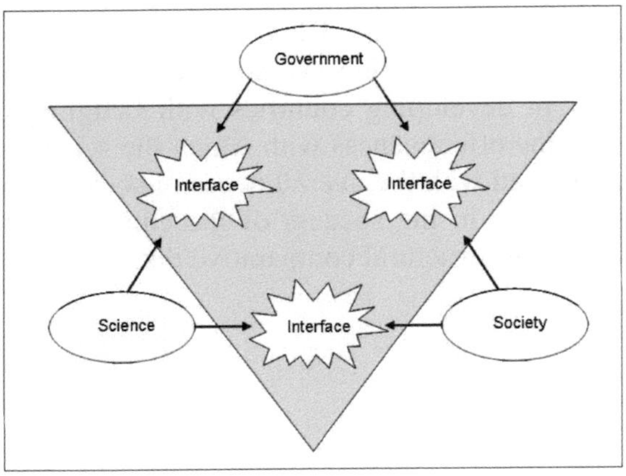

Fig. 1.3 *Opening up the black box of governance: a schematic representation of the Trialogue Model (adapted from Hattingh et al. 2005a). The shaded area represents the place where governance as process occurs*

Figure 1.3 shows the triangular relationship between the three actor-clusters along with their respective interfaces. The shaded area in the diagram spans the three interfaces, and it is this area in which *governance as process* functions. The quality of these three interfaces determines the extent to which government can generate the incentives needed to develop society by allowing science to inform the decision-making process. It can therefore be said that the quality of these three interfaces is an independent variable in determining the outcome of *governance as product.*

The Governance Interface

From the foregoing it can be seen that, in order to establish *governance as process,* all three of the actor-clusters have to engage one another via a series of interfaces. It is thus proposed that the management systems should also reflect the move from bi-lateral interfaces to the Trialogue, where all three actor-clusters meet.

An example of that three-way interface can be found in the way the new water policy in South Africa was introduced, when science, society and government engaged with equal vigour (DWAF 1997). The common purpose was motivated by an eloquently simple slogan, "*some for all, forever, together*", which embraced the principles of a finite resource, the need for equitable access to it, and for sustainable development. Ecologists, social scientists, lawyers, politicians and water resource managers could all align their expertise with this purpose. Similarly, the process for determining and implementing an ecological water reserve for aquatic ecosystems provided a common forum for researchers, planners and managers. These processes were feasible, and successful, only because all parties participated as partners in the adaptive development and implementation. This engagement was not without pain, but it did lead to invaluable learning and shared understanding (Postel and Richter 2003).

Applicability of the Trialogue Model to IWRM

To assess the applicability of the proposed Trialogue for effective ecosystem governance, it is instructive critically to examine the IWRM paradigm, as represented within the mainstream water resource management discourse. Arguably the best example of this is found in the work by Jønch-Clausen (2004), as this has been published by the GWP, as the custodian of that global discourse. The three pillars of IWRM presented in Figure 1.4 are taken directly from the mainstream GWP literature on the topic (Jønch-Clausen 2004:16).

The blocks with rounded corners to the right of the diagram presents the IWRM approach as standing on three pillars – Management Instruments, Enabling Environments and the Institutional Framework – with inputs provided by the three core values of economic efficiency, equity and environmental sustainability. The outcome of that process is seen to be a balance between water for livelihoods and water as a resource.

The square blocks to the left of the diagram show the clusters of issues that were isolated earlier in this chapter. Economic efficiency, equity and environmental sustainability can be thought of as representing *norms and values* because their quantification in a real sense is highly value-laden. Economic efficiency, as an example, is only something that is realistically possible once a high assurance of supply has been attained in a given society. This lies at the heart of the hydraulic mission of a country, with most development trajectories starting with relatively inefficient agriculture, and evolving over time to a mining-based and eventually an industrialised economy. Each of these evolutions improves efficiency in one sense by moving water away from the inefficient agricultural sector, thereby unlocking the gearing advantages inherent to inter-sectoral water allocative efficiency.

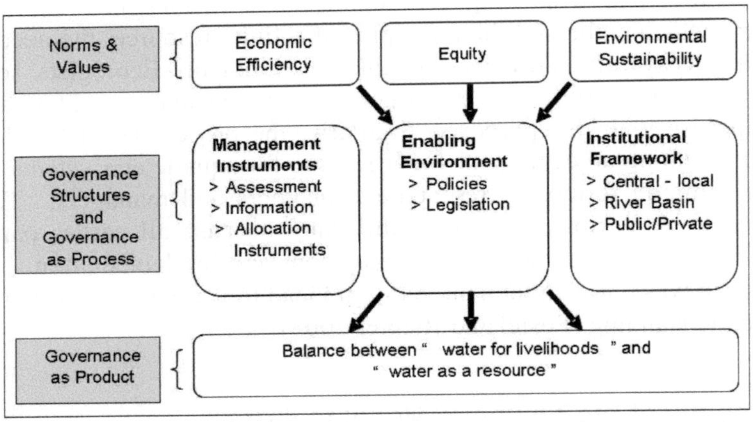

Fig. 1.4 *The three pillars of Integrated Water Resource Management, as depicted by the Global Water Partnership (redrawn from Jønch-Clausen 2004:16) showing linkages to the governance model developed in this chapter*

The same holds true with equity. It is only once social stability has been developed to the point of having core norms and values codified in the legal system that equity becomes a realistic objective.

Environmental sustainability is based on similar norms, because it can be argued that the developed (mature democracy) countries became developed by initially practicing unsustainable strategies, but once their development had resulted in a high degree of wealth accumulation in society, sustainable practices could be considered. It is hard to tell a peasant farmer that their practices are unsustainable, when they cannot get a fair price for their produce due to subsidies in the agricultural sector of the developed, mature democratic countries, and their families are in poor health with limited opportunities for breaking the poverty cycle.

The IWRM model presented in Figure 1.4 can also be understood in terms of *governance structures* by focussing on the institutional framework. *Governance processes* can be thought of as management instruments and the enabling environment. Achieving a balance between water for livelihoods and water as a resource seems a bit simplistic in the real world, but using this as an example, *governance as product* can be thought of as the degree of contestation of that balance (irrespective of what is being traded off to achieve a balance). Imbalance between these elements would produce a high level of social tension and would be regarded as poor governance. Whereas a more equitable balance would result in less social disruption and better prospects for sustained economic growth and social well-being of a nation. The IWRM model that was developed by the GWP (Jønch-Clausen 2004:16) is thus a useful one, but insufficient as it now stands if governance is to be added to the equation.

Lying at the very heart of governance is a form of unwritten contract between society and the government, which can be thought of as a hydro-social contract (Warner and Turton 2000; Turton and Meissner 2002). This hydro-social contract incorporates the norms and values of society that structure the relationships between key stakeholders. Once we speak of structuring relationships, we immediately need to focus on interfaces, because it is the quality of the interface that determines the nature and long-term viability of the relationship. Roux et al. (2006) refer to knowledge interfaces as the spaces where mutual understanding is developed. Interfaces are places of dialogue that facilitate the co-evolution of values, priorities, intent and action, which in turn provide robustness to decision-making. When scientists, resource policy-makers, managers and other stakeholders engage at the knowledge interface, they become a unified learning system in which new and shared experiences lead to new knowledge. In a functional interface the parties move beyond the traditional roles of knowledge provider and knowledge consumer, to that of partners that negotiate what is feasible, desirable and acceptable (Roux et al. 2006).

Structure of this Book

This book has been structured around different elements of the Trialogue Model which has been presented in this chapter. The book consists of five main parts. Part 1 sets the scene by presenting the Trialogue Model of governance. Part 2 deals with an overview of governance in general in an attempt to establish linkages where appropriate to external factors such as the drive towards sustainable development. Part 3 focuses specifically on the Trialogue as a concept by interrogating selected elements of the proposed model from the perspective of different case studies, each of which have been carefully selected to provide a spread of possible lessons to be learned. Part 4 teases out cross-cutting issues such as learning processes, communication and institutional dynamics of governance. Part 5 ties these together by reflecting on the book as a whole, suggesting areas of possible future research by revisiting the hypotheses presented in this chapter.

Part 2 provides an overview of governance and consists of Chapters 2–6. Alan Hall provides a global context for considering the ecosystem governance issues that are presented in the rest of the book. It looks at what governance is, which forms it takes and why more effective governance is important. It also deals with some of the fundamental principles that underpin governance, suggesting that this is part of the larger process of globalisation and the overall discourse on sustainable development. Alex Simalabwi shows that issues of scale are important. He focuses on power asymmetries, concluding that where governments have the power to legislate; our stakeholders have the power to legitimise. The need to balance ecosystems with global needs is dealt with in Chapter 4, where Malin Falkenmark shows that "good" governance consists of two distinct elements – what is being governed (human activities in the landscape) and how that process is to occur via a series of consecutive steps. This shows that a sound understanding is needed of the biophysical, social and governance systems, further suggesting that the Trialogue Model could be improved by replacing the corner labelled "science" with a box called "biophysical processes" instead. In Chapter 5 Peter Ashton shows that, while African countries are often tempted by the apparent advantages of the sustainable development concept, this is generally difficult to achieve in reality, due to shortages in social, technical and economic resources, generally known as second-order resource scarcities. He shows that the basic water management approaches in Southern Africa are well-aligned with general principles of good governance, though there is considerable variation in the degree of success achieved. He highlights the complex and multidimensional nature of integrated water resource management and good governance, both of which are crucial to the realisation of sustainable development. In Chapter 6 Ken Conca looks at international water law as a way of channelling interstate water disputes into potentially peaceful processes of bargaining, arbitration and conflict resolution. He notes that transnational social networks will also be crucial in the achievement of the twin aims of promoting effective ecosystem governance and stabilising the socio-ecological controversies that rage around these systems. This complex pattern of institutionalisation suggests both a validation and refinement of several of the core hypotheses linked to the Trialogue Model.

Part 3, consisting of Chapters 7–12, homes-in on the Trialogue Model of governance. In Chapter 7 Geoffrey Gooch focuses on the role of both formal and informal institutions in ecosystem governance, along with the roles of different forms of knowledge and civil society as key elements of the process. He then examines various organisational structures, their aims, norms and values, as well as the problems of cooperation arising from different institutional cultures. He suggests that there are a series of Trialogues and he discusses their relevance to ecosystem governance. In Chapter 8 Raya Stephan presents an assessment of the evolution from the traditional approaches of groundwater management in the context of international water law, to the latest tendencies in the International Law Commission and environmental treaties. This evolution shows the increased awareness of the need to understand groundwater, to integrate scientific knowledge and to establish adequate rules for its sustainable development. The role of law is highlighted in the context of the Trialogue Model. In Chapter 9 Michael Campana, Alyssa Neir and Geoffrey Kilse provide insight into governance processes between the USA, Mexico and Canada, with respect to the transboundary groundwater aquifers that are shared by those three countries. Eight specific case studies are presented to give a contextual assessment of considerable detail.

In Chapter 10 Nyambe Nyambe, Charles Breen and Robert Fincham state their view that public service agencies involved in managing the use of ecosystems have strong organisational cultures, partly because of their origins, which are rooted in certain ethical and moral precepts. They suggest that, by examining assumptions that lie at the core of organisational cultures, we can understand adaptation, responsiveness and management for change. They suggest that this plays an important role in harmonising the goals of society, science and government. In Chapter 11 Sandra Fowkes pays specific attention to the role of society in the processes of governance. Her work in fire management in the Cape Town area is particularly instructive, because it introduces a completely new dimension to ecosystem governance in the context of a fledgling democracy. She makes a point that science, or knowledge, may play a far more powerful role in shaping the governance decision space by making itself available to both government and society. In Chapter 12 Barbara Schreiner unpacks the notion of governance in a fledgling democracy by focusing on contemporary South Africa within a rapidly changing set of social phenomena. Given her senior management position within the Department of Water Affairs and Forestry of South Africa, and being responsible for policy-related aspects of the implementation of water to deepen democracy, her insights are particularly useful.

Part 4, comprising Chapters 13-15, develops more understanding of the cross-cutting themes and issues inherent to governance. In Chapter 13 Dirk Roux, Kevin Murray and Ernita van Wyk suggest that adaptation and the improvement of management practices are dependent on our capacity to learn, or to renew, our knowledge. The bridging of organisational, disciplinary, cultural and functional boundaries is central to promoting co-learning within the social system of the water institution. They propose nine principles that can guide organisations in developing good learning environments and practices. In Chapter 14 Wilma Strydom, Liesl Hill and Estee Eloff describe the role of communication between the different components of the Trialogue Model of governance. The quality of the various interfaces is assessed, using a South African case study known as the River Health Programme State of Rivers project. In Chapter 15 Jane Doolan gives an Australian example by focusing on the Victoria River Health Programme's evolution over the last 15 years. The Trialogue Model is supported in this chapter, but the author notes that it is crucial to include an addition dimension – time – which is not factored into the current thinking.

Part 5, consisting of Chapters 16–17, reflects back on the contents of the whole book. In Chapter 16 Linda Godfrey suggests that, if sustainable development is at the heart of good ecosystem governance, then a solid and productive relationship between government and society is the foundation of a successful outcome. The Trialogue Model is seen to be both complex and dynamic, being influenced by various factors such as the political system, the state of evolution of the democratic processes within that system, and the corporate culture within key government ministries. In Chapter 17 Anthony Turton and Hanlie Hattingh wrap the whole book up by revisiting the hypotheses presented in Chapter 1 against the background of what the various authors have presented in Chapters 2–16. This is done in the sincere hope that a concerted attempt by various research entities around the world, channelling their energies via a focussed research agenda, will yield additional insight into that complex concept known as governance.

Conclusion

This chapter has shown that the so-called "black box" of governance can be conceptually understood in terms of specific elements. The first of these are *elements of scale*, of which there are four: economic, political, administrative and international. The interaction of these is mediated by three key *structural aspects*: mechanisms, processes and institutions. Within that structural arrangement are four critical *process aspects*: interest articulation, the exercise of legal rights, the discharge of legal obligations and the mediation of potential disputes. Underlying all of this are the *norms and values* of a given society, some of which are codified as law, but many of which exist merely as perceptions in society.

A clear distinction can be made between *governance as product*, with qualitative aspects such as 'effective' or 'sustainability' being examples, and *governance as process*, with procedural and institutional arrangements being examples. Given the fundamental structure of the three major actor-clusters, and their dependence on communication with one another, their configuration can be described as a Trialogue. Within this, the effectiveness of governance becomes highly dependent on the quality of the interfaces and how well all three actors meet.

Finally, there is a natural tension that exists within the discourse on sustainable development that is translated into the discourse on IWRM, because the drivers of that discourse have a reflexive world view that is informed by the experiences of the developed countries with mature democracies. The needs of developing countries with fledgling democracies are largely based on infrastructural development with a distinct first-order focus, but the complexity associated with this is compounded by the fact that a second-order focus is also needed in the form of institutional development. It is the *problematique* arising from the tension between these two poles that is central to the GWP's view that the crisis in the water sector is a crisis of governance.

References

Allan JA (2000) The Middle East water question: hydropolitics and the global economy. London, IB Tauris

Biswas AK (1993) Management of international water: problems and perspective. Paris, UNESCO

Brauch HG (2005) Threats, challenges, vulnerabilities and risks in environmental human security. SOURCE Paper 1/2005, Bonn, United Nations University Press

Clark CW (1977) The economics of over-exploitation. In: Hardin G, Baden J (eds) Managing the commons. San Francisco, Freeman, pp 82–95
Cosgrove WJ , Rijsberman FR (eds) (2000) World water vision: making water everybody's business. London, Earthscan
DWAF (1997) White paper on a national water policy for South Africa. Department of Water Affairs and Forestry, Pretoria, South Africa
Falkenmark M (1989) The massive water scarcity now threatening Africa: why isn't it being addressed? Ambio **18**:112–118
Falkenmark M (1991) Rapid population growth and water scarcity: the predicament of tomorrow's Africa. In: Davis K, Bernstam M (eds), Resources, environment and population, Supplement to Population and Development Review **16**, New York, New York, The Population Council and Oxford University Press, pp 81–94
Gibbons M, Limoges C, Nowotny H, Schwartzman S, Scott P,Trow M (1994) The new production of knowledge: the dynamics of science and research in contemporary societies. London, Sage Publications
Giddens A (1990) The consequences of modernity. London, Polity Press
Gilman RT, Abell RA, Williams CE (2004) How can conservation biology inform the practice of integrated river basin management? International Journal of River Basin Management (2)2:1–14
Gleditsch NP (2003) Environmental conflict: Neomalthusians vs Cornucopians. In: Brauch HG, Liotta PH, Marquina A, Rogers P, Selim M (eds) Security and environment in the Mediterranean: conceptualizing security and environmental conflicts. Berlin, Springer-Verlag, pp 477-486
Gleick PH (1999) The human right to water. Water Policy **1**:487–503
GWP (2000) Towards water security: a framework for action. The Hague, Global Water Partnership
Hardin G (1968) The tragedy of the commons. Science 162:1243–48. Reprinted in: Shrader-Frechette KS (ed) (1981) Environmental Ethics. Pacific Grove, Ca, Boxwood, pp 242–52
Hattingh J, Kgomongoe M, Leaner J, Maree G, Strydom W, Turton AR, Van Wyk E (2004) Obstacles to successful implementation of government tools. Report ENV-P-1 2004-017. Pretoria, CSIR
Hattingh J, Claassen M, Leaner J, Maree G, Strydom W, Turton AR, Van Wyk E, Moolman M (2005a) Towards innovative ways to ensure successful implementation of government tools. Report ENV-P-I 2004-083. Pretoria, CSIR
Hattingh J, Claassen M, Viljoen P, Van Wyk J, Murray K, Roux DJ (2005b) Implementing resource directed water quality management: from policy to management instruments to operational adoption. Draft paper submitted for the 2005 SANCIAHS Conference, South Africa
Homer-Dixon TF (1994) Environmental scarcities and violent conflict: evidence from cases. International Security (19)1:5–40
Homer-Dixon TF (1995) The ingenuity gap: can poor countries adapt to resource scarcity? Population and Development (21)3:587–612
Homer-Dixon TF (1996) Environmental scarcity, mass violence and the limits to ingenuity. Current History **95**:359–365
Homer-Dixon TF (1999) Environment, scarcity and violence. Princeton, NJ, Princeton University Press
Homer-Dixon TF (2000) The ingenuity gap. London, Jonathan Cape

ICWE (1992a) International conferences on water and the environment: development issues for the 21st century. 26–31 January 1992, Dublin

ICWE (1992b) The Dublin statement on water and sustainable development. International conference on water and environment. Dublin, Ireland

Imperial MT (2005) Collaboration and performance management in network settings: lessons from three watershed governance efforts. Washington DC, IBM Center for the Business of Government

Jønch-Clausen T (2004) Integrated water resources management (IWRM) and water efficiency plans by 2005: why, what and how? TEC Background Papers 10. Stockholm, Global Water Partnership

Jones T, Phillips B, Williams C, Pittock J, Davis T (2003) Conserving rivers: lessons from WWF's work for integrated river basin management. World Wildlife Fund, Gland, Switzerland

Organisation for Economic Co-operation and Development (OECD) (1994) Main definitions and conventions for the measurement of research and experimental development (R&D): a summary of the Frascati manual 1993. Paris, France, OECD

Ohlsson L (1999) Environment, scarcity and conflict: a study of Malthusian concerns. Department of Peace and Development Research. Göteborg, Sweden, University of Göteborg.

Ohlsson L, Turton AR (1999) The turning of a screw. Paper presented: Plenary Session of the 9th Stockholm Water Symposium "Urban stability through integrated water-related management", 9–12 August by the Stockholm International Water Institute (SIWI). Sweden. Also available as MEWREW Occasional Paper 19, Available online at http://www.soas.ac.uk/Geography/Water-Issues/OccasionalPapers/home.html

Pallett J (1997) Sharing water in Southern Africa. Windhoek, Desert Research Foundation of Namibia

Pegram G, Mazibuko G, Von der Heyden C, Anderson A, Hollingworth, BH (2005) Strategic review of current and emerging governance systems related to water in the environment in South Africa. Final draft report, (WRC Project K5/1514)

Reisner M (1993) Cadillac desert: the American west and its disappearing water. Rev edn. New York, Penguin

Rogers P, Hall AW (2003) Effective water governance. TEC Background Papers 7. Stockholm, Global Water Partnership

Rosegrant M (1997) Water resources in the twenty-first century: challenges and implications for action. Food, Agriculture, and the Environment Discussion Paper 20. Washington DC, International Food Policy Research Institute

Roux DJ, Rogers KH, Biggs HC, Ashton PJ, Sergeant A (2006) Bridging the science-management divide: moving from unidirectional knowledge transfer to knowledge interfacing and sharing. Ecology and Society (11)1:4. Available online at http://www.ecologyandsociety.org/vol11/iss1/art4/

SIWI (2000) No freshwater security without major shift in thinking: ten year message from the Stockholm water symposia. Stockholm, Stockholm International Water Institute (SIWI)

Turton AR (1999) Water scarcity and social adaptive capacity: towards an understanding of the social dynamics of managing water scarcity in developing countries. MEWREW Occasional Paper 9. SOAS water issues study group. Also in: Proceedings of the conference "Sustainability, risk and nature: the political ecology of water in advanced societies", held at Oxford University on 15–17 April 1999. Available online at http://www.soas.ac.uk/-Geography/WaterIssues/OccasionalPapers/home.html

Turton AR (2002) WDM as a concept and a policy: towards the development of a set of guidelines for Southern Africa. Commissioned analytical paper for the IUCN Water Demand Management Programme for Southern Africa, phase II, Pretoria, IUCN. Available online at http://www.up.ac.za/academic/libarts/polsci/awiru.

Turton, AR, Meissner R (2000) Analysis of the influence from the international hydropolitical environment, driving the process of change in the water sector (Appendix F). In: Report of the institutional support task team – shared rivers initiative of the Incomati river. Scottsville and Harare, Institute of Natural Resources (INR) and Swedish International Development Agency (SIDA). Available online at http://www.up.ac.za/academic/libarts/polsci/awiru

Turton AR, Meissner R (2002) The hydro-social contract and its manifestation in society: a South African case study. In: Turton AR, Henwood R (eds) Hydropolitics in the developing world: a Southern African perspective. Pretoria, African Water Issues Research Unit (AWIRU), pp 37–60

Turton AR, Meissner R, Mampane PM, Seremo O (2004) A hydropolitical history of South Africa's international river basins. Pretoria, Water Research Commission

UNCED (1992) United Nations conference on environment and development. Agenda 21, Chapter 18, Freshwater, United Nations Publications

Vermillion G, Merrey D (1998) What the 21st century will demand of water management institutions. Journal of Applied Irrigation Science **33**:165–187

Walker B, Holling CS, Carpenter SR and Kinzig A (2004) Resilience, adaptability and transformability in social-ecological systems. Ecology and Society (9)2:5. Available at http://www.ecologyandsociety.org/vol9/iss2/art5

Walters CJ (1986) Adaptive management of renewable resources. Collier Macmillan, New York, New York, USA

Walters CJ (1997) Challenges in adaptive management of riparian and coastal ecosystems. Conservation Ecology (1)2:1. Available online at http://www.consecol.org/vol1/iss2/art1

Warner J, Turton AR (2000) Integrated management requires an integrated society: towards a new hydrosocial contract for the 21st century. AWIRU Occasional Paper 12. African Water Issues Research Unit, Pretoria University. Available online at http://www.up.ac.za/academic/libarts/polsci/awiru

Waterbury J (1979) Hydropolitics of the Nile Valley. New York, Syracuse University Press

Wilson EO (1998) Consilience – the unity of knowledge. Vintage Books, New York

Williamson OE (1988) Corporate finance and corporate governance. Journal of Finance (43)3:567–591

Wolf AT (1999) Water wars and water reality: conflict and cooperation along international waterways. In: Lonergan S (ed) Environmental change, adaptation, and human security. Dordrecht, Kluwer Academic, pp 251–265

Wolf AT, Yoffe SB, Giordano M (2003) International waters: identifying basins at risk. Water Policy (5)1:29–60

WWC (2000) Final report: second world water forum and ministerial conference. The Hague, Ministry of Foreign Affairs

Part 1

An Overview of Governance

Global Experience on Governance

Alan W Hall

Abstract

This introductory chapter provides a broad context for considering the ecosystem governance issues discussed in subsequent chapters. In an increasingly interconnected world, nations cannot alone solve many of the pressing social and environmental challenges. Establishing effective governance systems is gradually becoming recognised as fundamental to sustainable development although it remains a mystery to many. To provide some clarity to this esoteric subject this chapter examines what governance is, the various forms it can take, why more effective global governance is important, and some basic principles that underpin it.

Keywords: environmental governance, global governance, globalisation, Global Water Partnership (GWP), United Nations (UN), sustainable development,

An Interconnected World

We all live in an interconnected world. This manifests itself in many ways. Soon after hurricane Katrina had destroyed New Orleans, a Financial Times headline read: "Before deciding whether to rebuild New Orleans, President Bush should ask whether Premier Hu Jintao likes jazz." This surreal headline relates to the connection between the huge USA budget deficit that is partly financed from Chinese Central Bank surpluses. Increasing the budget deficit to rebuild New Orleans could strain the intricate global financial systems. Another disaster, the tsunami in Asia in December 2004, solicited the biggest-ever response for aid. Globalisation is about much more than trade or the Internet.

Globalisation is a long-term process that transforms the spatial organisation of social relationships. Driven mainly by transcontinental or interregional interactions and exercise of power, globalisation is not a new phenomenon. Throughout history, military conquest and territorial expansion have been driving forces behind globalisation. Today, economic trade has become the primary driver, and the ease with which trading partners can communicate, and the speed with which goods and services can be moved or provided globally, has led to truly dramatic increases in the pace of globalisation.

This growing interconnectedness gives rise to the need for effective governance at a global scale. That is, the need to harmonise individual and group actions both within societies and, increasingly, between different societies.

Understanding Governance

There is no single, all-encompassing definition of governance. The term is a rather awkward one that refers to the general and very complicated political atmosphere in which we all live. It is important to distinguish between several terms that are commonly used when discussing governance. For example, *politics* is a process by which individuals and/or groups establish a society's agenda or vision; politics is not governance. *Management* refers to the implementation of actions aimed at achieving a society's vision; management is not governance. *Government* refers to the form or system of rule by which the actions of the members of a society are controlled; government is not governance. *Governance* relates to the broader social systems of governing, systems that enable society to accept or reject alternative political agendas or societal visions. Governance refers to the manner in which power is balanced in the administration of a country and embraces the traditions and institutions by which authority is exercised.

There are many books that consider the meaning of governance[1] and, while there is no "one size fits all" definition of governance, the following useful description has been developed by the United Nations Development Programme:

"Governance is the exercise of economic, political and administrative authority to manage a country's affairs at all levels...it comprises the mechanisms, processes and institutions through which citizens and groups articulate their interests, exercise their legal rights, meet their obligations and mediate their differences."

In short, governance sets the rules of the game and the systems in which we operate. Governance is important because: there is a strong causal relationship between better governance and economic growth, social equity, and sustainable development; there is ample evidence that the bulk of foreign direct investment goes to countries with good governance systems;

[1] See Pierre J (ed) (2000) Debating Governance, for a comprehensive discussion on all aspects.

and that better governance leads to poverty reduction. Normally, governance has been considered at the national level as in the above definition, but the effective governance of globalisation is becoming as important as governance at the national level.

There are three principal types of governance. Traditional **hierarchical governance systems** were put in place so that the state could steer society. Under this approach, the state directs, controls and cares for its citizens and the market. This model is increasingly challenged due to the fiscal crisis of the state, deregulation of financial markets, technological advances that facilitate networks and decentralisation, increased sub-regional autonomy, and excessive pressures and workloads on declining state bureaucracies. It is now generally accepted that the state, acting alone, cannot solve societal problems.

Market-led governance systems came to the fore at the end of the Cold War era. The Reagan/Thatcher doctrine reversed the post-World War II philosophy of a strong central state. In its place, the market was proposed as the primary mechanism for allocating resources. A period of deregulation, privatisation and managerialism followed that was characterised by more individual and less collective decision-making. Many people now question the effectiveness of market-led governance systems because markets are generally not seen as being truly representative of societal values.

There is now a greater reliance on **distributed governance systems** – through informal and voluntary sharing of authority and responsibility – as a means for societies to manage themselves and coordinate with others. The distributed governance model is characterised by networks and other forms of public-private alliances, which derive their legitimacy from the efficacy of their efforts to resolve complex national or transboundary problems. Distributed governance attempts to provide a balance between government efforts to control society, the dictates of the market, and the interests of society represented by community groups. Contemporary governance systems are thus centred on the expressed and empirically-verified needs of society; government steers but does not control; and there is considerable interdependence at all levels of authority. The fear is that this is slow, costly, easily distorted by special interest groups, and can lead to stalemate (Rogers and Hall 2003).

As a particular society develops, the intricate web of policies, institutions, laws, customs and practices that comprise its governance system also evolve. Historically, these evolutionary processes were shaped primarily by the needs and opportunities that emerged within the borders of each nation state. Increasingly, however, governance systems are shaped by regional and global considerations.

Global Governance in a Globalising World

Increasingly rapid globalisation gives rise to a pressing need for more effective global governance systems (Held and McGrew 2003). Those opposed to globalisation point out the distortions in society, but to oppose globalisation on that basis is to be looking backwards, not forwards. The future belongs to those who face the challenge of our interconnectedness and recognise that this challenge can be met if we can get global governance systems right. In fact, improved global governance should be seen for what it is: a positive response to governing globalisation.

Having made that point, we have a long way to go to develop good global governance systems. Indeed, one of the critical challenges of the 21st century is how to shift from exclusive sovereign rule over finite territories and their geopolitical mechanisms to a complex multilateralism for global governance. We must resolve the fundamental question of how world affairs are to be governed, and in so doing determine how we will achieve the elusive but increasingly essential balance between economic growth, social equity and environmental sustainability. This is illustrated by the global debates on the Kyoto Protocol and climate change, and regional debates on transboundary waters in the Danube, Nile and other river basins.

State sovereignty is still the principal driver in world affairs, but governments are increasingly constrained in their actions by international treaties and conventions negotiated through the United Nations or other international organisations. In fact, sovereignty is often voluntarily sacrificed when issues are seen to require collaborative or collective action. Working in isolation, nation states simply cannot regulate and control the global forces that impact on the state. Issues impinging on economic growth, social equity and environmental sustainability often cross frontiers and are not amenable to resolution without the kind of transboundary coordination and cooperation that can be engendered by effective global governance.

In essence, then, global governance is all about coordinating the development of coherent and effective solutions to shared problems among multiple national and transnational entities in a complex and interconnected world. Movement in this direction is readily apparent. Since the beginning of the 20th century – and especially since the early 1990s – the world has witnessed a huge increase in regional and global institutions, all seeking to intervene, promote and regulate the affairs of humanity. At the start of the 20th century there were some 37 Inter-Governmental Organisations (IGOs) and 176 International Non-Governmental Organisations (INGOs). At the start of this century there were about 5,200 IGOs, and a whopping

25,500 INGOs, (Held and McGrew 2003) plus many other forums such as the G8 (Group of eight most-industrialised countries), G20, etc., and a host of regional and sub-regional entities. Even the terms to use for many organisations are unclear.

Prior to the 1960s there was little United Nations interest in the environment. The Stockholm Conference in 1972 was a landmark event that called for help to less-developed countries to forestall environmental problems. It also led to the creation of the United Nations Environment Programme (UNEP). In the 1990s there was a proliferation of organisations and treaties on environmental matters. This has led to turf wars, jurisdictional gaps and overlaps, as well as overload for national level institutions. Global environmental governance is presently weak with a lack of resources and little authority. For example, despite considerable concern and public pressure on water issues, it still remains fragmented within the UN system, as well as in many government institutional set-ups.

Does this growing assortment of organisations, conventions and treaties constitute an evolving global governance complex? There is considerable disagreement about that, but two things seem clear: governments, especially in developing countries, are being overwhelmed, and globalisation in a number of domains is driving the proliferation of multilateral initiatives aimed at promoting more coherence and equity in governance on a global scale. Moreover, the environmental dimensions of interdependence are increasingly recognised as being inextricably linked to the issues of security, economic development and human welfare.

Need for Effective Global Environmental Governance

Population growth, increasing wealth, climate change and other indicators have long suggested the existence of limits to natural resources and to growth. This led to the conceptualisation of sustainable development, and this in turn needs to lead to global environmental governance, that is, global governance that extends beyond the current primary focus on peace, security, international trade and global finance. Clearly, economic growth is the strongest force underpinning the trend towards global governance, but people are increasingly concerned that, without social equity and environmental sustainability, conflict and short-term perspectives will dominate. Global governance systems are needed in order to achieve a much-needed balance – and also to ensure the sharing of information and science and technology in achieving sustainable development.

The growing visibility and mainstreaming of environmental issues in the context of globalisation has led to a multitude of international and regional conferences, treaties, declarations and calls for new institutions (although rarely for the reform of existing ones). This comprises a very fragile and not very efficient multilateral framework in which political changes can easily derail advances and agreements that can take decades to negotiate. Excessive and sometimes utopian demands can result in government or societal backlashes that set progress towards solutions back and reduce confidence in global governance. Moreover, many issues – perhaps too many – are pushed towards the global level for resolution. We must be stricter in applying the principle of subsidiarity in problem solving; otherwise global environmental governance may collapse before it really gets started.

Indeed, most environmental issues are local or regional, rarely inherently global. That said, the principles and concepts of sustainable development are global and should be enshrined in global, rule-based systems (Esty and Ivanova 2002). Moreover, apart from obvious concerns such as climate change, a number of specific environmental issues are amenable to global governance:

- Spillovers – dealing with the effects of environmental degradation in one state or region on others, for example, coping with refugees fleeing drought;
- Regulating transboundary pollution and dealing with the degradation of the global commons, for example acid rain and river/lake/sea pollution;
- Displacement – trade in natural resources that leads to the exploitation of one country's natural resources for the sake of preserving the resources of another country; for example, China endangering sustainable development in Africa and elsewhere by importing huge volumes of natural resources whilst conserving its own;
- The international transportation of toxic wastes;
- Transportation of species that may become invasive and damage other ecosystems, for example, wildlife, some aquatic plants; and
- The need to develop global and regional environmental regulatory institutions, treaties, conventions and laws, for example, the Kyoto Protocol and European Union Water Framework Directive.

As noted earlier, territorially-delimited political communities are not able to handle such challenges, even if they enjoy extensive control within their own borders.

Idealism can be a threat to effective global environmental governance. Some people are fervently anti-science or anti-technology, for example, yet many solutions to pressing environmental problems can only come from science and technology. Some people tend to be fervently anti-private sector, yet business can be very effective in providing technological solutions. The key is to develop the right regulatory systems that control the excesses of commercialism and reduce the risk of market failures. Such systems need to be well designed so that obstacles are not placed in the way of development or efficient decision-making, for example, the World Commission on Dams Report (2000) was felt by many governments in developing countries to be too restrictive and, rather than ensuring sustainable development, could in fact prevent development.

In general, scientific research can be critical to progress in the environmental arena. Unfortunately, research has sometimes been devalued by a lack of objectivity – real or perceived – poor quality, and political correctness. Universities are increasingly seen as being too commercial. Research on the potential environmental effects of genetically modified organisms that is funded by agribusiness is naturally perceived as suspect. Research on the effectiveness of private sector water services that is funded by trade unions is also unlikely to be objective. Raising public awareness can change political will and direction, but overstating the case can be a serious problem in the long term, diverting resources from truly pressing issues and reinforcing a growing negative attitude towards science and technology. There is a fear that researchers are becoming advocates for causes, or mercenaries producing research results suited to those funding the work.

In addition to more objectivity, researchers need to agree on a few key issues that are beyond question of global importance, and in so doing help politicians to focus on and resolve them, and not hide behind their complexity.

Principles for Effective Global Environmental Governance

There is no single model for effective global environmental governance. There are, however, some basic principles or attributes that are considered essential components of effective governance systems, and that pertain as well to global environmental governance systems.

Institutions should work in an open and transparent manner. They should use accessible language that can be easily understood by the general public so as to increase confidence in deliberations. Good governance requires that all policy decisions are transparent so that both insiders and outsiders can follow the steps taken in policy formulation.

Institutions should be inclusive and communicative. The quality, relevance and effectiveness of government policies depend on ensuring wide participation throughout the policy chain – from conception to implementation. Governance institutions and systems need to communicate with stakeholders in very direct ways, and in so doing enhance the integration of civil society into governance systems.

Policies and action must be coherent. The need for harmony and coherence in governance increases as the range of issues and tasks grows and becomes more diverse. Such environmental challenges as climate change cross the boundaries of sectoral policies on which government traditionally rests. Coherence requires political leadership and a strong responsibility on the part of governance institutions at different levels to ensure consistent approaches to global environmental issues.

Governance systems must be equitable. Equity, between and among, all stakeholders must be carefully monitored throughout the process of policy development and implementation. Legal and regulatory frameworks must be fair and enforced impartially.

Accountability is critical to good governance. The rules of the game need to be clearly spelled out, as should the consequences for violation of the rules, and arbitration mechanisms need to be built into the governance system to ensure that satisfactory solutions can still be reached when seemingly irreconcilable conflicts arise among stakeholders.

Governance systems must be efficient. Political, social and environmental efficiency needs to be balanced against the dictates of economic efficiency. Transaction costs should be minimised so as to not impede action, but the views of all stakeholders need to be obtained and considered in order to achieve efficient governance in the broader sense.

Governance systems must be responsive and sustainable. Effective governance will be demand-driven, based on clear objectives, and on an evaluation of likely future impacts. Policies should be incentive-based and decisions should be taken at the most appropriate levels. This will contribute to empowerment, a sense of ownership, and clarity regarding social and economic benefits. This in turn will contribute to the sustainability of the governance system.

Conclusion

Global environmental governance has been discussed for the last 10 to 15 years and is now at a crossroads (Ayre and Callway 2005). Sovereign states are increasingly enmeshed in an array of complex, multi-layered governance structures and networks with overlapping mandates, and many countries – especially in the developing world – find it progressively more difficult to cope, often delegating authority to INGOs or IGOs. There is a malaise, because people do not feel the results match their ambitions.

Degradation is moving faster than debate. To overcome this malaise there is a need to conclude the debate and strengthen the global environmental governance system, even if it does not satisfy every need. Establishing an effective system would need to take account of the role of existing bodies and to create new entities to fill the gaps. Any new global entity should have a clear mandate that focuses only on global environmental issues and not on those that can, and should, be resolved nationally, or regionally. This would help to give more confidence that the potentially negative impacts of globalisation can be mitigated. There is also a pressing need to scale down the rhetoric and solve a few key practical problems, rather than trying to solve the entire world's problems at one stroke. For example, the Global Water Partnership (2004) is currently working with governments on practical steps towards good water governance, by preparing integrated water resources management action plans, which provide a roadmap that converts principles into practice. The focus has thus shifted to working on the elements that make up a good governance system, rather than advocating for a perfect holistic system (Rogers and Hall 2003). It is hoped that taking bite-sized chunks in a structured way will advance change and lead to better water management for the benefit of all.

A starting point may be to establish a high-level task force, under the auspices of the UN Secretary General, to prepare a proposal for a global environmental body that would act as a watchdog and dispute-settlement mechanism for any cross-border environmental issue. It is unfortunate, but likely, that any separate body, related purely to the environment, will remain weak. Also, it is unlikely that any existing international environmental entities would be suitable for such a role. To be effective, such a body should be closely linked to an economically powerful institution. For example, it could be established under the umbrella of the World Bank, similar to the International Centre for the Settlement of Investment Disputes. This would avoid the need to establish a completely new institution whilst giving the new entity strong legal powers.

References

Ayre G, Callway R (eds) (2005) Governance for sustainable development. Earthscan, London, UK

Global Water Partnership (2003) Effective water governance: learning from the dialogues. GWP, Stockholm, Sweden

Global Water Partnership (2004) Catalyzing change: a handbook for developing integrated water resources management and water efficiency strategies. GWP, Stockholm, Sweden

Held D, McGrew A (2003) Governing globalization. Polity Press, Cambridge, UK

Esty EC, Ivanova MH (eds) (2002) Global environmental governance: options and opportunities. Yale Center for Environmental Law and Policy, USA

Pierre J (ed) (2000) Debating governance: authority, steering and democracy. Oxford University Press, Oxford, UK

Rogers P, Hall AW (2003) Effective water governance. TEC background paper 7, GWP, Stockholm, Sweden

World Commission on Dams (2000) Dams and development: a new framework for decision-making. Earthscan, London, UK

National Perspectives on Water Governance: Lessons from the IWRM Planning Process in Malawi and Zambia

Alex Simalabwi

Abstract

At the Second World Water Forum in 2000, the problems relating to water around the world were described as a consequence of the lack of good governance in water. The Global Water Partnership defines water governance as the range of political, social, economic and administrative systems that are in place to develop and manage water resources, and delivery of water services, at different levels of society.

As a contribution to good water governance, the GWPSA has been facilitating the development of Integrated Water Resources Management and Water Efficiency (IWRM/WE) Plans in Malawi and Zambia. Effective water governance is crucial for the implementation of IWRM. While the process is still ongoing, various lessons can be drawn in relation to water governance.

This chapter argues that, while governance may be seen to be dependent on three key clusters; Government, Society and Science and the interactions among them, there are no distinct boundaries among the three clusters. Further, lessons from the IWRM/WE process highlight the importance of scale and power relations to water governance. IWRM Plans are being developed for river systems and natural resources at the national scale, confined to national boundaries. However, the transboundary nature of water resources requires effective interactions between and across the different scales. International conventions, protocols, declarations and targets such as the 2015 Millennium Development Goals (MDGs) are some of the factors at the international scale that have a strong influence on the IWRM Planning process at the national scale. Another important issue is that of power relations among players at a given scale, and also between different scales. The way decisions and information is communicated from central government to local government and vice-versa or from the catchment to the sub-catchment scale is crucial to good water governance. These interactions and process are highlighted in this chapter.

Keywords: governance, IWRM Plans, scale, Trialogue, Integrated Water Resources Management (IWRM)

Introduction

At the Second World Water Forum in 2000, the problems relating to water around the world were described as a consequence of the lack of good governance (GWP 2000). This fact was also confirmed in a presentation to the United Nations Secretary General High Level Panel at WSSD in 2002 by HRH Prince Willem-Alexander of the Netherlands (WSSD 2002). The Prince echoed earlier sentiments at the Second World Water Forum that the global water crisis is a crisis of water governance. At the 2002 World Summit on Sustainable Development (WSSD) in Johannesburg, South Africa, the international community agreed that all countries should prepare IWRM/WE Plans by the year 2005, with support to developing countries (WSSD 2002) as a contribution to good water governance and ultimately sustainable development. The Plans are an important milestone to achievement of the 2015 MDGs and thus reduction of poverty and improvement of peoples' livelihoods. The IWRM/WE Plans are thus an important precursor to improved water governance, with the overall goal of sustainable development, through better management of water resources.

In response to the WSSD call, the Canadian Government is supporting the development of IWRM plans in five African countries through CIDA: Mali, Malawi, Zambia, Kenya and Senegal through an initiative called Partnership for Africa's Water Development (PAWD). The Global Water Partnership Southern Africa (GWPSA) is coordinating the development of these Plans in Malawi and Zambia. The support is being facilitated through the Malawi Water Partnership (MWP) and Zambia Water Partnership (ZWP) respectively.

Through GWPSA facilitation, both Malawi and Zambia are designing strategies, policies and institutional frameworks that are socially acceptable and capable of assisting national governments to mobilise sufficient resources in support of their implementation. To ensure that the implementation of these strategies is achievable, the formulation process of IWRM/WE Plans involves a broad range of key stakeholders. The process involves a dynamic partnership between government, the market and civil society, and has generated experiences and lessons that can advance the theory of water governance.

Turton et al. (2005) have proposed a hypothesis that states that the degree to which governance is successful depends on six essential elements:

- Science Actor-cluster;
- Government Actor-cluster; and
- Society Actor-cluster.

These three actors are linked by:

- An interface between society and science;
- An interface between government and society; and
- An interface between government and science.

This chapter shares the IWRM planning process experience in Malawi and Zambia. It highlights the importance of scale and power relations as other factors important to water governance, but not fully explicit in the proposed CSIR Water Governance Trialogue Model.

Governance Defined

Various definitions for water governance exist. The Global Water Partnership defines water governance as the range of political, social, economic and administrative systems that are in place to develop and manage water resources, and delivery of water services, at different levels of society. The United Nations Development Programme (UNDP) defines water governance as the exercise of economic, political and administrative authority to manage a country's affairs at all levels. It comprises processes and institutions through which citizens and groups articulate their interests, exercise rights, meet their obligations and mediate their differences (UNDP 1997).

While many other definitions exist on water governance, it is generally agreed that good governance requires the establishment of an enabling environment (GWP 2003). Putting IWRM into practice depends on effective governance. Weak governance leads to government failure, market failure and system failure. Good water governance forms an important pillar of the IWRM/WE Plans and is about local change and reform, and strategies to ensure good water governance need to be developed as part of the IWRM Plans required under the WSSD Target.

Elements of Governance

In both Malawi and Zambia, government and private stakeholders are actively engaged in the process of developing IWRM Plans. The interests and needs of the various stakeholders involved are largely dictating the course of the IWRM/WE planning process. As mentioned above, the CSIR proposes three main clusters as elements of ecosystem governance; Government, Science and Society (Turton et al. 2005).

Lessons from the IWRM/WE planning process show that these clusters, though not exclusively separate, are present. GWPSA has facilitated the establishment of responsive IWRM planning structures at country level that involves various stakeholders. Through participation at the high-level steering committee, government ensures that constitutional and statutory regulations are observed and adhered to. The water strategies must be consistent with long-term policy and address priorities. Participation by private sector representatives such as sugar estates, commerce, and so forth, ensures that society's perspective is taken into consideration. The academic institutions also active in the planning process bring in an important scientific perspective.

Two key issues not so obvious from the proposed Trialogue Model, are the issues of scale and power. The way in which decisions and information is communicated from central government to local government and *vice-versa* or from the catchment to the sub-catchment scale is crucial to good water governance. This highlights the importance of scale and power. At the regional and international scale, various international agreements, protocols related to transboundary water management and commitments such as achievement of the MDGs, play an important role in the IWRM/WE planning process.

This chapter thus argues that water governance is not only dependent on the six elements of the Trialogue Model, as scale as an important element to governance. While the three main clusters exist they are not exclusively separate, as in the Trialogue Model, but overlap in various structures at different scales. The IWRM/WE Planning process being facilitated by GWPSA is taken as a case in point.

IWRM/WE Planning Process

A National IWRM Plan is a road map to guide a country on the changes needed to move from fragmented to integrated ways of developing, managing and using water resources, and to accelerate actions towards those ends (GWP-TAC 2004). It clearly outlines the actions and resources required by a country to move from managing and developing water resources in a disjointed and uncoordinated manner, to well-coordinated and harmonised manner, in order to maximise the socio-economic benefits and improvement of people's livelihoods.

An IWRM/WE Plan provides guidance to national governments in addressing a country's key water-related development problems such as water for agricultural productivity, people, energy, and the environment. Furthermore, an IWRM/WE Plan helps to strike a balance between the use of resources for livelihoods and conservation of the resources to sustain its ecological functions for future generations. It seeks to avoid the lives lost, the money wasted, and the natural capital depleted because of fragmented decisions that did not take into account the larger ramifications of sectoral actions. The plan provides practical guidance of how institutions involved in the development and management of water resources; government agencies, civil society, private sector and cooperating partners, can operate in a coordinated and integrated manner for economic efficiency, environmental sustainability and social equity. Water as an integral part of the ecosystem is crucial for sustenance of biodiversity, and thus good water governance is important in ensuring ecosystem integrity. IWRM advocates for environmental efficiency. One of the objectives of IWRM/WE plans is to contribute to environmental sustainability and thus good ecosystem governance.

In order to secure the co-ordination of water management efforts across water related sectors, and throughout entire basins, formal mechanisms for cooperation and information exchange need to be established. Ideally, such coordination mechanisms should be created at the highest political level and put in place in all relevant levels of water management. An IWRM planning process creates an environment where such coordination mechanisms can be forged and consensus reached among various stakeholders on appropriate governance structures.

The Process

IWRM/WE planning is a cyclic process. This entails continuous review of the status at regular intervals in order to deal with new or additional priority water resources issues, management requirements and infrastructure requirements. The IWRM process is illustrated in Figure 3.1 as the "Integrated Water Resources Management Cycle." The cycle starts with the planning processes and continues into implementation of the frameworks, action plans and monitoring of progress. Stakeholders have to evaluate from whether new reform needs have appeared, or whether the reform process has led to the expected improvements. If the latter is not the case then the cycle must be repeated.

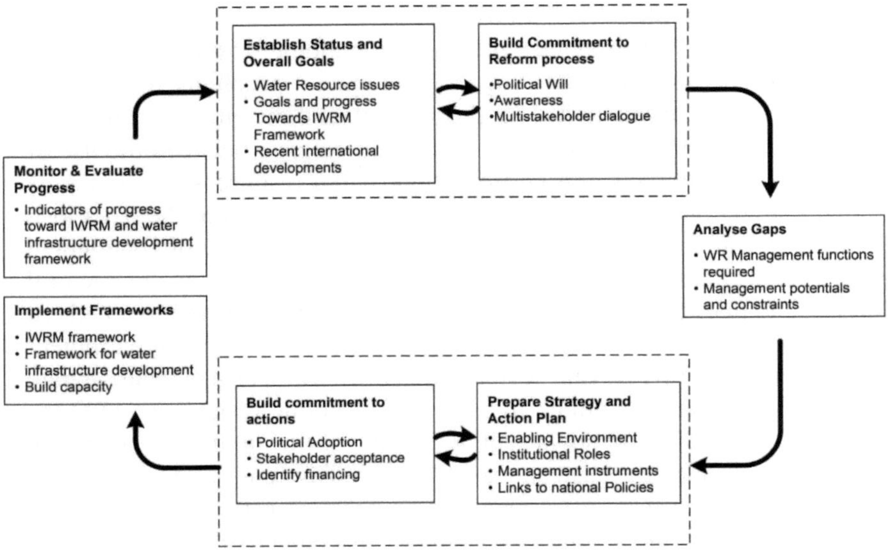

Fig. 3.1. *The Integrated Water Resources Management Cycle (Source: GWP-TEC 10 Background Paper, IWRM/WE plans by 2005, Why, What and How)*

Feedback Loops in the Process Cycle

Active stakeholder involvement is a key element, as is the commitment and practices of managing the process cycle, and should at any stage feedback to repeat some of the steps in the light of new developments. However, two feedback loops are particularly important in the planning cycle.

The first deals with prioritisation of the water resource issues and the status of the present water resource management system, including taking stock of those recent international developments of importance for the national water resource management process. Priority setting and commitment to reform requires political will, awareness to be raised and an active stakeholder dialogue.

The cycle illustrates that before priorities for reform can be agreed there may be need for reviews, extended dialogue and including new stakeholders.

The second feedback loop deals with the process of preparing the strategy and, in particular, the "plan." This requires extensive policy consultations and stakeholder involvement. It illustrates that the final action plans need political agreements at the highest political level, acceptance from the main stakeholders and raising the necessary financial means from domestic and international resources.

IWRM Planning Process in Malawi and Zambia: Lessons Learnt

The ongoing IWRM/WE planning process in Malawi and Zambia has generated considerable information and experience, which can contribute to the current body of knowledge on the issue of water governance. This is presented below. An attempt has been made in this chapter to link the lessons learnt in the process to the Trialogue Model (Turton et al. 2005).

Critical analyses of the lessons from the IWRM/WE Planning process also brings in other dimensions to proposed elements of water governance; that of scale and the interface between the various scales with the Science, Government and Society processes; and that of power relations. Who is eligible to make decisions at each scale and how these decisions are made and communicated are other important factors to water governance? The role of the media in water governance is also highlighted.

Government Process

Good governance requires the establishment of an enabling environment. IWRM is a political process and involves conflicts of interest that must be mediated through a good governance structure. A system with clear rules of accountability, participatory mechanisms and respect for law and obligations are prerequisite for a good governance structure. An appropriate institutional structure needs to be developed. Governments play a key role in the establishment of an enabling environment. They must also be the main regulators and controllers in the water sector with its associated infrastructure. Further, governments promote improvements in the public sector, regulate the private sector involvement, and decide on market mechanisms. However, in keeping with internationally accepted principles of subsidiarity, water is a resource to be managed at the lowest appropriate level. In terms of this paradigm, it is government working with civil society that must raise awareness of the importance of improved water resources management among policy makers and the general public.

Figure 3.2 shows the process management structure in place for the IWRM/WE planning in Malawi and Zambia. Progress to date confirms that political support is critical to the IWRM Plan development process.

Political support is crucial not only for political legitimisation, but also for ensuring that government, being the custodian of national development, takes ownership of the Plan, and financing for implementation of the Plan. Large-scale mobilisation of resources is required for implementation and this is only possible if the process has sufficient support from government.

In the IWRM/WE planning process, senior officials from government are participating actively in the project steering committees and some are also involved in the project management teams. In Malawi, tremendous political support exists with the Country Vice President, who was also Minister of Water Development at the start of the project in 2004. The new Minister of Water Development is also enthusiastic and has been closely following up on developments in the project.

In Zambia, a high level Cabinet Inter-Ministerial Committee is responsible for overall policy guidance of the water reforms and IWRM planning process. The Cabinet committee, chaired by the Minister of Water Affairs and Energy Development (MEWD), was constituted by Cabinet to provide cross-sectoral policy guidance to the Water Resources Action Programme (WRAP) funded by World Bank, NORAD and Dannida. By anchoring the planning process at this highest policy planning level, this has ensured that the strategies being defined are not only consistent to national policy but, also address national priorities.

The IWRM/WE process management structure includes directors and senior government officials from the water line ministries. These are part of the PAWD Core Team (PCT) whose primary responsibility is to ensure that the various components of IWRM in different sectors are integrated in the IWRM plan being developed. The PCT is basically the integrating committee of the planning process. The project core team ensures that sector plans from their respective line ministries are taken into consideration. In Zambia, the PCT is chaired by the Director of Water Affairs while in Malawi the Permanent Secretary in the Ministry of Energy and Environmental Affairs chairs the PCT. The PCT also involves representatives of civil society and academic institutions and this provides for interaction between Government, Science and Society as proposed in the Trialogue Model.

Contrary to the boundaries supposedly created by the Trialogue Model, there are no distinct boundaries between government, society and science. The three clusters interact and overlap in the different project structures. Government is present in the stakeholder catchment fora, the PCT and also in the PMT, and so is science and society, suggesting that the interfaces are sound.

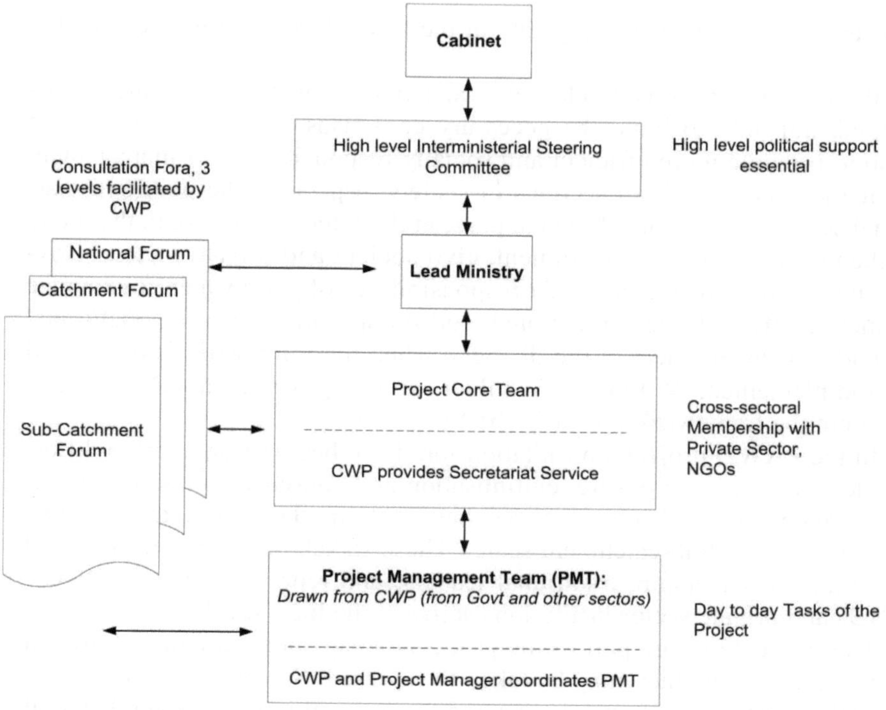

Fig. 3.2. *PAWD process management structure (Source: GWPSA PAWD project document)*

Stakeholders at the national scale dominate representation in the PAWD structures. To facilitate participation by stakeholders from the lower scales, both Malawi and Zambia have set up project structures at catchment and sub-catchment scales, and have put in place mechanisms of communication among these different scales. This is shown in Figure 3.2.

Society Process

Effective stakeholder participation depends on a conducive governance regime at the national level, which allows for equitable participation of society in the governance of water resources. Laws need to set out clear understanding between National, Basin and State level responsibilities as well as promote integration across sectors. According to Manzungu (2004), in South Africa a water law with devolved responsibilities could not be implemented effectively as the required decision-making powers and resources and capacity at the lower levels did not accompany it.

Participation by stakeholders in water management programmes is required, not only to build the necessary consensus for policy reform, but also to promote more efficient and socially responsible water management strategies that benefit all sections of society, especially the poor and marginalised. Dialogues need to take place at the interfaces between the many stakeholders involved, government, civil society and private sector. Governments can only exercise their responsibilities of good water governance if they involve all relevant national (and if appropriate also regional/transboundary) stakeholders in the dialogue when the framework is developed and implemented. Without stakeholder support, government efforts to implement the framework can easily be frustrated.

In the PAWD project, consultation fora have been set up at three different levels in order to ensure legitimisation and consensus on issues affecting stakeholders. Both countries have set up these fora at National, Catchment, and Sub-catchment scale. These stakeholder fora include local government leadership, traditional leaders, and representatives of NGOs, CBOs and private sector institutions active at the local level.

These structures are part of the process management structure shown in Figure 3.2. While they may be taken to represent Society in the Trialogue Model, society is not homogenous and the issue of power among different players is critical. Stakeholders hold different interests largely determined by the stake they hold and this may translate into power differences. The GWP sees its role as facilitating participation among different stakeholders in society, and this involves mediation and negotiation as conflicts among stakeholders arise. Operational guidelines are in place and highlight who is eligible to make certain decisions (power) and how such decisions should be communicated.

Government-Society Interface

Stakeholder Participation

Lessons from the IWRM/WE planning process confirm that stakeholder participation is critical to the IWRM plan development process as it provides for consensus and legitimisation of the process.

A multi-stakeholder platform comprising of stakeholders, beyond water practitioners, is important for effective stakeholder representation. Effective stakeholder participation depends on a conducive governance regime at the national level, which often lies outside the purview of water practitioners (Manzungu 2004). GWP's country water partnerships have been instrumental in bringing various stakeholders to neutral platforms. The multi-stakeholder platforms in both countries have established multi-sectoral steering committees to spearhead the development of the plan (Figure 3.2).

The membership of these steering committees includes line water ministries such as agriculture, environment, local government, finance, natural resources, etc. Other members include NGOs, academic and research institutions and community representative organisations such as farmer associations. The value of the CWP platform lies in the fact that stakeholders have equal status, which enables open discussions on issues that affect them.

The membership profile in the CWPs and project structures highlight the lack of separate clusters as proposed in the Trialogue and reinforces the fact that the clusters are not distinct but overlap, indicating an effective interface.

Engaging the Media

Another key lesson is the critical role that media can play in the water sector. Awareness rising has been a crucial part of the project. In Malawi, the MWP organised a media workshop on the PAWD project prior the launch of the PAWD project. Journalists from both print and electronic media attended the workshop. This enabled them to aptly discuss the value of water in economic development in their publications and the critical role of multi-stakeholder platforms such as the MWP. The MWP has since received widespread coverage with several institutions requesting to join the MWP and be part of the PAWD programme (V Chipofya 2005, pers. comm). The media is thus a critical partner in the government-society interface and plays the role of "messenger" in transmitting and promoting the flow of information from government to society and vice-versa, and across the different scales. The media also plays a fundamental role in facilitating debate among stakeholders on the different issues related to governance such as accountability, transparency, inclusiveness, and also critiquing the consequences of certain strategic alternatives for water governance. The media therefore enhances the quality of the interfaces in the Trialogue Model.

Science Process

A critically important element of IWRM is the integration of various sectoral views and interests in the development and implementation of the IWRM framework (GWP-TAC 1999). Integration should take place within:

- The *natural system*, with its critical importance for resource availability and quality; and
- The *human system*, which fundamentally determines the resource use, waste production and pollution of the resource, and which must also set the development priorities and control associated infrastructure.

Integration within the natural system concerns the integration of land and water management, surface and groundwater, upstream and downstream water related interests, recognising the full hydrologic cycle. Public pressure caused for example by a lack of safe and affordable drinking water and basic sanitation, pressure from national economic sectors like energy and agriculture due to lack of water for development, transboundary conflicts and crises and international agreements on water, all justify the importance of science and research for new and innovative and of improving water resource management. Water scarcity and deteriorating water quality have been, or will soon become, critical factors limiting national economic development, expansion of food production and/or provision of basic health and hygiene services to the population. The recognition of the need to redress these weaknesses in their water governance structures has convinced many countries that a new water management framework, which recognises the important contribution of science, is needed. Other common critical issues include:

- *Inappropriate pricing structures* and hence limited cost recovery resulting in inefficient operation and maintenance of water systems, as well as in misallocation and loss of water;
- *Inadequate information and data* to support sound management of water;
- *Water degradation, health and loss of productivity*;
- *Soil degradation and loss of productive land.* The way water is managed in coordination with land management has significant effects on agricultural production; and
- *Risk management, floods and droughts.* Economic losses from floods, droughts and climate variability are globally experienced at a very large scale. The drought in Zimbabwe in the early 1990s entailed a 45% decline in agricultural production and an associated 11% decline in

gross domestic product (GDP) (GWP-TEC 10, 2004). El Niño floods (1997-98) caused an estimated economic loss exceeding US$1.7 billion in Kenya and US$2.6 billion in Peru. Mozambique suffered a 23% reduction in GDP following the floods in 2000. In Zambia, agriculture sector share in GDP decreased from 16% in 1991 to 8% in 1992 mainly due to drought (Hunt et al. 1994) while in Malawi, GDP declined by 7.9% as result of the 1992 drought (Malawi Economy 2006).

Adopting an integrated approach takes into account the science process including risk management and the prudent coordinated management of land and water. Monitoring, forecasting and contingency planning is important to inform decisions aimed at alleviating grave economic consequences.

In IWRM, the role of science is critical in providing planning guidance material through the GWP Technical Committee, GWP-Tool box, Cap-Net and Water-Net. These are key players in the planning process. Water-Net comprises about 40 academic institutions within the SADC region and these provide a body of scientific knowledge to tap from.

Science-Society Interface

There are no distinct boundaries among the three clusters proposed in the Trialogue Model because the quality of the interfaces is good. Membership of PCT includes academic and research institutions as well. In fact the IWRM/WE process is hosted by academic institutions, the Malawi Polytechnic and the University of Zambia. Other academic institutions are involved at various levels of the project management and these are key in providing science based guidance and contribution to the planning process.

Government-Science Interface

The PCT comprising of government, civil society and academic institutions provides a platform for interaction between the government and academic institutions. This clearly demonstrates that the governance Trialogue comprising of Government, Science and Society is at play in the GWPSA IWRM/WE planning facilitated process. The IWRM/WE process management structure in place (Figure 3.2) provides for the interface of the three clusters: Government, Science and Society. The boundaries, as proposed in the model, are not explicitly there due to the quality of the interfaces and the issue of power as an implicit element is important.

Discussion

This section argues that, while governance may be seen to be dependent on three main clusters; Government, Society and Science and the interactions among them, scale and power are other important elements that should be factored in the Trialogue Model. IWRM Plans are being developed for river systems and natural resources at the national scale, confined to national boundaries. However, the transboundary nature of water resources implies that external factors, existing at different scales within and beyond national boundaries, are important aspects to be considered. Interactions across the different scales is important and so is the issue of power relations among players at a given scale and also between different scales. The way decisions and information is communicated from central government to local government and vice-versa or from the catchment to the sub-catchment scale is crucial to good water governance. International conventions, protocols, declarations and targets such as the 2015 MDGs are some of the factors beyond the national scale that have a strong influence on the IWRM Planning process.

Importance of International Scale

The SADC Protocol on Shared Watercourses is one case in point that both Malawi and Zambia have taken under consideration. The issue of harmonising these national plans with the plans among riparian states of the Zambezi Basin has often come up. The countries are cautious of the obligations of riparian states as espoused in the SADC Protocol (SADC 2000). The Plans were also born out of the 2002 World Summit on Sustainable Development (WSSD) in Johannesburg, South Africa, where in the international community agreed that all countries should prepare IWRM/WE plans by the year 2005. While the 2005 target has generally been accepted to be an ambitious target, both Malawi and Zambia maintained that they wanted to develop the plans by 2005, even when it was clear that the target was ambitious. Both countries have managed to develop draft plans by 2005. The insistence by these countries to meet the 2005 target as called for by the WSSD international community, is an example of how processes at the international scale influence governance at the national level. The IWRM plans are by themselves international commitments that seek to accelerate efforts towards achievement of the 2015 MDGs. Consequently both Malawi and Zambia have proposed to structure the IWRM plan towards the 2015 target in accordance with the SADC Vision for Water, Life and the Environment (Hollingworth et al. 2005).

Other examples of process at the global/international scale that influence water governance at the national scale include Agenda 21 that gave birth to IWRM. The World Water Forums have contributed to calls for water and environmental planners and managers to promote IWRM approaches (Varady 2003). These processes have contributed to national water and ecosystem governance programmes by influencing national water and environmental policies.

In Malawi, stakeholders at a national consultative meeting called for the revision of the 1969 Water Resources Act so as to, among other things, align it with current international developments for water resources management (MWP 2005). This is being pursued in the Malawi PAWD process.

The influence of international process on national processes shows the complexity of ecosystem governance and the fact that governance involves interaction of processes across different scales, both vertically and horizontally. Further, within the PAWD management structures in GWP, linkages with the key partners at a global, regional, and national level, have been extremely useful to the development of the PAWD project. Project Management Teams for the entire programme have been set up from global to country level (Figure 3.3).

Being part of the GWP international global family has also added value and benefits to the availability of critical information whilst developing the Plan. Guidance is critical and at global level the GWP Technical Committee (TEC), through active interaction with the PAWD teams from the countries involved in the project, has produced a handbook titled "Catalyzing Change: A Handbook for Developing Integrated Water Resources Management (IWRM) and Water Efficiency Strategies" (GWP-TEC 2004). Such guidance ensures that countries interacting with the larger global family benefit from global knowledge. Also, as new lessons and experiences are captured in the process, these are shared with the global family through the GWP IWRM Tool Box, and made accessible to the teams involved in the project. Cap-Net, a GWP capacity building Associated Programme (AP), has developed training materials for IWRM Planning based on the experiences of countries in the PAWD project, and in conjunction with Waternet, conduct training courses on IWRM planning (Cap-Net 2005). Gender and Water Alliance (GWA), working with GWP at the global level, has also developed training guidelines for mainstreaming Gender in IWRM (CAP-Net, GWA 2006). This information is also being used by the teams in both Malawi and Zambia.

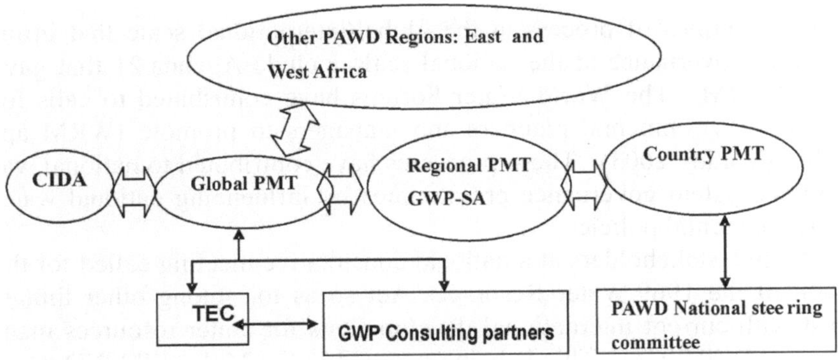

Fig. 3.3. *Schematic PAWD management structure, global to country level*

Importance of Power Relations

The importance of power in governance cannot be overemphasised. One school of thought on governance advocates that governance is fundamentally about power, relationships and accountability: who has influence, who decides, and how decision-makers are held accountable (Graham et al. 2003). Within the PAWD project in Malawi and Zambia, operational guidelines have been put in place to primarily regulate power relations among different stakeholders involved in the planning process. The guidelines serve as control mechanism on who has the power to make decisions. These guidelines also articulate how stakeholders are to be chosen and who makes the choices. When decisions are made at country level, the guidelines stipulate that such decisions should be communicated to stakeholders through the CWP coordinator who is also responsible for communicating to the regional GWPSA office. At the regional scale, GWPSA regional office is part of the national project steering committees. The regional office represents the project management team in Sweden that is responsible for facilitating the process globally. Both Malawi and Zambia are part of a global family (international scale) that has strong influence on the development of the Plans. Lessons emerging from the planning process have confirmed the importance of accounting for power relations. In both Malawi and Zambia, government have applauded the CWPs for being able to bring stakeholders from different backgrounds to dialogue on important water governance matters (MEWD 2006, MoD 2004). While the CWPs have this responsibility, and seem to ostensibly have the power to bring stakeholders together, governments take the lead in matters of policy and thus have the power to provide strategic direction. The dynamics of power are not static as stakeholders ultimately want to feel that government is making correct decisions and is accountable to the stakeholders. In Zambia, several stakeholder consultations have been conducted on the Water Resources Bill, and while government has the ultimate responsibility of legislating the bill into law, stakeholders feel that the bill cannot be legislated until they have been adequately consulted.

Another example on accounting for power relations can be seen from the negotiations on signing the Zambezi River Basin Commission (ZAMCOM) agreement for management of the basin among the eight riparian states; Angola, Botswana, Malawi, Mozambique, Namibia, Tanzania, Zambia and Zimbabwe. The Government of Zambia was the only government of the eight riparian states that did not sign the ZAMCOM agreement in 2005, citing the need to further consult with national stakeholders (P Chola 2005 pers. comm). In Malawi, a national workshop on IWRM sensitisation for Permanent Secretaries failed to take off at first, because the invitation letter to other Permanent Secretaries had initially been written by the Permanent Secretary in the Ministry of Water and Irrigation Development. When the same letter was written and signed off by the secretary to Cabinet, 29 out of 33 Permanent Secretaries in the country attended and participated in the workshop (V Chipofya 2004, pers. comm).

The above clearly shows the dynamic nature of power in water governances and in this instance, while governments have the power to legislate; stakeholders have the power to legitimise.

Conclusion

Management in an effective governance system would need an integrated approach where the needs and demands of all water users are provided for, and consequences of the water resources taken into account in decision-making process in all sectors of society. While the main clusters of Government, Society and Science as proposed in the Trialogue, provide an important framework for such an approach, the model is simplistic in that the fundamental interplay across scales is not fully explored. Vertical scales are "hidden" and yet interaction of processes across scales is a fundamental component of water governance as highlighted in the IWRM planning process. The transboundary nature of water resources requires effective interactions across different scales within and beyond national boundaries. International conventions, protocols, declarations and targets, such as the 2015 MDGs, are all factors at the international scale that have a strong influence on the IWRM Planning process. The issue of scale is thus an important element that should be factored in the Trialogue Model.

Power is another important factor that needs to be understood in the Trialogue. The way in which decisions are made, and who has the power to do so, relationships and accountability: who has influence, and how decision-makers are held accountable, are important elements not clear in the Trialogue.

From the IWRM/WE planning process, it is clear that issues of scale and power are important elements of governance. Further, while the main clusters (Government, Society and Science), and their corresponding interfaces arise in the ongoing IWRM/WE planning process, they are not exclusively separate and in some cases, these are mutually intertwined.

References

CAP-Net (2005) Integrated water resources management plans, training manual and operational guide. Netherlands Commission on Sustainable Development. Available online at www.un.org/esa/sustdev/csd/csd12/

CAP-Net, GWA (2006) Why gender matters: a tutorial for water managers. Multimedia CD and booklet. CAP-Net International Network for Capacity Building in IWRM, Delft

Chipofya V (2005) GWPSA regional capacity building workshop. Personal communication with Professor Victor Chipofya, coordinator of MWP

Chola P (2005) ZWP/PAWD national consultative workshop. Personal communication with Mr Peter Chola, Deputy Director, Ministry of Water and Energy Development (MEWD), Zambia

Graham J, Amos B, Plumper T (2003) Principles for good governance in the 21st Century. Policy Brief 15, Institute on Governance, Ottawa

GWP-TEC (2003) Effective water governance. TEC Background Paper 4, Elanders Novum Press, Sweden

GWP-TAC (2004) Integrated water resources management (IWRM) and water efficiency plans by 2005. GWP Technical Committee, Stockholm, Elanders Inforlogistics vast Press, Sweden

GWP-TEC 10 (2004). Integrated water resources management (IWRM) and water efficiency plans by 2005. Why, what and how? Stockholm, Elanders Infologistics vast Press, Sweden

GWP-TEC (2004) Catalyzing change: a handbook for developing integrated water resources management and water strategies. Stockholm, Elanders Infologistics vast Press, Sweden

GWP (2000) Global water dialogues for water governance. Elanders Infologistics vast Press, Sweden

GWP-TAC (1999) Integrated water resources management. GWP TEC Background Paper 4. Available online at www.gwpforum.org

Hollingworth BE, Chiramba T (eds) (2005) Implementing the SADC regional strategic action plan for integrated water resources management (1999-2004): lessons and best practice. SADC, GTZ, Inwent

Hunt S, Kaite C, Kasonde T, Campbell E (1994) Nutrition and food security. analysis of the Zambia priority survey I. Food Studies Group, Oxford University

Malawi Economy (2006) The encyclopedia of the nations. Available online at www.nationsencyclopedia.com

Manzungu E (2000) Towards effective stakeholder participation for improved governance of water resources in Southern Africa. CASS, Harare, Zimbabwe

MEWD (2006) Letter to the GWP executive secretary Mr Emilio Gabrealli from the Government of Zambia. GWP continued support to IWRM planning in Zambia

MoD (2004) Letter to the GWP executive secretary Mr Emilio Gabrealli from the Government of Malawi. GWP support towards the WSSD target in Malawi

SADC (1998) Regional strategic action plan for integrated water resources management. Maseru.,SADC

SADC (2000) The revised protocol on shared watercourses. Gaborone, SADC

SADC (2000) Protocol on shared watercourses systems in Southern Africa Development Community (SADC). Signed in Windhoek, August 2000. SADC, Gaborone

Turton AR, Hattingh J, Claassen M, Roux DR, Ashton PJ (2005) Towards a model for ecosystem governance: an integrated water resource management example. Paper presented at the International Symposium on ecosystem governance, 10-13 October 2005, Kwa Maritane Bush Lodge, Pilanesberg, South Africa

Varady RG (2003) Global water initiatives: observations on their evolution and significance. Presented at the Third Annual Meeting of the International Water History Association. IWHA, University of Arizona, Udall Center for Studies in Public Policy

UNDP (1991) Status of desertification and implementation of the United Nations plan of action to combat desertification. UNEP, Nairobi

UNDP (1997) Governance for sustainable human development. UNDP Policy Document, New York

WSSD (2002) World Summit on sustainable development plan of implementation. Johannesburg, 2000. Available online at www.johannesburgsumit.org

Good Ecosystem Governance: Balancing Ecosystems and Social Needs

Malin Falkenmark

Abstract

The overarching problem behind the need for good ecosystem governance is the fact that human needs for water, food, energy, etc., generally demand manipulations of landscape components. Due to water's role as the bloodstream of the biosphere, with many parallel functions in the landscape, and ecosystems' water-dependence, ecosystems tend to get impacted by those manipulations. Societal activities that have to be incorporated in good ecosystem governance include land use change, water use, flow control, waste production, and alien species. Different distinctions have been highlighted: i.e. between avoidable and unavoidable manipulations, and between local ecological landscape components as opposed to the whole catchment as a composite ecosystem. There are also two contrasting time perspectives to keep in mind: repairing of already-manifested ecosystem degradation versus avoiding foreseeable future ecosystem degradation in a world living with change in response to strong societal driving forces. Local-scale ecosystems have to be protected by addressing their key water determinants; catchment-scale ecosystems by benefiting from water's function as an integrator through efforts to orchestrate society-driven manipulations for internal compatibility. The latter involves trade-off striking and balancing of different interests, and will demand both well organised stakeholder participation, and the definition of bottom lines and resilience criteria to protect key ecosystems. Good ecosystem governance has been characterised as follows: WHAT to govern, i.e. human activities in the landscape, HOW to govern involves an array of consecutive steps: fact finding and problem analysis; strategic plan of action; tools to make such action possible, such as legislation, financing, competent institutions, stakeholder participation etc.; and tools to secure its implementation, such as incentives/sanctions, capacity building, media campaigns, etc. The road towards good ecosystem governance will be demanding due to the dominance, at present, of partial reality-conceptualisation. A shift in thinking is absolutely essential to get out of this trap. Good understanding will be needed in the three different systems: natural biophysical system, social system, and governance system. The Trialogue hypothesis could be improved by changing the 'Science process' corner of the triangular model into a 'Biophysical process' corner. Science processes will be needed for all the three components of the Trialogue.

Keywords: ecosystem governance, science processes, social system, governance system, water cycle, catchment management, water use, Millennium Ecosystem Assessment

Introduction

The overarching problem behind the need for good ecosystem governance is the interaction between the global network of nested ecosystems and the social system. Humans try to manage the complexity of living organisms and their non-living surroundings – the ecosystem – since it provides essential goods and services on which the development of human welfare is based. Human needs are driven by population growth and expectations for wealth increase. In trying to meet those human needs, waste is introduced and various biophysical disturbances are produced, together degrading the life supporting web of ecosystems (Figure 4.1).

Human interaction with the life-supporting web of living matter has recently been assessed in the Millennium Ecosystem Assessment (MA 2005) project. To be as effective as possible in reaching decision-makers with the resulting recommendations for action, the conceptual framework used for the assessment was based on the concept of ecosystem services. The idea of using this concept is to demonstrate the fundamental dependence of human well-being on the web of ecosystems at all scales.

The outcome of the assessment was summarised in the following four findings:

- Over the past 50 years, ecosystems have been changed more extensively than ever before in human history;
- These changes have contributed to substantial net gains for society, achieved at growing cost, however, in terms of the degradation of many ecosystem services and the increased risk of sudden non-linear changes;
- The degradation could grow significantly worse in the first half of the present century; and
- Reversing the degradation, while meeting increasing human demands, will involve significant changes in policies, institutions and practices.

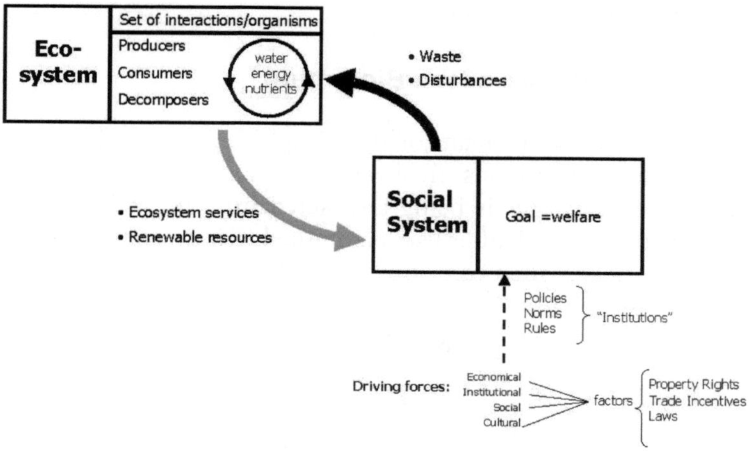

Fig. 4.1. *Humanity depends critically upon ecological links between nature and society. Because driving forces act on the social system, ecosystem management is a question of living with change while securing long-term ecosystem productivity*

In the MA conceptual framework, water is seen both as a resource and as part of the non-living surroundings of the global web. This interest in the links between the water cycle and ecosystem-related phenomena is the background to the International Symposium on Ecosystem Governance held at KwaMaritane South Africa.

This chapter will build on the author's studies on the links between water and ecosystems (Falkenmark and Folke 2002, 2003; Falkenmark 1997, 2003a, 2003b; Falkenmark and Rockström 2004) and on the experiences from the discussions in the Task Force on Environmental Sustainability of the Millennium Project (TF 6). It will try to dissect the environmental degradation phenomenon and to identify principal action opportunities in terms of ecosystem governance, what to govern, and how to govern. This chapter will first examine how water is involved in the causal chains linking human activities with impacts on ecosystems, in order better to understand WHAT to govern more exactly. This will be followed by an analysis of HOW to govern. Some fundamental governance problems of the past will be highlighted and three main components of an overall ecosystem governance are identified by benefiting from water's deep involvement.

Water and Ecosystem Change

Water – the Bloodstream of the Biosphere

Water constitutes the bloodstream of the biosphere and thereby links atmosphere, terrestrial ecosystems, freshwater flows and aquatic ecosystems (Falkenmark 2003b). Water is being abstracted from the water cycle for use in the social system and, after use, is returned to it by consumptive water use and return flows, often carrying pollutants (Figure 4.2).

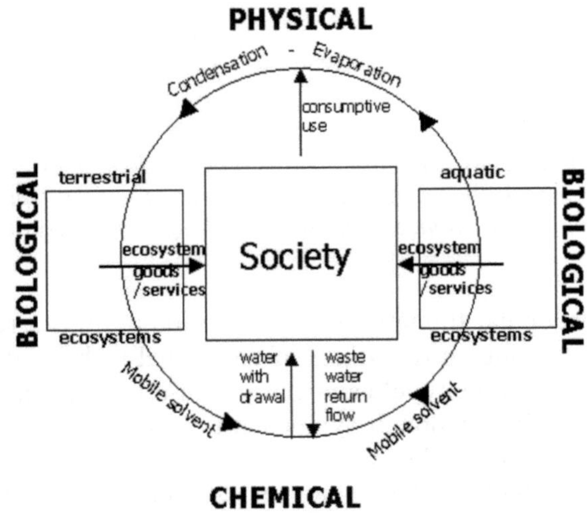

Fig. 4.2. *Water– the bloodstream of the biosphere. Schematic illustration of linkages between cycling freshwater, the terrestrial ecosystems feeding on it, the aquatic ecosystems thriving in the habitats formed by it, and human society withdrawing water from it, and after use returning it either as polluted return flow or as a vapour flow/consumptive use. The diagram also shows in which phase of the water cycle the respective set of balancing processes are most active*

Three pairs of balancing processes are involved in the water cycle system (Ripl 2003):

- Physical processes: evaporation/condensation;
- Chemical processes: dissolution/crystallisation; and
- Biological processes: photosynthesis/respiration.

Human welfare development problems are currently highlighted through the Millennium Development Goals (MDGs) and related activities. Within the Millennium Project, water is directly targeted only in terms of safe water supply and sanitation within the environmental sustainability goal (MDG 7). Water is, however, indirectly deeply involved in many of the other MDGs, especially MDG 1, the alleviation of poverty and hunger. Many of the MDGs are linked internally as a consequence of water's many parallel functions: as a resource, as a mediator, and as an integrator (Figure 4.3).

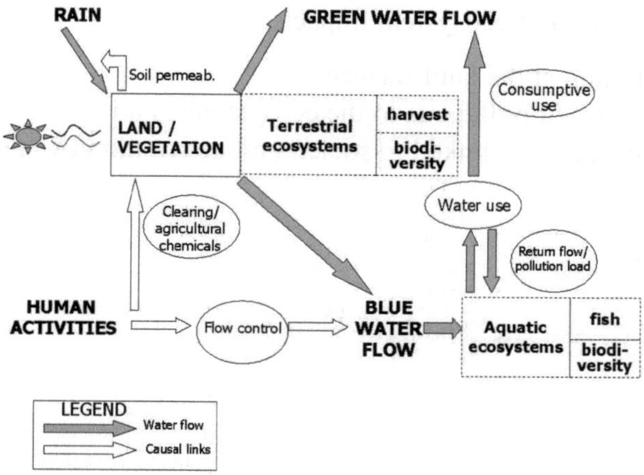

Fig. 4.3. *Water's multifunctionality makes it play many different roles in linking activities related to different Millennium Development Goals: societal water supply, income generation to alleviate poverty, food production to alleviate hunger, environmental sustainability etc.*

In the Millennium Project, the environmental sustainability aspect of human interactions with the life support system was discussed by Task Force 6 (Melnick et al. 2005). This group defined environmental sustainability as: *meeting human needs without undermining the capacity of the environment to provide for those needs and support life over the long term.*

Why do Ecosystems Change?: Looking Through the Fresh-water Lens

Human needs for food, water, energy, minerals, etc., cannot be met without manipulating the life-support system with its natural resource base and its incessant biomass production processes (Falkenmark 1997, 2003b). These manipulations interfere with three key water processes that are crucial for the generation of ecosystem impacts:

- Water partitioning at the land surface;
- Water as a carrier of pollutants to the ecosystems; and
- Consumptive water use linked to societal modes of direct water use.

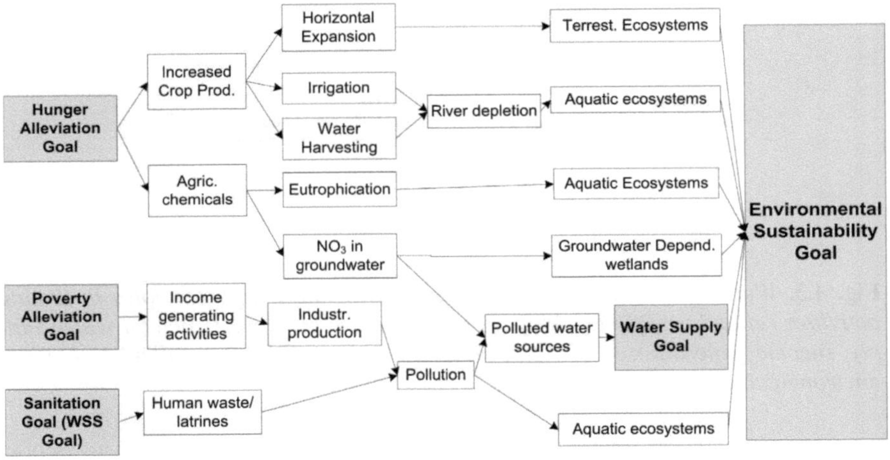

Fig. 4.4. *Ecosystems may be impacted from three different societal entry points: by land use activities, by water use and introduced pollution load, and by flow control measures. The dark arrows denote water flow and the open arrows denote causal links*

As seen from a water cycle perspective, four main types of manipulations interact with the water cycle:

- Land cover change altering soil infiltration, vegetation and therefore consumptive water use, groundwater recharge and runoff generation;
- Water withdrawal to supply society in different ways including municipal supply, industrial supply, and irrigation and, during use, partitioned between consumptive water use and return flow, the latter often loaded with pollutants;

- Waste-production that tends to follow wealth-development. Soluble waste products are caught by the circulating water and carried into water-dependent ecosystems;
- Water storage to overcome seasonality problems, and flow regime changes through stream flow regulation; and
- Invasive alien species that may impact water partitioning by altering consumptive water use.

Since ecosystems are water-dependent, they are easily impacted when water's activities in the life-support system are disturbed (Figure 4.4). Thus, although the aim of the manipulations mentioned is to meet human needs of food, water, energy and goods, the changes generated in the biosphere bloodstream system translate themselves into ecological side-effects, most of them mediated by water's multifunctionality. The consequence is environmental degradation, influencing the capacity of ecosystems to produce goods and services.

This water-based conceptual approach suggests that the manipulations that will have to be addressed by ecosystem governance are primarily the following:

1. Land-cover change, which implies alteration of rainwater-partitioning at the land surface between consumptive water use/evapotranspiration, surface runoff and groundwater recharge/dry season flow.
2. Direct water use, which involves water withdrawal from rivers or aquifers, followed by a partitioning between consumptive water use and return flow. The consumptive use components tend to cause river depletion, which can now be seen over large areas of the planet and has reached a level where many rivers in irrigation-dependent areas are now closed in the sense that no more water can be withdrawn without it causing even more serious effects on aquatic ecosystems.
3. The massive waste-production that has gone hand-in-hand with socio-economic development in the past and resulted in a build-up of water pollution all over the world, more serious in certain hot spots (Falkenmark 2005). This build-up is the result of a willful neglect of water pollution that is quite difficult to bring under control, due to its close links to industrial activities and employment.

As shown i.a. by South African experiences, alien species also have to be kept in mind, since they may cause serious problems if natural enemies are lacking. Terrestrial aliens are exemplified by the eucalyptus in South Africa, planted by forest companies but which, lacking natural enemies, have spread all over the country. This problem is now addressed by the large-scale Working for Water program by which an additional 10% of runoff is going to be gained (Van Wilgen et al. 1998). In lakes and rivers aquatic species may unbalance the ecosystem, one example being the Nile perch introduced into Lake Victoria.

Countermeasures Available

When analysing the countermeasures required to minimise ecosystem change, an analogy with medical treatments may be useful. In the medical world there are basically three types of treatment used in trying to minimise the problems of a particular manifestation of disease:

- Elimination of the cause (medicine against infections, operation against tumours);
- Reduction of the symptoms (pain reduction); and
- Build-up of immunity (vaccination).

The most obvious way to reduce the ongoing undermining of the life-support system by ecosystem degradation is to address the causes, in particular the manipulations which cause unacceptable ecosystem degradation as side effects – see example in Figure 4.5, related to the special perspective of agricultural production (SIWI et al. 2005).

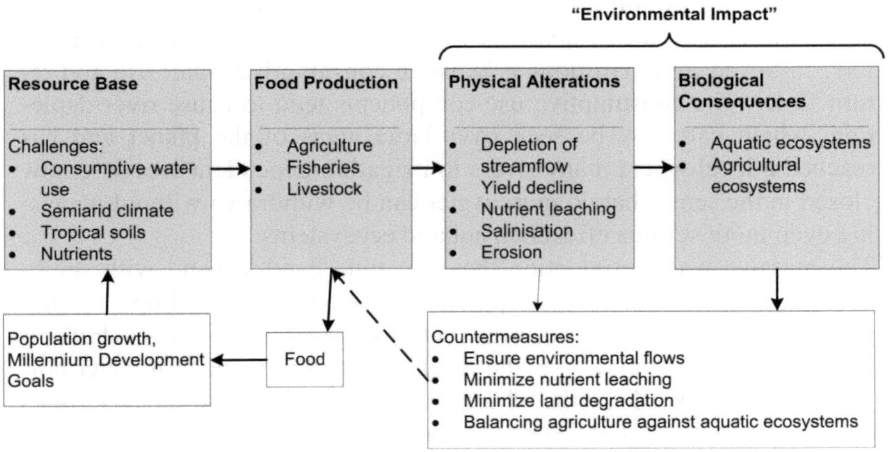

Fig. 4.5. *Links between food production to meet global food needs, environmental impacts generated by agricultural practices, and possible countermeasures towards environmental sustainability*

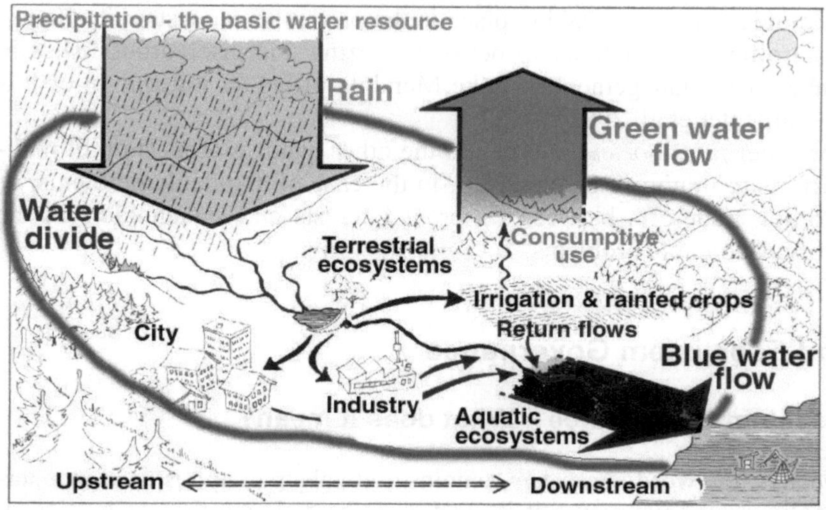

Fig. 4.6. *The catchment requires an integrated approach to all water-related phenomena at work inside its boundaries. All the rain falling within the catchment is partitioned between humans and nature, between land use/terrestrial ecosystems and water use/aquatic ecosystems, and between upstream and downstream uses and phenomena*

Some Important Distinctions

In analysing the possible countermeasures against environmental degradation, it is also important to identify the causes, in terms of whether the manipulations are avoidable or unavoidable. The ones which are principally *avoidable* include i.a. incautious land use changes, containable waste loads, use of toxic chemicals that will escape to the life-support system, etc. The *unavoidable* manipulations include i.a. consumptive water use which is part of the photosynthesis process by which biomass is produced (Falkenmark and Lannerstad 2005). For the latter type of manipulation, trade-offs will have to be struck.

There are also two contrasting time-perspectives involved: on the one hand, repairing already-manifested ecosystem degradation and, on the other, avoiding foreseeable future ecosystem degradation in response to the strong driving forces.

The *repairing* type may try either to intervene in the causal chain that has resulted in the unwanted phenomenon, for instance eutrophication of a lake, by controlling fertilisation and nutrient flows, or try to intervene in the ecological processes taking place in the particular ecosystem itself: e.g. by controlling the relationship between grazing and piscivorous fishes in the lake. The management of Lake Mendota offers an illustration of this case (Carpenter et al. 2001).

The latter *future-oriented* type, on the other hand, is a question of making efforts to minimise carelessness, in the case of avoidable problems, or of striking trade-offs between different interests and stakeholders in the case of unavoidable problems.

Good Ecosystem Governance

Ecosystem Governance – What does it mean?

Nowadays, the word *ecosystem* remains quite abstract and is used in a surprisingly amorphous way – it is used more as a slogan than as a science-based concept. Even the MA did not really explain the concept in its summary for decision-makers, but preferred to use the ecological service approach in their advocacy for action.

First of all, there are two basic types of land-based ecosystems:

- Terrestrial ecosystems, which depend on and interact with soil moisture, vapour flow and groundwater; and
- Aquatic ecosystems, which depend on water in rivers.

Forests belong to the former group; lakes, deltas and stream ecosystems to the latter. Wetlands may be of either type, or even have mixed origins.

Second, ecosystems may be of very different scales, which of course influence governance solutions. On the one hand, the concept may refer to local landscape components of particular interest from societal or scientific perspectives. Examples in terms of terrestrial ecosystems may include a certain local forest with high biodiversity, or a groundwater-dependent wetland with particularly rich biodiversity due to the shifting mix of groundwater discharge and inundating surface water.

On the other hand, ecosystems can be thought of as meso-scale conglomerates of local ecosystems, internally linked by water flows into a catchment ecosystem. This is the way the Global Environment Facility (GEF) has used the concept when highlighting the need for "land-water integration in a catchment-based ecosystem approach" (GWP 2000). When thinking at this scale, we might think of the catchment as a biological fabric linked by flows of water and nutrients.

The governance implications of these different aspects of the concept ecosystems evidently differ:

- Local-scale ecosystem protection is basically an issue of protecting the key water determinants, e.g. stream flow seasonality, groundwater table, aquifer outflow, chemical water composition, etc.; and
- Catchment-scale governance is a question of orchestrating society-driven landscape manipulations for compatibility with minimum requirements of the biological fabric. For instance, upstream water alterations have to be in balance with bottom lines for downstream aquatic ecosystems. Proper attention has to be paid to biodiversity and ecosystem resilience against creeping or sudden changes in the surroundings.

Finally, since for demographic and other reasons, societal needs evidently keep changing with time, the overarching challenge is to *live with change*. This follows from the abovementioned four findings of the MA, with the focus put on the major changes expected in the first half of the present century.

What will be needed is a radical shift in thinking and a conceptual development away from the sheer protection/conservation mode towards a balancing mode as will be discussed below.

A Key Challenge will be the Striking of Trade-offs

Taking the above approach, the basic challenge of ecosystem governance develops into the task of *controlling the side-effects of something wanted.* Since some of the side-effects of the biophysical alteration of landscape components needed to meet societal needs may be avoidable (e.g. erosion and pollution) and others unavoidable (e.g. consumptive water use), there are two basic modes involved:

- For the former type, the minimisation of carelessness, for instance in terms of pollution or flow-control rules; and
- For the latter type, the striking of trade-offs between the manipulation necessary for the wanted ecosystem service (e.g. food, timber, etc.) and the unavoidable side-effects produced.

The latter task varies, depending on whether the trade-offs are known or unknown:

- *Known* trade-offs call for analysis, the ability to balance different interests, and access to tools for both of these steps (such as EIA for the former; environmental flow and flood mimicking for the latter), stakeholder involvement, etc.; and

- *Unknown* trade-offs have to be met with preparedness (e.g. clarifying of resilience conditions) and adaptive management (e.g. a preparedness to modify decisions as the understanding advances and monitoring to watch out for unwanted ecosystem changes).

Successful governance should also seek multiple-use solutions and the benefit from possible synergies through win-win solutions.

Ability to Orchestrate for Compatibility

Good governance will depend on process-understanding within three systems – society, government and the biophysical system – and on properly functioning interfaces between government at different levels and society, in line with biophysical realities.

A basic criterion for good governance has to be the ability to ensure that societal action stays within the potential of the surrounding ecosystems to absorb the unavoidable consequences without switching into an unwanted state (e.g. from a clear lake to a turbid lake, or from a savannah grassland to a scrubland). Governance activities involve four main phases or steps: 1) fact-finding and problem-analysis; 2) strategic plan of action; 3) making those actions possible; and 4) securing their implementation.

Table 4.1. *Differences in focus in ecosystem governance*

SCALE	***REPAIRNG***	***AVOIDING***
LOCAL	*Improving key ecosystem determinants* (e.g. key pollutant in Lake Mendota)	*Protection against change of determinant of core catchment ecosystems* (e.g. groundwater recharge area protection)
CATCHMENT	*ILWRM on basin level for matching upstream/ downstream interests* (e.g. CAP for irrigation water withdrawal in Murray Darling basin)	*Planning phase striking of trade-offs* • *known trade-offs*: stakeholder participation • *unknown trade-offs*: secure resilience of core catchment ecosystems

Getting from Here to There

Moving from today's confused situation to more successful ecosystem governance will primarily call for a fundamental shift in thinking.

Past efforts towards ecosystem governance have evidently not been very effective, for a number of reasons. The present situation is characterised by a great confusion, wherein different groups are addressing different realities. A type of "technical fix" approach is frequently used, where the public discourse is directed towards finding a "silver bullet fix" that will "solve" the ecological problem at hand.

The tools developed have had their main focus on aquatic ecosystems, such as environmental flow, flood flow mimicking, etc. At the same time, terrestrial ecosystems have been surrounded by myths, such as the one incorporated in the Millennium Ecosystem Assessment instructions, claiming that forests are water *providers*. In reality, it is the other way round: global forests are generally water-consuming – they consume altogether 40,000 km^3/yr out of the 70,000 km^3/yr of continental rainfall returning to the atmosphere (Rockström et al. 1999). The main exception here is cloud forests, since they capture wind-driven horizontal rain and thereby tend to regulate water flow throughout the year and thus increase dry season flow (DFID 2005).

Dominance of Partial Reality Conceptualisation

The era of environmental degradation build-up has been an era of an unfortunate conceptual standstill. Without proper concepts, theories cannot be developed, problems cannot be correctly defined and can therefore not be solved. The serious failures, over three or four decades, of attempts to come to grips with environmental degradation can be seen as linked to several phenomena:

- Lack of overarching population/environment/development theory;
- Coarse or misleading concepts;
- Absent bridging between hydrology and ecology;
- Partial-reality oriented conventions; and
- A low level of public understanding.

The slow development that has taken place has been dominated by a biological angle, since that is how observable effects became evident, following the approach of Rachel Carson, the first observer of environmental degradation. Even today, in 2005, the community of concerned ecologists tends to limit itself to *advocacy* rather than to *guidance* on HOW to protect ecosystems and biodiversity. In the absence of proper guidelines, conservation of the existing bio-system and the development of desirable biodiversity evidently becomes difficult. What is needed to promote biodiversity is tangible definitions and implementable action on the ground. This remaining advocacy orientation is illustrated by the joint statement of the heads of the five global biodiversity-related conventions to the 2005 World Summit (the High Level Plenary Meeting of the UN General Assembly 14–16 September 2005) in which they called upon the world's leaders "to recognise that to make the MDGs a reality on a densely-populated planet, biological diversity needs to be used sustainably and its benefits more equitably shared." The statement says nothing, however, about HOW to do it, only WHY.

Efforts to explain the observations by bridging biologists' and biophysicists' understanding of the causal chains have been extremely slow, which has allowed misinterpretations to enter policies and institutions (Calder 2005). One example is the above-mentioned idea that forests are water providers and that the planting of forests would secure dry season flow in degraded areas – a view that neglects the increased consumptive use that might go with the altered vegetation. South African scientists realised long ago that forest ecosystems are water-consumptive, rather than water-providing. Hence, forestry is seen as a stream-flow-impacting activity, reflected both in the South African government organisation where water and forestry are joined in one Ministry, and in the new water legislation.

The conceptual development during the last three decades has to a certain degree been crisis-driven, with the African drought in the 1970s as an early component, generating the later much-criticised concept of desertification and reflecting a lack of transdisciplinary bridging. This crisis-driven development of a piecemeal understanding, combined with a high level of frustration and concern, is the background also to a set of unidimensional international conventions: desertification, biodiversity, climate, international waters, forest principles etc. Due to the fact that the life-support system functions as a closely woven fabric of interlinkages, many of which are water-mediated, none of these conventions would be effective if applied in isolation. A better approach would be to implement them in combination. This is the way tried within the Global Environment Facility (Duda 2003).

Critical Moves towards Good Ecosystem Governance

Looking at good ecosystem governance from a more pragmatic angle, with attention given to the need to find ways to balance humans and ecosystems, my suggestion is that it has to include the following three main aims (Falkenmark and Rockström 2004):

- *Secure* the life-support system goods and services: water security, food security, and ecological security;
- *Avoid* difficulties in terms of pollution and silting, linked to water's lift up/carry away functions and its mobility, to be achieved by water pollution abatement and soil protection; and
- *Foresee* unavoidable conflicts and difficulties linked to climatic variability (e.g. floods, droughts), driving forces at work (e.g. population growth, improved quality of life), and to water's multiple functions in non-negotiable natural processes in the landscape.

The goal for good governance can be described as reaching an *eco-hydro-solidarity*, whereby the rainwater input to a catchment is wisely organised between all different water-dependent and water-impacting activities and ecosystems. An introductory analysis will have to be undertaken in order to clarify the water-related links between major land-uses, water-uses and the ecosystem. The crucial resilience capacity of the ecosystem to absorb change, without loss of stability, must be established. There has also to be a broad realisation of the fact that a land-use decision is also a water decision, and that all ecosystems are in fact water-dependent. Without water there would not be any ecosystems (Ripl 2003).

Summarised Findings

The basic *problematique* behind the need for more successful ecosystem governance is the fact that, in order to meet societal demands of natural resource based goods and services, physical landscape manipulations are unavoidable. Besides making it possible to meet human needs, the physical alterations produce biological consequences, many of which are unavoidable and are the price paid for the services achieved. The task of controlling these biological consequences involves the minimisation of carelessness, especially in the case of avoidable problems (e.g. pollution, erosion), and the striking of trade-offs in the case of unavoidable problems (e.g. consumptive water use in terrestrial ecosystems).

It has been shown that ecosystem governance is an amorphous concept. There are two main interpretations of ecosystem governance: on the one hand, local-scale protection of highly-appreciated landscape components (e.g. a lake, a wetland or a forest), on the other hand, catchment-scale compatibility control efforts. The basic challenge is that of controlling the side-effects of something wanted by society.

The current situation is still dominated by partial reality approaches, almost a conceptual stand-still. In other words, definition of the problem remains biological, effect orientated, and with a poor understanding of the involvement of water in many different functions in both the causal chains behind these effects and, as key determinants of the terrestrial and aquatic ecosystems.

A shift in thinking is absolutely essential if we are to get out of this trap, especially in view of the strong driving forces of demographics, human rights and globalisation. This shift in thinking has to have its focus on a whole set of science, societal and government processes.

Conclusions

Is the Hypothesis Valid?

This chapter sees the task of good ecological governance as taking place at the government-society interface and as being based on an understanding of biophysical realities. This means that, in order to be successful, it will be essential to achieve a good understanding of three different systems:

- The natural biophysical system, which is the basic scene in which humans and ecosystems interact;
- The social system, where needs and wants originate, where frustration develops and where societal acceptance of trade-offs and limitations will have to be achieved; and
- The governance system, with all its intricacies, which will have to be changed in such a way as to host the task of orchestrating human activities for catchment-scale compatibility.

In terms of the validity of the hypothesis, a doubtful component is the third corner of the triangular model: "science processes". This corner will have to be replaced by "biophysical processes", which are the ones for which the governance processes will have to secure adaptation. In my understanding, science processes are needed at all three corner-processes, and therefore contributes a fourth overarching dimension to the model.

References

Calder IR (2005) The blue revolution: integrated land and water resources management. Second edn. Earthscan, London

Carpenter S, Walker B, Anderies JM, Abel N (2001) From metaphor to measurement: resilience of what to what? Ecosystems 4:765–781

DFID (2005) From the mountain to the tap: how land use and water management can work for the rural poor. Department for International Development, UK

Duda AM (2003) Integrated management of land and water resources based on a collective approach to fragmented international conventions. In: Freshwater and welfare fragility: syndromes, vulnerabilities and challenges. Theme issue, Philosophical transactions, biological sciences 358:2051–206. The Royal Society, London

Falkenmark M (1997) Society's interaction with the water cycle: a conceptual framework for a more holistic approach. Hydrological Sciences 42:451–466

Falkenmark M (2003a) Water management and ecosystems: living with change. TEC Background Papers 9. Global Water Partnership, Stockholm

Falkenmark M (2003b) Freshwater as shared between society and ecosystems: from divided approaches to integrated challenges. In: Freshwater and welfare fragility: syndromes, vulnerabilities and challenges. Theme issue, philosophical transactions, biological sciences 358:2037–2050

Falkenmark M (2005) Water usability degradation: economist wisdom or societal madness. Water International 30:136–146

Falkenmark M, Folke C (2002) The ethics of socio-ecohydrological catchment management: towards hydrosolidarity. Hydrology and Earth System Sciences, 6:1–9

Falkenmark M, Folke C (2003) Introduction. In: Freshwater and welfare fragility: syndromes, vulnerabilities and challenges. Theme issue, philosophical transactions, biological sciences 358:1917–1919. Falkenmark M, Rockström J (2004) Balancing water for humans and nature: the new approach in ecohydrology. London, Earthscan Publications, Falkenmark and Lannerstad 2005

Falkenmark M, Lannerstad M (2005) Consumptive water use to feed humanity – curing a blindspot. Hydrology and Earth System Sciences

Global Water Partnership. GWP 2000 Towards water security: a framework for action. Global Water Partnership, Stockholm

Millennium Assessment. MA 2005 Millennium ecosystem assessment synthesis report. In press

Melnick D et al. (2005) Environment and human well-being: a practical strategy. UN Millennium Project, Earthscan, London

Ripl W (2003) Water: the bloodstream of the biosphere. In: Freshwater and welfare fragility: syndromes, vulnerabilities and challenges. Theme issue, philosophical transactions, biological sciences 358:1921–1934. Rockström J, Gordon L, Folke C, Falkenmark M, Engwall M (1999) Linkages among water vapor flows, food production and terrestrial ecosystem services. Conservation Ecology 3:5

SIWI et al. (2005) Let it reign: the new water paradigm for global food security. Final Report to CSD–13. Stockholm International Water Institute

Van Wilgen BW, Le Maitre DC, Cowling RM (1998) Ecosystem services, efficiency, sustainability and equity: South Africa's working for water programme. Trends in Ecology and Evolution 13:378

The Role of Good Governance in Sustainable Development: Implications for Integrated Water Resource Management in Southern Africa

Peter J Ashton

Abstract

The philosophy and principles of sustainable development offer African countries the alluring promise of being able to develop prosperous societies and economies without exhausting the natural resource-base on which these are based. However, urgent needs for social and economic relief, coupled with shortages of social, technical and economic resources, make it almost impossible for many of these countries to achieve this ideal in the short- to medium-term. Because sustainable development depends on the choices that society makes and the participative processes by which these choices are made, there is a clear need to ensure wide acceptance of the governance processes that are used in decision-making. Here, 'good governance' is recognised as a complex and multi-dimensional concept that incorporates a guiding philosophy or set of operating principles, a preferred process or way that people interact with each other, and a desired situation or outcome. The 'Trialogue Model' of the partnership between government, civil society and science offers useful insights into the attributes of good governance and the way that this underpins and facilitates prudent resource management. An examination of the approaches used by southern African countries to manage their water resources reveals that some SADC countries have entered a development phase that is characterised by shared systems of values and a growing alignment of national and regional policies, statutes and plans. This is well aligned with the philosophy of good governance, which requires full commitment from stakeholders at all levels of decision-making. Ultimately, the extent to which a governance system can be regarded as 'good' or 'weak' depends on whether or not the five key principles of good governance (openness, participation, accountability, effectiveness and coherence) are explicit in every decision-making process that affects the livelihoods of stakeholders.

Keywords: sustainable development, Integrated Water Resource Management (IWRM), good governance, water use, Southern African Development Community (SADC), NEPAD, Millennium Ecosystem Assessment, African Union

Introduction

Africa has been generously endowed with an enormous array of natural resources that underpin the livelihoods of societies across the continent (Scholes and Biggs 2004). However, this endowment is unevenly distributed and the rates at which natural resources are used have often exceeded their replenishment rates (SARDC 1996). Many areas of Africa are now characterised by a vicious downward spiral of resource-depletion that is accompanied by increasing poverty amongst progressively larger numbers of people (Scholes and Biggs 2004). Diminishing stores of key resources such as water, arable land, food, fuel, medicines and building materials, emphasize the urgent need to find long-term solutions for these problems.

The worrying prospects of rising poverty and further economic decline have forced many African governments to seek short-term assistance for communities or regions in dire need of support. Here, various forms of overseas development aid have been seen to offer countries the opportunity to overcome their most pressing social and economic problems (World Bank 2005). However, while many of these short-term initiatives achieve their objectives, they inevitably provide purely cosmetic solutions to the underlying problems, and seldom halt or reverse the adverse conditions (Ashton 2002). Instead, many African countries have become progressively more dependent on these external sources of aid (World Bank 2005), many of which are also linked to unfavourable terms of trade that entrench economic indebtedness (GWP 2000; Gleick 2002).

These unfavourable trends cannot easily be diverted or reversed unless African governments and societies work together to transform the ways that they view, value and use their resources. However, this will be problematic for those African countries that are characterised by insufficient institutional and legal frameworks, poor governance systems, a weak infrastructure base, and insufficient scientific, technical and educational capacity (Shela 1996; Scholes and Biggs 2004). Taken as a whole, African countries face a series of daunting challenges that will require bold displays of commitment, concerted action and political will if success is to be achieved.

It is here that adherence to the principles of sustainable development, first articulated by the World Commission on Environment and Development (WCED 1987), offer African countries the greatest potential for success in their efforts to reverse or ameliorate the adverse social and economic prospects facing them. The concept of *sustainable development* reflects the fundamental interdependence between economic development, the natural environment, and people (WCED 1987), and seeks to ensure that social and economic development follow a path that enhances the quality of life of humans whilst ensuring the long-term viability of the natural systems (resources) on which that development depends (UNCED 1992).

Importantly, the relative simplicity of the words used to describe sustainable development masks a great deal of underlying complexity that must be accepted, rather than ignored, if African governments and societies are to realise their development aspirations. Attempts to improve natural resource management practices are more likely to succeed if all segments of society are able to co-operate within a multi-layered governance system that reflects their values, principles, aspirations, imperatives and objectives (Folke et al. 2002). Adherence to the guiding ethics and values that characterise 'good governance' will help to ensure that such a governance system is effective, efficient and socially relevant (Ashton et al. 2005).

This chapter examines the central role that good governance can play in supporting the principles of sustainable development, with a particular emphasis on prudent water resource management. The conceptual 'Trialogue Model' of governance that has been proposed as a possible mechanism to link government, science and civil society, is used as a framework to interrogate and interpret the role of 'good governance' in water resource management approaches. Examples of southern African development priorities and water resource management approaches are used to illustrate regional challenges and opportunities, highlighting the central role of good governance in meeting the long-term development needs of southern African societies.

Africa's Development Challenge

The Scale and Severity of the Problems Facing African Countries

The main challenges to socio-economic development that face many African governments are revealed by a comparison of the available social, economic and health status data provided by Gleick (2002), FAO (2005) and CIA (2005), and data on water availability and water dependency (FAO 2005), for each country in Africa. Data for 2004 indicate that some 38% (329 million) of Africa's total population of 866 million have no access to an improved water source, while 40% (346 million) lack access to appropriate sanitation systems (CIA 2005; FAO 2005). These numbers are a stark reminder of how difficult it will be for these countries to achieve their Millennium Development Goals. Given the relatively high levels of prevailing poverty in most African countries (37 of the 47 African countries have a per capita GDP below US$10/day; six of these countries have a per capita GDP of under US$2/day), and annual population growth rates of around 2% (CIA 2005), countries will be hard-pressed merely to match the growing demand for basic water and sanitation service provision. Attempts to deal with the accumulated backlog of people who need these services simply accentuate the severity of these problems.

The Southern African Situation

The twelve mainland African countries comprising the Southern African Development Community (SADC) represent a microcosm of the African situation, having similar levels of poverty and low levels of service provision, though the SADC states have far higher adult prevalence rates of HIV/AIDS (Table 5.1). This table illustrates the scale of the dilemma posed by the HIV/AIDS pandemic, as well as widespread poverty and health issues. Life-expectancy at birth has declined from 50 to 41 since 1990 (CIA 2005), mainly due to high infant mortality rates. This has been exacerbated by the high prevalence rates of HIV/AIDS and tuberculosis (Ashton and Ramasar 2002). The provision of social services such as education, health, transport, water and sanitation is improving in urban areas but remains weak and unreliable in many rural areas (Table 5.1).

Prolonged periods of civil war, or wars of liberation, have complicated the situation in some SADC countries (Christie and Hanlon 2001). With the cessation of these conflicts there is an even more pressing need for the countries concerned to reconstruct weakened economies, rebuild shattered infrastructure, reinstate fragmented governance systems and redress the iniquities of past political dispensations (Porto and Clover 2003). These pressures on the respective governments are accentuated by the need to ensure that good governance processes characterise their decisions and actions (Turton et al. 2006).

Highlighting the Need for Sustainable Development

Global thinking and international experience have been pivotal in shaping the international debate around the need for sustainable development in Africa and the principles that promote effective public participation in decision-making processes (IUCN-UNEP-WWF 1980; WCED 1987; UNCED 1992; WCD 2000). Now there is rapidly-growing agreement amongst African governments that socio-economic well-being and a healthy biophysical environment are inseparable (UNECA 1991, 1992). In addition, there is also a growing awareness of the importance of international collaboration and the vital role that the citizens of a country should play in both local- and national-scale environmental management, especially with regard to alleviating the problems linked to poverty and resource degradation (Hirji et al. 2002; Turton et al. 2006).

In Southern Africa, acceptance of these concepts and approaches has been marked by the development of new national policies and laws, as well as regional protocols, accords and strategies, which formalise the need to engage stakeholders in appropriate decision-making processes (e.g. SADC 1998; IUCN-ROSA 2001). However, shortages of economic, technical and human resources, plus differences in managerial systems within specific countries, as well as the diverse array of rights, obligations and practices of stakeholders, have made it difficult to implement these instruments in practice (Ashton and Chonguiça 2003). In addition, the presence of dual legal systems in most Southern African countries has often led to the trivialisation of customary or traditional laws and practices (Ashton and Neal 2005). As a result, many individuals and communities feel alienated from planning and management processes because they believe that their beliefs and value systems have been discounted in the new national approaches (IUCN-ROSA 2001).

Southern Africa's pressing need for social and economic development has prompted the governments of the SADC countries to focus on broader issues of social equity and resource stewardship. National development goals have changed from their original primary focus on resource extraction, and now focus instead on broadening the participation of society in each country's economic base (SADC 2005). In the SADC countries, specific objectives focus on improving the social and economic equity of the poor majority, the prudent and rational management of natural resources, the eradication of pervasive poverty, the development of the human population, and the improvement of economic growth.

In terms of the SADC vision for sustainable development (SADC 2005), the region must:

- Accelerate economic growth with greater equity and self reliance;
- Improve the health, income and living conditions of the poor majority; and
- Ensure equitable and sustainable use of the environment and natural resources for the benefit of present and future generations.

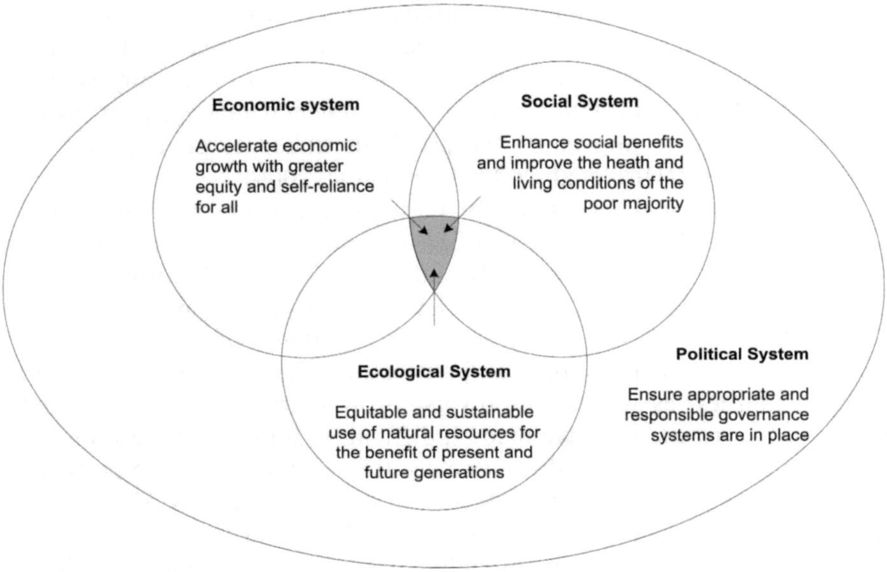

Fig. 5.1. *Conceptual diagram showing the four interacting systems and their principles that promote movement towards the ideal of sustainable development in SADC countries (shaded). Figure redrawn from Ashton and Chonguiça (2003)*

On their own, the attainment of these three goals cannot ensure the success of sustainable development initiatives and an important fourth dimension, governance, needs to be added (Figure 5.1). SADC governments have recognised that an increased emphasis on policy implementation depends on the creation of institutions of governance that can meet stated objectives. Ideally, these governance systems need stakeholders to engage transparently throughout the process, from local to national levels, and for partnerships to be formed between governments and private individuals or organisations. These partnerships have become increasingly important in countries whose governments lack sufficient resources to discharge their responsibilities effectively (IUCN-ROSA 2001).

Table 5.1. *Selected demographic, economic and health statistics for the twelve mainland African countries comprising the Southern African Development Community (SADC), plus equivalent statistics for the entire SADC region*

Statistic	Units	Country Statistics												SADC Totals
		Angola	Botswana	DRC	Lesotho	Malawi	Moz bique	Namibia	S. Africa	Swaziland	Tanzania	Zambia	Zimbabwe	
Population														
Total area of country	km^2	1 246 700	600 370	2 345 410	30 355	118 480	801 590	825 418	1 219 912	17 363	945 087	752 614	390 580	9 293 879
Total population (June 2005 estimate)	Millions	11.919	1.640	60.086	1.867	12.159	19.405	2.031	44.344	1.174	36.766	11.262	12.747	215.399
Population growth rate (June 2005 estimate)	%	1.90	0.00	2.98	0.08	2.06	1.48	0.73	-0.31	0.25	1.83	2.12	0.51	1.58
Percentage of population urbanized	%	31	64	29	25	14	35	37	49	32	25	43	33	34
Life expectancy at birth (June 2005 estimate, includes HIV)	Years	36.6	33.9	49.4	36.7	37.0	40.3	43.9	43.3	35.7	45.2	39.7	36.7	41
Adult percentage of population (aged 15+ years)	%	53.7	57.5	44.9	61.0	50.4	54.1	57.7	64.5	55.6	53.4	51.1	57.1	53.5
Adult literacy rate	%	42	78	63	84	62	47	84	86	81	78	79	89	71
Economy														
Country GDP (as PPP$ in 2004)	US$ Billions	23.2	15.1	42.7	5.9	7.4	23.4	14.8	491.4	6.0	23.7	9.4	24.4	687.3
GDP growth (from 2003 to 2004)	%	11.7	3.5	7.5	3.3	4.0	8.2	4.8	3.5	2.5	5.8	4.6	-8.2	4.3
Per Capita GDP (PPP$ in 2004)	US$ / P	1 946	9 177	711	3 156	609	1 205	7 268	11 082	5 127	645	835	1 912	3 191
Population below poverty line (US$1/day 2002-2003)	%	50	24	51	43	42	38	35	15	19	20	64	59	41
GINI Coefficient (gap - highest and lowest 10%)	-	0.54	0.54	0.63	0.56	0.62	0.40	0.70	0.59	0.51	0.38	0.53	0.63	0.58
Health														
Proportion of total population malnourished	%	51	24	58	27	39	51	26	8	8	44	45	39	40
Under-5 mortality rate	%	2.60	1.10	2.05	1.32	1.83	1.97	0.67	0.71	1.49	1.65	2.02	1.23	1.55
HIV/Aids prevalence (June 2004)	% of adults	3.9	37.3	4.2	28.90	14.2	12.2	21.3	21.5	38.8	8.8	16.5	24.6	7.1
Malaria prevalence	%	8.80	4.87	37.50	0.15	27.68	18.12	1.50	0.14	2.84	1.21	34.20	5.41	16.55
Tuberculosis prevalence	%	0.17	0.51	0.19	0.29	0.23	0.23	0.13	0.32	0.33	0.21	0.32	0.21	0.24
Improved water source (Urban)	%	34	100	79	88	90	81	98	96	81	90	80	85	28
Improved water source (Rural)	%	44	90	26	74	44	41	67	73	64	57	48	73	32
Appropriate sanitation (Urban)	%	38	92	62	72	92	66	95	90	86	95	86	71	25
Appropriate sanitation (Rural)	%	30	53	25	50	70	36	47	80	57	86	64	57	36

Social, economic and health status data sourced from CIA (2005), FAO (2005) and UNAIDS (2005); data on access to improved water supplies and appropriate sanitation taken from Gleick (2002). SADC totals are weighted by population

The Importance of Governance

Four key African initiatives have emphasised the importance of strong political commitment and good governance in all African countries, to ensure that development does not destroy the resource base on which it is based (OAU 1985; UNECA and UNEP 1989; UNECA 1991, 1992). The central theme that emerged from these meetings was the recognition that sustainable development was essential, but would only be achieved if African countries had sufficient capacity in terms of viable institutions, appropriate and relevant technology, as well as adequate human and financial resources (SARDC 1996). Significantly, it was also recognised that achievement of this ideal required all sectors of society (government, civil society – or the lay public – and scientists or technology providers) to cooperate closely and share a common vision of the future. This provides strong support for the 'Trialogue' model of governance that links government, civil society and science, and which promotes close collaboration and interactions between these three sectors. Because government and science are both components of society, it is perhaps more useful to indicate the third component of the 'Trialogue' as the wider, or lay, public (Figure 5.2).

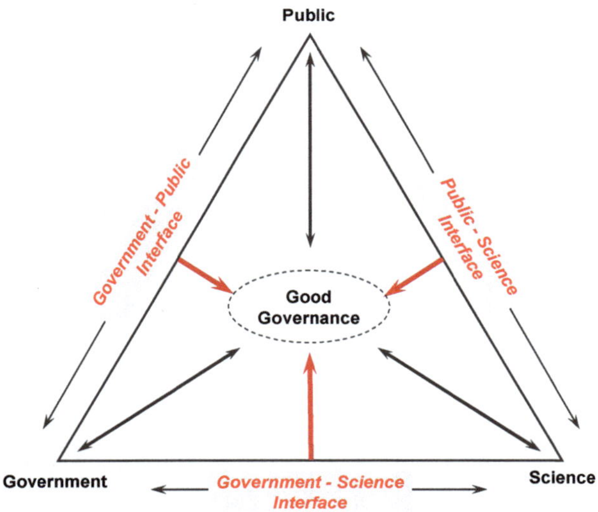

Fig. 5.2. *Conceptual diagram illustrating the linkages and interfaces between the lay public, government and science, and their collective contribution to 'good governance'*

Increased support for this inter-dependent, 'Trialogue' concept of governance has arisen in recent years from the rapidly growing awareness of the need to promote good governance at all levels of government and across all sectors of society (Ashton et al. 2005). Additional impetus has been provided by the fact that many African countries share similar visions and face comparable problems, while also sharing several geographic, historical, cultural and linguistic ties that supersede political boundaries (Ashton and Chonguiça 2003). This has prompted the formation of a variety of regional institutional structures such as SADC, and continental initiatives such as the New Partnership for Africa's Development (NEPAD). All of these initiatives aim to improve the engagement of stakeholders from every sector of society in decision-making processes that affect their livelihoods and well-being, and are a direct response to the growing demands from society for governments to be more accountable and transparent, and to respect human rights. Taken together, these initiatives reinforce the need for countries to broaden and strengthen their systems of governance at all levels, so that they can accommodate and comply with decision-making processes and management structures that now extend beyond national boundaries (MacKay and Ashton 2004).

In practice, the contextual components of governance are sometimes misunderstood, or there is a tacit assumption that every person shares a common set of ideals and goals. In reality, despite similar displays of patriotism or other outward signs of unity, people living within the same community often have wide differences in their goals, ideals and aspirations; equally important differences exist between different communities. A similar type of problem hampers the effectiveness of governance systems that assume all stakeholders can be engaged and informed in a uniform way. Again, the reality is quite different; stakeholders vary widely in their ability to understand and adopt governance processes or instruments that they are not familiar with (Turton et al. 2006). Ultimately, an ideal governance system needs to ensure that the participation of stakeholders at all levels is carefully balanced and integrated so that the best and most sustainable outcomes can be agreed to and achieved.

The preceding discussion has illustrated two key principles of governance systems. First, the context of a governance system that is related to the use of natural resources will be shaped partly by social values and imperatives, and partly by the constraints and opportunities afforded by the natural resource system around which the governance system has evolved (Ashton et al. 2005). Secondly, the effectiveness of a governance system will be determined by the extent to which stakeholders are informed and engaged in decision-making processes, and benefit from the outcomes of these processes.

In the context of natural resource management, the ideal governance system should include all the mechanisms and processes for managing the natural resource in question according to goals that reflect the agreed aims, objectives and values of society. Such an idealised governance system would include all three sectors of society (government organisations, non-government organisations and civil society), and would be stratified from international through to local scales (Ashton et al. 2005). This governance system would also consist of several levels, from principles, through policies and legislation, to regulations and practice (MacKay 2003). The complete 'net' of governance is three-dimensional, spanning level of governance, geographic scale and responsible agents (Ashton et al. 2005). In essence, therefore, the term 'governance' is a complex and multi-dimensional concept, since it incorporates three key components, namely:

- A guiding 'philosophy', or core set of agreed operating principles;
- The preferred 'process' that guides the way that people interact with each other and with institutions; and
- A desired set of 'products' or outcomes.

Understanding the Dimensions of 'Good Governance'

Despite clear evidence of the benefits to be gained by ensuring that the broader public are correctly engaged in resource management decisions, there is little guidance on how best to achieve this. Experience has shown that not all approaches are equally effective, and an approach that succeeds in one situation may not work as effectively in another. Where acrimonious disputes arise between stakeholders and authorities, the authorities may be prompted to avoid future stakeholder involvement (Bruch et al. 2005). In the absence of an agreed system or set of standards whereby states can ensure effective public participation, and with no guarantee that a system will have the desired outcome, decisions on how best to engage the public in decision-making often remain ad hoc in nature. In democratic societies, the ballot box allows citizens to hold their governments accountable for their commitments. If this mechanism fails to bring about the desired changes then the problem rapidly increases in complexity and significance, and accentuates public perceptions of the need for greater transparency and participation in decision-making processes (Bruch et al. 2005).

Where public involvement has been constructive, the process has proved to be a collectively enriching experience of consensus-building that has enhanced the substantive content of the final decisions and improved the working relationships among the various parties (Bruch et al. 2005). Ultimately, this develops greater feelings of 'ownership' and 'accountability' amongst participants and a far better chance that the agreement reached will be successful in the longer term. This is important since it is the community members 'on the ground', rather than state ministries or institutions, that will determine whether resource management principles, policies and programmes are effective (Ashton and Neal 2005).

Given the complex and multi-dimensional nature of governance, all participants should clearly understand their roles and responsibilities, and adhere to a set of common principles that define "good governance" if such a system is to be effective, efficient and socially relevant (Ashton et al. 2005). In this context, good governance needs to be based on, and to incorporate, the attitudes, values and practices of society, while also giving focus and meaning to the aspirations and objectives of society. The European Union has presented a set of five principles of good governance – as the basis for their attempts to improve performance (European Union 2001). These principles, and their supporting rationales, are:

- **Openness**, where governance institutions work in an open and transparent manner, communicating freely about what they do and the decisions that are taken, using language that is accessible to and understandable by the general public. All stakeholders are fully informed of their rights, roles and responsibilities, they are provided with sufficient relevant information in a form that is appropriate for them to take a balanced decision, and their ability to make decisions is not hampered by an inability to articulate their positions due to literacy levels or financial circumstances;
- **Participation**, where the quality, relevance and effectiveness of policies, legislation, regulation and practice depend on wide public participation throughout the policy chain, from conception to implementation, to create greater confidence in both the institutions of governance and in the outcomes of policy. There is an explicit demonstration that democratic values underpin stakeholder representation, participation and power sharing, the processes of decision-making are visibly fair and equitable to all parties, and it is clear that decisions are just and are based on the outcomes of consensus-seeking and joint fact-finding processes;
- **Accountability**, where every role in the legislative, administrative and executive processes is made clear; where every institution takes full responsibility for what it does, and where there is appropriate clarity and responsibility from everyone who is involved in developing and implementing policy at every level. There are clear and explicit demonstrations of honesty by participants and leaders at all levels, human rights

and freedoms are protected, and the processes and decisions adopted are aligned with, and supported by, the information provided to stakeholders;
- **Effectiveness**, where policies are both timely and appropriate, delivering what is needed, based on clear and agreed objectives that have taken account of past experience and potential future impacts. Effectiveness also depends on implementing policies in a proportionate manner and on taking decisions at the most appropriate level; and
- **Coherence**, where both policies and implementation actions are clearly aligned and well understood by all participants, and are consistent with other related initiatives within a complex system.

While close adherence to these five principles is seen to be all that is necessary to provide the necessary impetus for maintaining good governance systems in Europe (European Union 2001), the first three of the principles listed above need to be given special attention in the pursuit of good governance in the African context. More specifically, these problems relate to: low levels of literacy and a lack of familiarity with technical terminology; widespread poverty that was initiated during previous administrative dispensations and has been sustained by continuing inequalities in terms of access to resources and finance; and a lack of familiarity with democratic processes, often accompanied by mistrust of unfamiliar representatives and self-appointed 'leaders'.

Here it is important to note that good governance requires a systems approach that is based on the inclusion and inter-dependence of all its components. All the principles listed above need to be included and integrated into a coherent system; none of them should be ignored, avoided or diminished. The effectiveness of a governance system does not relate to the extent to which one or more of these principles is included. Instead, governance systems become progressively impaired where these principles are ignored, or where insufficient time is allowed for stakeholders to be informed or helped to participate on equal terms with their peers. In cases where stakeholders were unable to participate properly, this led to a situation that is commonly referred to as "the illusion of inclusion" (Ashton and Chonguiça 2003). A similar situation arises where there is an implicit assumption that English is the universal language of development and no provision is made to ensure that non-English-speakers are equally well-informed or able to participate in decision-making processes. In such situations, the resulting outcomes seldom meet the requirements of good governance.

The Role of Good Governance in Integrated Water Resource Management (IWRM)

Setting the Scene: The Availability of Water in Africa

On average, Africa has some 4,645 km^3 of freshwater, about 70% of which consists of annually renewable surface water resources (FAO 2005). Approximately 30% of Africa's water resources are groundwater, a large proportion of which consists of so-called "fossil water" located under the Sahara Desert and which has not been recharged for several thousand years (Kuwairi 2004). Africa's 63 shared river basins cover approximately 63% of the surface area of the continent, contain 78% of the human population and, more significantly, hold over 90% of the continent's surface water resources (Ashton and Turton in press; Turton et al. 2006). While far less information is available on the water contained in the 38 shared aquifer systems (UNESCO-ISARM 2004), or the precise extent to which countries in the more arid regions of Africa rely on these systems, they represent critically important sources of water for the countries located in Africa's desert and semi-desert areas (Kuwairi 2004). Similarly, many major cities, towns and smaller communities across Africa rely wholly or partially on groundwater resources to meet the needs of domestic and industrial water users (UNESCO-ISARM 2004).

In terms of the water crowding index devised by Falkenmark (1986, 1989), 13 countries in Central and West Africa have ample water available for each person and enjoy relative "water security". In contrast, 16 African countries experience occasional water shortages, nine countries experience frequent water shortages and droughts, while nine North and East African countries experience "chronic water scarcity". The extent to which African countries rely on shared surface water resources can be seen in their water dependency ratio values, i.e. the percentage of a country's renewable water resource that is generated outside its borders (FAO 2005). Interestingly, five of the nine African countries that experience chronic water scarcity receive no inflows from neighbouring states; the remaining four countries (Egypt, Eritrea, Kenya and Libya) obtain between 29,7% (Eritrea) and 96% (Egypt) of their water from their neighbours. Collectively, the SADC countries of southern Africa contribute some 26 km^3 of water to their northern neighbours; the majority of this water is derived from high rainfall regions within Angola and the Democratic Republic of Congo (FAO 2005).

The Critical Role of Water in Sustainable Development

Reliable and sustained access to water supplies is a strategic issue for developing countries situated in arid and semi-arid regions, because the availability of water is fundamental to the economic growth and security of all states. In Southern Africa, the significance of water is clearly illustrated by the fact that the first cooperation protocol signed within the SADC region was the Protocol on Shared Watercourse Systems (Ramoeli 2002). Also, Heyns (2002) noted that one of the major development challenges facing the SADC region in the near future is the need to implement regional water sharing and transfer schemes that can alleviate the economic limitations imposed by looming water scarcity in some countries. This is a strong call for a cooperative regional programme to develop the water supply infrastructure needed for future economic growth.

Several social and economic studies (e.g. Gleick, 1998; Hirji et al. 2002) have shown that shortages of infrastructure, institutions, money and skills worsen the problems caused by water shortages, and impede the ability of countries to achieve sustainable social and economic development. So-called 'poor' states seldom achieve sustained security of water supply because they cannot mobilise sufficient economic, human and technological resources. Conversely, relatively 'rich' states can more easily deploy a variety of economic and technological resources to resolve their water supply problems (Ashton 2004).

The Importance of Integrated Water Resource Management (IWRM)

Modern approaches to water resource management acknowledge that water resources can only be managed effectively and efficiently when the entire river basin or catchment forms the basic management unit. Furthermore, surface water and groundwater are closely interlinked, and they should be managed together as a single resource. These principles form the foundation for integrated water resource management (IWRM), and have been widely accepted throughout the world. An equally important component of the IWRM philosophy, and one that is often conveniently forgotten or ignored, is that effective implementation of IWRM requires the engagement of all stakeholders in decision-making processes (Acreman 2004).

Several southern African countries have recognised the need for IWRM approaches and have already drawn up policies, implemented the required legislation, and initiated actions designed to achieve these objectives within their territories (Hirji et al. 2002). Additional evidence is provided by the increased numbers of joint river basin commissions or other institutions on shared rivers (Ashton and Turton in press). Though several of these institutions are still in the early stages of development, they offer promising evidence of a common commitment to fair and equitable management of shared water resource systems.

In their ideal form, IWRM approaches to catchment (or river basin) management provide both the guiding philosophy and a practical framework for actions that promote cooperative decision-making and responsible management of water (and other) resources. If these two factors are implemented effectively, the outcome is prudent water resource management within the river basin. Because IWRM comprises a guiding philosophy, a set of agreed operating procedures and a desired outcome, it clearly shares these characteristics with the concept of good governance. Consequently, it should come as no surprise that these basic similarities are complementary and reinforce each other at every level, thereby extending the suite of advantages to be gained from ensuring that good governance is an integral part of IWRM.

Importantly, while effective and efficient water management institutions are usually regarded as 'technocratic', they rely on good governance processes to ensure that all government and civil society stakeholders are engaged effectively in decision-making processes. A critically important element of such partnerships is the awareness that the rights and obligations of each party are mutual and reciprocal, rather than unilateral (Wolf 1999).

Where is 'Good Governance' Working in Practice?

The answer to this important question is: "partly". Many Southern African countries, most notably Malawi, South Africa, Zambia and Zimbabwe, have made significant steps to improve their governance systems for natural resource management (SADC 1998). These countries have forged strong partnerships between different government departments, technology providers in the private and public sector, and stakeholder groups, which share common objectives and appreciate the need to work closely together. This has enabled these groups to translate their closer alignment into remarkable successes in terms of the development and implementation of new policies, statutes and plans (Bruch et al. 2005). While these results are very promising, and occur at a relatively small-scale, it is still too soon to evaluate the ultimate outcomes of these initiatives. Nevertheless, these programmes provide valuable and very visible examples and incentives to other Southern African countries.

Elsewhere in the SADC region, countries such as Botswana, Lesotho, Mozambique and Swaziland are working to align their policy frameworks more closely with the environmental sustainability principles advocated by regional environmental and water sector coordinating units (SADC 1998; Hirji et al. 2002; Acreman 2004). However, despite these welcome policies and their associated legislative and institutional reforms, some SADC countries still retain elements of their original segmented focus (Ashton 2004). Clearly, these collaborative efforts need to be supported and extended if all Southern African countries are to meet their Millennium Development Goals and attain their individual and collective aspirations of sustainable development.

Charting the Road Ahead for IWRM in Southern Africa

Despite the fact that the SADC countries have accepted that IWRM approaches offer the best prospects for prudent long-term management of their water resources, and have publicly supported the need for good governance, there are still several 'hurdles' to be overcome before success can be achieved.

Full cooperation between countries in matters regarding the management of their own and shared water (and other) resources will require alignment and harmonisation of their various resource management policies and legislation. This is not a trivial undertaking, particularly where the countries may have differing social, political, economic, and judicial systems. The private sector, civil society and various technical specialists (i.e. 'science') will also need to support these initiatives because they would be most affected by any changes that occurred; similarly, governments would be unlikely to achieve the required degree of compliance if they worked in isolation from their constituents. This will require all participants to maintain close working partnerships based on the principles of good governance. While this may be difficult to achieve, the potential benefits that could accrue to each country far outweigh the potential difficulties in realising this objective.

Five key steps or processes can be identified where Southern African countries need to reach agreement so that the necessary alignment and harmonisation of policies, legislation and procedures can be achieved.

These steps are listed and then discussed, briefly, below:

- Creating a shared vision for the future;
- Aligning policies and legislative frameworks;
- Creating sufficient human (professional) capacity;

- Creating appropriate institutions of governance; and
- Designing processes for conflict prevention, mediation and compensation.

Creating a Shared Vision for the Future

The launch of the African Union and the NEPAD strategy provide clear evidence that Africa's leaders share a broad vision for the future development of the continent that extends beyond any social, economic and political differences they might have (Hirji et al. 2002). In particular, NEPAD represents a firm commitment by the African Heads of State to the principles of sustainable development and good governance as essential prerequisites for reducing poverty and ensuring social and economic upliftment. These guiding principles are also founded on a shared realisation that greater regional integration is an essential requirement for sustainable development in the region (SADC 1998, 2005).

Aligning Policies and Legislative Frameworks

Interestingly, the differences in the policies and legislative frameworks of African countries that share a water resource are related more to the ways in which they are implemented, than to the ways in which they are structured (Pallett 1997; Heyns 2002). These inconsistencies have arisen because of differing priorities in the countries concerned, making it difficult for the respective authorities to achieve comparable levels of management efficiency and control over the resources in their care. However, it is important to note that the central components of the policy and legislative frameworks in each country share several similarities and were drawn up from identical or very similar principles (SADC 1998).

Creating Sufficient Human (Professional) Capacity

The chronic shortage of trained technical and scientific personnel and well-informed stakeholders is a perennial problem for resource management institutions throughout Africa (GWP 2000; Kakonge 2002). However, the devastating effects of the HIV/Aids pandemic are particularly visible in southern Africa (Table 5.1) where up to 38% of adults may be HIV+ in some countries (Ashton and Ramasar 2002). This has enormous implications for all capacity-building efforts and constrains the development of national and regional efforts to manage resources.

Creating Appropriate Institutions of Governance

The effective, efficient and integrated management of a water resource within a country, or one that is shared by two or more countries, requires explicit adherence to the principles of good governance at all levels. At the highest level, governments must trust each other and there must be a firm commitment to inter-state collaboration and co-operation (Lundqvist 2000). None of these responsibilities are easily incorporated into existing institutional (government) structures within participating countries and many of the policies, priorities and strategies that are needed extend well beyond the line function boundaries of conventional government departments (Wolf 1999). At lower levels, stakeholders must display a common commitment to participating fully and effectively in decision-making processes, being responsible and accountable for their decisions and actions (Bruch et al. 2005).

Dealing with Conflict Prevention, Mediation and Compensation

For stakeholders, one of the most controversial aspects of development is the seeming inevitability with which approval is granted for development projects to proceed, whilst local communities and individuals feel that they are the ones who bear the real 'costs' in terms of lost 'ownership' or access to resources. Part of this problem centres on the perception that the natural resource-bases of a country are 'public goods', with the government as custodian (Christie and Hanlon 2001). This is particularly true of scarce natural resources such as water or arable land, where private ownership of the resource has been replaced with the right of equitable use of the resource (Hirji et al. 2002). Stakeholder perceptions need to be addressed very sensitively and, where real loss of access or use has occurred, there may be grounds for considering some form of compensation. The process of identifying and eliciting these concerns and helping to prevent possible conflict requires input from skilled mediators and facilitators (Ashton and Chonguiça 2003). Close adherence to the principles of good governance will help to avoid potential disputes or conflicts.

Conclusions

The 'emergence' of the African Union (AU) and the declaration that NEPAD will provide the framework for Africa's future development trajectory offers exciting opportunities for future regional-scale and continental-scale development projects. The individual and collaborative efforts of all African states will be needed in order to realise the objectives of the AU and NEPAD. This is equally true for individual countries and for multi-national situations such as the management of a shared river basin. The same holds true for the SADC countries, which have already formed several institutions for the collaborative management of their shared river systems (Hirji et al. 2002; Turton et al. 2006).

The growing awareness of the urgent need for sustainable development throughout Africa provides the greatest stimulus for national governments to work together to develop shared, regional approaches to resource management (GWP 2000). Once again, the SADC countries represent a microcosm of the rest of Africa and these countries have acknowledged that the principles of sustainable development must underpin both national and regional development initiatives. However, while important lessons can be learnt from attempts to adopt sustainable development approaches elsewhere in the world, it is important to remember that many of Africa's problems have unique local characteristics that may prevent or hinder the direct 'transplantation' of a solution that has been generated elsewhere (Heyns 2002). In every case, it is essential to understand fully the characteristics and needs of the African situation so as to ensure that, whatever solution is proposed, it is able to meet these requirements over the long-term.

The proposed Trialogue Model of governance highlights the need for government, science and civil society or the lay public to form lasting partnerships that will promote responsible decision-making and shared responsibility for prudent water resource management. These partnerships also require each group to accept the need for formal governance structures, processes and instruments that can complement and strengthen the underlying philosophy of cooperation. In turn, for these to be truly effective, all stakeholders need to understand the multidimensional nature of governance and their individual roles and responsibilities. Taken together, the five principles of good governance provide a suitable 'blueprint' for building and guiding effective and responsible interactions between stakeholders.

The acceptance and implementation of IWRM approaches in Southern Africa has highlighted the critical need for skilled personnel and informed and committed stakeholders; this is particularly important for the prudent and effective management of shared water resources. This problem can only be addressed effectively through the concerted and collaborative efforts of all the SADC states (GWP 2000). The individual and collective responses of SADC countries to the HIV/AIDS pandemic will determine whether or not they are able to develop and retain the necessary skilled personnel. The suite of national and regional responses will also influence the extent to which individual stakeholders will be able to contribute to the required partnerships between government, science and the lay public.

A closely related issue is the need for appropriate trans-national institutional structures that can effectively manage shared resources on behalf of the states concerned (Pallett 1997; Heyns 2002). These institutions would help to demonstrate good corporate and public governance practices, promote genuine regional cooperation amongst the states concerned, and help to enhance the social and economic development of the region and the continent as a whole. There is good evidence to indicate that the early attempts to form joint institutional structures to manage shared water resources in Southern Africa have provided valuable insights and learning that can be used as a firm basis for future actions. These initiatives now need to be extended and expanded to formally embed the principles of good governance, and then demonstrate these in practice, so that the SADC region and its component states can attain their development goals and aspirations.

Acknowledgements

Three anonymous referees are sincerely thanked for their constructive suggestions on how best to improve this paper.

References

Acreman M (2004) Water and ecology. In: Dooge J, Selborne J (eds) Water and ethics. Paris, United Nations Educational, Scientific and Cultural Organization (UNESCO)

Ashton PJ (2002) Avoiding conflicts over Africa's water resources. Ambio **31**:236-242

Ashton PJ (2004) Water and development: a Southern African perspective. In: Trottier J, Slack P (eds) Managing water resources past and present: the 2002 Linacre lectures. Oxford, Oxford University Press, pp 149-174

Ashton PJ, Chonguiça E (2003) Issues and trends in the regional harmonization of EIA processes in Southern Africa, with special reference to trans-boundary and cumulative impacts. In: Chonguiça E, Brett R (eds) Assessing the need for a regional approach to environmental impact assessment in Southern Africa. Harare, IUCN Regional Office for Southern Africa, pp 26-59

Ashton PJ, Neal MJ (2005) Public involvement in water resource management in the Okavango river basin. In: Brusch C, Jansky L, Nakayama M, Salewicz KA (eds) Public participation in the governance of international freshwater resources. Tokyo, United Nations University Press, pp 169-195

Ashton PJ, Patrick MJ, MacKay HM, Weaver AvB (2005) Integrating biodiversity concepts with good governance to support water resources management in South Africa. Water SA (31)4:449-456

Ashton PJ, Ramasar V (2002) Water and HIV/AIDS: some strategic considerations for Southern Africa. In: Turton AR, Henwood R (eds) Hydropolitics in the developing world: a Southern African perspective. Pretoria, African Water Issues Research Unit (AWIRU), pp 217-235

Ashton PJ, Turton AR (in press) Water and security in Sub-Saharan Africa: emerging concepts and their implications for effective water resource management in the Southern African region. In: Brauch H-G, Grin J, Mesjasz C, Krummenacher H, Behera NC, Chourou B, Spring UO, Liotta PH, Kameri-Mbote P (eds) Facing global environmental change: environmental, human, energy, food, health and water security concepts – volume II. Berlin, Springer-Verlag

Bruch C, Jansky L, Nakayama M, Salewicz KA, Cassar A (2005) From theory to practice: an overview of approaches to involving the public in international watershed management. In: Brusch C, Jansky L, Nakayama M, Salewicz KA (eds) Public participation in the governance of international freshwater resources. Tokyo, United Nations University Press, pp 3-18

Christie F, Hanlon J (2001) African issues: Mozambique and the great flood of 2000. Bloomington, Indiana University Press

CIA (Central Intelligence Agency) (2005) The world factbook 2005. Washington DC, Central Intelligence Agency of the United States of America. Available online at http://www.cia.gov/cia/publications/factbook/geos [Accessed November 2005]

European Union (2001) European governance: a white paper. Brussels, Commission of the European Communities. Available online at http://europa.eu.int/-comm/governance/white_paper/index_en.htm [Accessed May 2003]

FAO (Food and Agriculture Organisation of the United Nations) (2005) Aquastat information service on water in agriculture and rural development. Rome, Food and Agriculture Organization of the United Nations. Available online at http://www.fao.org/waicent/FaoInfo/Agricult/AGL/AQUASTAT/afric.htm [Accessed June 2005]

Falkenmark M (1986) Freshwater: time for a modified approach. Ambio (15)4:192-200

Falkenmark M (1989) The massive water scarcity now threatening Africa: why isn't it being addressed? Ambio (18)2:112-118

Folke C, Carpenter SR, Elmqvist T, Gunderson LH, Holling CS, Walker BH, Bengtsson J, Berkes F, Colding J, Danell K, Falkenmark M, Gordon L, Kasperson RE, Kautsky N, Kinzig AP, Levin S, Mäler K-G, Moberg F, Ohlsson L, Olsson P, Ostrom E, Reid WV, Röckstrom J, Savenije HHG, Svedin U (2002) Resilience and sustainable development: building adaptive capacity in a world of transformation. Scientific background paper for the World Summit on Sustainable Development. Stockholm, Environmental Advisory Council

Gleick P (ed) (1998) The world's water 1998-1999: the biennial report on freshwater resources. Washington DC, Island Press
Gleick P (ed) (2002) The world's water 2002-2003: the biennial report on freshwater resources. Washington DC, Island Press
GWP (Global Water Partnership) (2000) A vision of water for food and rural development. The Hague, Global Water Partnership
Heyns PSvH (2002) The inter-basin transfer of water between countries within the Southern African Development Community (SADC): a developmental challenge of the future. In: Turton AR, Henwood R (eds) Hydropolitics in the developing world: a Southern African perspective. Pretoria, African Water Issues Research Unit (AWIRU), pp 157-176
Hirji R, Johnson P, Maro P, Chiuta TM (eds) (2002) Defining and mainstreaming environmental sustainability in water resources management in Southern Africa. Maseru, Harare, Washington DC, SADC, IUCN, SARDC, World Bank
IUCN-UNEP-WWF (International Union for the Conservation of Nature – United Nations Environment Programme – World Wildlife Fund) (1980) World conservation strategy. Gland, International Union for the Conservation of Nature
IUCN-ROSA (International Union for the Conservation of Nature – Regional Office for Southern Africa) (2001) Dialogue on transboundary natural resource management in Southern Africa. Report of the workshop held in Harare, Zimbabwe, 20-21 February 2001. Harare, IUCN-ROSA. Available online at http://www.iucnrosa.org.zw/tbnrm/publications_content.html [Accessed June 2005].
Kakonge JO (2002) Water scarcity and related environmental problems in parts of Sub-Saharan Africa: the role of the transboundary environmental impact assessment convention. Impact Assessment and Project Appraisal (20)1:49-59
Kuwairi A (2004) The great man-made river project. Paper presented at the Pan African Water 2004 Conference. Sandton, South Africa, 29-30 November 2004
Lundqvist J (2000) Rules and roles in water policy and management – need for clarification of rights and obligations. Water International (25)2:194-201
MacKay HM (2003) Water policies and practice. In: Reed D, De Wit M (eds) Towards a just South Africa: the political economy of natural resource wealth. Washington DC, WWF Macroeconomics Programme, pp 49-83
MacKay HM, Ashton PJ (2004) Towards co-operative governance in the development and implementation of cross-sectoral policy: water policy as an example. Water SA (30)1:1-8
OAU (Organisation for African Unity) (1985) First African ministerial conference on the environment (AMCEN), Cairo, 16-18 December 1985. Cairo, Organisation for African Unity. Available online at http://www.unep.org/roa/amcen/-decisions_1.pdf [Accessed July 2005]
Pallett J (1997) Sharing water in Southern Africa. Windhoek, Desert Research Foundation of Namibia
Porto JG, Clover J (2003) The peace dividend in Angola: strategic implications for Okavango basin cooperation. In: Turton AR, Ashton PJ, Cloete TE (eds) Transboundary rivers, sovereignty and development: hydropolitical drivers in

the Okavango river basin. Pretoria, African Water Issues Research Unit and Geneva, Green Cross International, pp 65-81

Ramoeli P (2002) SADC protocol on shared watercourses: its history and current status. In: Turton AR, Henwood R (eds) Hydropolitics in the developing world: a Southern African perspective. Pretoria, African Water Issues Research Unit (AWIRU), pp 105-111

SADC (Southern African Development Community) (1998) Regional strategic action plan (RSAP) for integrated water resources development and management in the SADC countries (1999-2004). Summary report. Maseru, SADC Water Sector Co-ordinating Unit

SADC (Southern African Development Community) (2005) SADC review and country profiles. Harare, Southern African Development Community. Available online at http://www.sadcreview.com/country%20profiles%202005/ [Accessed June 2005]

SARDC (Southern African Research and Documentation Centre) (1996) Water in Southern Africa. Harare, Southern African Research & Documentation Centre

Scholes RJ, Biggs R (eds) (2004) Ecosystem services in Southern Africa: a regional assessment. Pretoria, Council for Scientific and Industrial Research

Shela ON (1996) Water resource management and sustainable development in Southern Africa: issues for consideration in implementing the Dublin Declaration and Agenda 21 in Southern Africa. In: Proceedings of the global water partnership workshop. Harare, Global Water Partnership

Turton AR, Earle A, Malzbender D, Ashton PJ (2006) Hydropolitical vulnerability and resilience along Africa's international waters. Chapter 3, in: Wolf AT (ed) Hydropolitical resilience and vulnerability along international waters. Nairobi, United Nations Environment Programme, Report UNEP/DEWA, 58 pp

UNAIDS (Joint United Nations Programme on HIV/AIDS) (2005) Epidemiological fact sheets on HIV/AIDS and country profiles – country data for 2004. Geneva, United Nations AIDS Organisation and World Health Organisation (UNHCR, UNICEF, WFP, UNDP, UNFPA, UNODC, ILO, UNESCO, WHO, World Bank). Available online at http://www.unaids.org/countryprofiles [Accessed July 2005]

UNCED (1992) Agenda 21 – Earth summit agreement. United Nations Conference on Environment and Development (UNCED), Rio de Janeiro, Brazil. Available online at http://www.un.org/esa/sustdev/agenda21.htm [Accessed July 2005]

UNECA (United Nations Economic Commission for Africa) (1991) Report on the Pan-African conference on environment and sustainable development, Bamako, 23-30 July 1991. Addis Ababa, United Nations Economic Commission for Africa, 11 pp. Available online at http://www.uneco.org/ECA/-Documents.html [Accessed June 2005]

UNECA (United Nations Economic Commission for Africa) (1992) African common position on the African environment and development agenda, Abidjan, 11-14 November 1991. Addis Ababa, United Nations Economic Commission for Africa. Available online at http://www.uneca.org/-ECA/Documents.html [Accessed July 2005]

UNECA and UNEP (United Nations Economic Commission for Africa and United Nations Environment Programme) (1989) Report of the first African regional conference on environment and sustainable development, Kampala, 12-16 June 1989. Addis Ababa, United Nations Economic Commission for Africa, and Nairobi, United Nations Environment Programme. Available online at http://www.uneca.org/ECA/NRD/ARCED/MC3 [Accessed July 2005]

UNESCO-ISARM (United Nations Educational Scientific and Cultural Organisation – International Shared Aquifer Resource Management) (2004) Managing shared aquifer resources in Africa. IHP-VI, series on groundwater 8. CD plus accompanying maps. Hanover, Germany, BGR – Fachgruppe Internationale Zusammenarbeit, Grundwasser

WCD (World Commission on Dams) (2000) Dams and development: a new framework for decision-making. London, Earthscan Publications

WCED (World Commission on Environment and Development) (1987) Our common future. New York, Oxford University Press

Wolf AT (1999) Criteria for equitable allocations: the heart of the international water conflict. Natural Resources Forum **23**:3-30

World Bank (2005) African development indicators 2005. Washington DC, Development Data Centre, World Bank. Available online at http://www4.worldbank.org/afr/stats/adi2005/default.cfm [Accessed September 2005].

Transnational Dimensions of Freshwater Ecosystem Governance

Ken Conca

Abstract

This chapter examines the diverse array of processes of institution-building that have emerged in response to the increasingly transnational challenges of freshwater ecosystem governance. Although some of these processes reflect the traditional confines of interstate diplomacy, based on state-centred authority, others transcend the confines of a narrowly statist framework. Expert networks, based on a new knowledge paradigm of Integrated Water Resources Management (IWRM) and advocacy networks engaging controversies such as water infrastructure projects and water marketisation, have become increasingly important mechanisms for the generation of new water norms and the shaping of water-related governance practices.

Keywords: institutions, transnational networks, ecosystem governance, Integrated Water Resource Management (IWRM), globalisation, knowledge, international law, Millennium Ecosystem Assessment

Introduction: Water, Globalisation, and Governance

Water flows across social boundaries, and people who seek to use water in various ways pursue their interests across those boundaries. Along with other socially-constructed boundaries such as culture, race, gender, or region, the boundaries of nation-states create difficult challenges for ecosystem governance. This chapter examines formal and informal processes of institution building that seek to address the governance challenge across national borders. Central themes are the multiplicity of emergent institutional forms; the limits of traditional interstate mechanisms to manage social controversies around water; and the growing importance of transnational networks that facilitate stakeholder dialogue on those controversies.

The Clash of Water's Multiple Meanings and the Tensions of Globalisation

The centrality of water to life goes without saying. Given its importance as a basic human need, water generates social cooperation on scales ranging from the local to the global – from municipal water supplies and local irrigation cooperatives to international river-basin treaties and global financing mechanisms.

Beyond its status as a basic human need, water has other important meanings. It is a critical element of life-sustaining ecosystems. It is an anchor of local communities, deeply integrated into landscapes, livelihoods and cultures. Water is also a scarce resource in many parts of the world and, increasingly, a marketable commodity. These multiple meanings make freshwater ecosystems – including rivers, lakes, watersheds, estuaries, and floodplains – a source of widespread social conflict. Manifestations of conflict include, but are not limited to, disputes over scarce supplies; struggles over whether, where, and how to build water infrastructure; disagreements over the proper balance between extractive and instream uses of water; disputes over development and preservation in and around freshwater ecosystems; and controversies over water pricing and allocation mechanisms.

Both the cooperative and conflictual dynamics surrounding water play out in increasingly transnationalised ways. Water is often viewed as primarily a local concern, best handled at the scale of watersheds, municipalities, and communities. Yet water reflects the social dynamics of globalisation in several ways:

- *Legally*, water has become the subject of increased diplomatic energy toward international treaty making. The UN Committee on Economic, Social and Cultural Rights (2002) recently declared water to be a human right;
- *Conceptually*, water reflects the knowledge dimension of globalisation. The past few decades have seen robust development of expert networks promoting the science and values of integrated water resources management. This trend has had important ramifications for how water resources are governed, how water infrastructure is financed and constructed, and how freshwater ecosystems are managed;
- *Economically*, water has, for over a decade, been a prime target of international structural-adjustment and privatisation pressures; and
- *Politically*, water has become the subject of increasingly transnationalised controversies. As alliances among environmentalists, human rights activists, affected local communities, labour unions, and indigenous

peoples' organisations have grown wider, deeper, and stronger, these emergent transnational networks have turned state versus society conflicts over large dams or privatisation schemes into matters of global discussion.

One ramification of these developments is that the evolutionary trajectory of national water governance systems depicted in Chapter 1 must be understood to occur in a setting of increasingly strong transnational influences. Given the many ways that water reflects the dynamics of globalisation, it should be no surprise that policies, laws, and practices related to water, rivers, watersheds, and freshwater ecosystems have been pushed and pulled simultaneously by several incongruous, even contradictory, transnational forces. In particular, this chapter contrasts two different ways of thinking about the emergence of global water governance. One perspective on global governance views it as a shift of rule-making to a higher, more aggregate, level, either through formal processes of international law or through softer processes by which governing norms are articulated, disseminated and diffused. Viewed from this perspective, the central mechanisms of emergent global governance are what international relations scholars refer to as "international regimes", or issue-specific bundles of rules that prescribe acceptable behaviour and proscribe undesirable actions (Krasner 1982; Rittberger 1993). In the water sphere, the most advanced process of international regime formation is the attempt to create global rules for the governance of shared river basins.

A second perspective on globalisation and global governance focuses less on intergovernmental cooperation and centralised processes of rule-making, stressing instead cross-border networks of social relations (Conca 2006). Here, two observations are central. First, we are seeing a multiplicity of emergent transnational networks that shape how water is governed locally. Second, given the great social controversies that infuse the activities in which these networks engage (including tensions over water infrastructure projects and the debate over water pricing and marketisation, discussed in greater detail below), rulemaking about water becomes as much a challenge of conflict resolution as it is of cooperative coordination. As the effects of globalisation make it more difficult for governments to manage complex ecosystems exclusively through formal institutions, these parallel processes of transnational networking have become increasingly important mechanisms of water governance.

This chapter examines both interstate diplomacy and transnational networks as emergent forms of global water governance. After discussing each of these domains in turn, the discussion takes up the specific hypotheses related to the Trialogue Model of governance presented in Chapter 1.

Particular attention is paid to hypotheses about the different "communities of practice" in the Trialogue Model, the context of values and norms in water-related decision making, and the way in which we understand the role of democracy in good governance.

Sovereign Diplomacy for International Rivers

Some of the emergent transnational forces shaping water governance fit comfortably within the domain of international law and interstate diplomacy. These include basin-specific efforts to hammer out cooperative international agreements on shared rivers and global-scale efforts to craft soft-law principles and "best practices" frameworks for such cooperative arrangements. Viewed through this lens, the "international" dimension of the problem is the physically border-crossing aspect of flowing water resources, and international water governance becomes a classic problem of international cooperation. A classic problem, and also a difficult one: it is well known that the upstream-downstream power asymmetries surrounding shared rivers can make political cooperation difficult (Beach et al. 2000). In the absence of other forces, when downstream states are the more powerful they will typically use that power to protect their basin-level interests. When upstream states are more powerful, they are likely to be more comfortable living in a Hobbesian state of nature, such as is seen, for example, in China's reluctance to be engaged by the states of the lower Mekong, or in Turkey's resistance to the entreaties of its downstream neighbours Syria and Iraq.

How successful has international law and organisation been at 'swimming upstream' against these power-dynamics and basin-level realities to create a rule-based framework for shared river governance? Efforts to establish soft international law for shared river basins have a longstanding history. In 1966 the International Law Association issued the Helsinki Rules on the Uses of the Waters of International Rivers, which stressed the concept of the equitable use of shared waters and the importance of peacefully negotiated dispute resolution (McCaffrey 2001). Shortly thereafter, the UN International Law Commission (ILC) took up the question. After more than two decades of often tumultuous debate, the ILC issued draft articles for a convention on the principles of cooperative governance that should apply to international rivers, which moved to the General Assembly for discussion. The result was the 1997 UN Convention on the Law of the Non-Navigational Uses of Internationally Shared Watercourses (hereafter, Watercourses Convention) (United Nations 1997). The Watercourses

Convention articulated several core governing principles for shared rivers, including sovereign equality, participation by all basin states, equitable use of shared waters, avoidance of significant harm, information sharing, environmental protection, and peaceful dispute resolution (see below).

The convention offers a framework, or "best practices" approach, in which general principles ensconced in the convention are to be translated by governments into specific practices at the level of individual river-basin agreements. The real test, therefore, is whether basin-level agreements have been spreading and deepening the embrace of these practices across time and space. A research project conducted with colleagues at the University of Maryland sought to answer this question (Conca et al. 2003; Conca et al. 2006). The project identified 62 international river agreements for the period 1980-2000 that dealt with significant aspects of river governance, but omitted a somewhat larger number of accords on narrow matters such as boundary demarcation, ice-breaking, navigation, or fishing rights. The project asked two questions: First: Are governments converging on common governing principles for shared river basins? And, second: To the extent that they are, has the global-level effort to create a principled framework for shared rivers shaped the content of individual international accords at the basin level?

To answer these questions, we looked for the presence in these basin-specific agreements of several core principles articulated in the ILC draft articles and the Watercourses Convention. Specifically, we looked for evidence of the following stipulations being followed:

- *Participation.* According to Article 4, every watercourse state is entitled to participate in negotiations over the entire watercourse and consult on lesser agreements affecting that state;
- *Equitable use.* Article 5 stipulates that states must exercise "equitable and reasonable use" of international watercourses within their territories;
- *Significant harm.* Article 7 obligates states not to cause "significant harm" to other states sharing a watercourse;
- *Sovereignty.* Article 8 affirms that states are obligated to cooperate on the basis of "sovereign equality, territorial integrity, mutual benefit and good faith.";
- *Information exchange.* Article 9 says that watercourse states shall regularly exchange information and data;
- *Consultation.* Article 11 calls on states to consult with other watercourse states on the effects of any "planned measures.";

- *Prior notification.* Article 12 requites states to give prior notification for any "planned measures which may have a significant adverse effect" on other watercourse states;
- *Environmental protection.* Articles 20-23 create obligations to protect and preserve ecosystems, control pollution, prevent the introduction of alien species; and protect the marine environment; and
- *Dispute resolution.* States are obliged to resolve disputes peacefully; the convention endorses arbitration and mediation and identifies procedures for fact-finding commissions (Article 33).

The list below summarises the main findings of this research. The data reveal a complex process of soft-law development. A few core principles emanating from global legal efforts have shown significant growth, diffusion, and deepening (a term we in the US use to refer to increased precision, specificity, or obligation); others are common in basin agreements but show no particular pattern of diffusion or deepening; still others are weakly represented in the data throughout the entire study period. Specifically, we found rapid growth of concepts of environmental protection and peaceful dispute resolution, and the emergence of international basin commissions as a specific mechanism of river management and dispute resolution. More contentious principles, such as equitable use of water or a responsibility to avoid significant harm to other basin states, show less dramatic growth or, in some instances, little development at all.

Findings from a content analysis of 62 international river basin agreements for the period 1980-2000[1]:

1. **International agreements are relatively infrequent:** We were able to identify 62 formal and informal public international agreements on shared river basins for the period 1980-2000. During this period, agreements were created in only 36 (14%) of the world's 263 internationally-shared river basins. Three-fourths of these agreements were in basins where states had a prior history of cooperation on that watercourse.
2. **The rate of agreement formation has dropped off:** A large pulse of international agreements occurred in the immediate wake of the 1992 UN Conference on Environment and Development (UNCED). Formation of new agreements tailed off substantially after the post-UNCED peak of the mid-1990s.
3. **Multilateral initiatives, but bilateral governance:** Most international rivers are shared by two countries, and yet nearly four-fifths of the agreements in the study period occurred in multilateral basins. However, within those multilateral basins, a bilateral agreement is twice as

[1] From Conca et al. 2003

common as a multilateral one. Shared governance is fragmentary, in that few agreements in multilateral basins include all basin states. There is no discernible trend toward greater inclusiveness of basin states in more recent accords.

4. **Only some UN Convention principles are well-established at basin level:** Of the core principles in the 1997 UN Watercourses Convention, those most frequently invoked at the basin level include environmental protection, peaceful dispute resolution, consultation, and information exchange. Less common are the principles of prior notification, equitable water use, and avoiding significant harm to other parties. Explicit reservations of sovereign rights were also relatively uncommon, though strongly increasing in the post-UNCED period.
5. **Both environmental protection and sovereignty are increasingly invoked:** Only two principles showed a statistically significant increase in frequency over time: environmental protection and the sovereign rights of the parties. There is also an increase of borderline statistical significance in the principle of avoiding significant harm. Recent agreements are not significantly more likely than earlier ones to incorporate the other core principles of the UN Watercourses Convention: equitable use, information exchange, peaceful dispute resolution, consultation, or prior notification of measures likely to cause adverse effects.
6. **Are two distinct regimes emerging?** Correlation analysis reveals two bundles of commonly-associated principles. One is centred on the idea of protecting a state's water rights and includes equitable use, water allocation mechanisms, recognition of sovereign rights, and provisions exempting domestic waters from international obligations. A second bundle is centred on dealing with the potentially harmful consequences of water use and includes prior notification, information exchange, consultation, regular meetings, peaceful dispute resolution, formation of basin commissions, and environmental protection.
7. **Little evidence of normative deepening:** For most of the core principles articulated in the 1997 UN Watercourses Convention, we see no evidence of normative "deepening" over time, in the sense of a trend toward greater specificity, broader scope, or greater intrusiveness in the responsibilities or obligations created for states. The exception is the principle of consultation, in the sense that it becomes more likely over time that an agreement will form a permanent basin commission as the specific consultative mechanism.

There are, of course, limits to the interpretation of these data. We looked at joint articulation of principles by governments, not at actual implementation. We did so on the grounds that joint articulation was a readily-observable and necessary, though clearly not sufficient, condition for effective cooperative governance. The results suggest that, while we are seeing processes of norm-development and dissemination on a worldwide scale, the path has hardly been a smooth, continuous or unidirectional one. The growing norm of environmental protection has been met with a backlash reinforcing the state's sovereign rights. Rather than seeing a single bundle of principles, we seem to be seeing two conflicting clusters, one stressing a state's right to water and the other stressing being a good water neighbour. Perhaps most importantly, the debate within the United Nations on how to reconcile core principles of using water equitably and avoiding harmful effects to other countries when using water, has not been easily reconciled at the basin level, either.

Limits to Conventional International Institution Building: Authority, Knowledge, and Territory in Ecosystem Governance

I have argued elsewhere (Conca 2006) that the barriers to effective international environmental diplomacy around local ecosystems are not limited to conflicting interests, weak institutionalisation, or the asymmetric distribution of global power – although these problems are both daunting and well known. A more deeply-rooted problem is that the basic conditions required for international cooperation are difficult to create, given the political dynamics surrounding localised ecosystems. Specifically, creating the necessary conditions of authority, knowledge, and territoriality that underpin conventional formal-legal processes of institutionalised international cooperation is a daunting, perhaps even impossible, challenge.

The Hybridisation of Authority

One central problem involves authority. The sovereign underpinnings of international relations impose two foundational assumptions on most international cooperative initiatives: that states are, by definition, authoritative actors with the legitimate power to enter into and implement international cooperation, and that non-state actors, be they corporations, private voluntary organisations, local communities, professional associations, or social movements, are not authoritative actors. Both of these assumptions make conventional forms of international cooperation difficult to sustain around watersheds, forests, coastal zones, and other socio-ecological systems subject to the multiple forms of meaning discussed at the outset of this chapter. The presumption that states, and only states, are the authoritative agents of governance is a polite fiction that often cannot be sustained in the real world – and particularly so in these cases. It is an assumption that holds constant, at one extreme end of the spectrum, a key bundle of variables related to the constitution, distribution, and legitimisation of authority.

National governments are generally taken to be both the subjects and objects of international environmental cooperation: their bargaining, agreement, and ratification determine whether a legitimate agreement exists, and they assume responsibility, and thus authority, for compliance. States are also taken as the primary objects of cooperation: governmental compliance is the presumed key to effectiveness, and governmental implementation is the primary task as a means to that end (Victor et al. 1998; Haas et al. 1993; Weiss and Jacobson 1998; Bernauer 1995). Non-state actors may nip at the heels of authority, but are not constructed in traditional cooperative frameworks as authoritative in their own right.

The problem with this formulation will be apparent to anyone who has worked on socio-ecological dynamics at the scale of a watershed. The presumption of state authority common to sovereign diplomacy misreads what is often a very different distribution of authority in a particular local context. Frequently, "the state" lacks the uncontested authority to control local access to nature or uses of nature, and efforts to exert such control become part of larger struggles for legitimate authority or material power. Historically, the ability to control rules of access to the environment and natural resources – to define who may alter, and to what extent, which specific natural materials, systems, and processes – has been a central component of state authority (Lipschutz and Mayer 1993). Conversely, where that power is lacking, authority is not fully consolidated.

At the same time, there are also many circumstances under which non-state actors have at least some authority to establish such controls. The growing invocation of the need to incorporate "stakeholders" into institutional frameworks for environmental and resource governance (Hemati 2002) is in part recognition of the growing power of actors traditionally left outside such frameworks to throw a wrench into the machinery of business as usual. An important corollary must also be recognised, however: local societal organisations are often better positioned than the state to mediate the process by which people gain access to nature (Dolšak and Ostrom 2003; Ribot and Larson 2004).

The Destabilisation of Knowledge Frameworks

In parallel with the problem of authority is the problem of knowledge. Environmental governance is often characterised as an institutional response to inherently uncertain threats and problems, and for this reason environmentalists often champion the precautionary principle. However, in practice, institutions of environmental governance are typically rooted, not in precaution, but rather in what Sheila Jasanoff (1998) termed "knowledge stabilisation." This is particularly the case for international environmental cooperation mechanisms, which emerge from interstate bargaining dynamics and which tend to coalesce around an authoritative problem statement, the establishment of a body of official knowledge, an optimistic stance toward the reducibility of uncertainty, and the embrace universalistic, scientific knowledge frameworks as the most important way of knowing about environmental problems and solutions. Uncertainty, local contingency, and other unstable aspects of our knowledge are acknowledged, but are not the central organising principles for governance.

Environmental problems are typically approached in this manner for several reasons, including the general hold of modern science upon imagination. However, there is also a crucial political dimension: the legitimacy of governance initiatives in this domain derives largely from their ability to pose as a way to solve a specified problem by linking cause-and-effect analyses of that problem to an optimised array of solutions. Knowledge is never fully stabilised, of course, but successful international environmental diplomacy typically does involve some important aspects of closure: A dominant problem construction becomes embedded; an officially sanctioned body of universal, technical knowledge begins to emerge; the boundary between official truths and acknowledged uncertainties is defined; a quest to reduce uncertainty, in particular ways and specific directions, crystallises. As Jasanoff (1998:86) suggests,

Contingency...is only half the story.... Equally important is the conclusion that, in spite of all indeterminacy and uncertainty, knowledge and social order are not perceived by human societies to be fluid at all points. Both can be made to hold still through institutions, material technologies, and shared norms and practices. The stabilisations brought about by international technical standards and transnational epistemic communities are particularly significant for environmental regimes.

Viewed in this light, it should be no surprise that conventional international environmental diplomacy has not fared particularly well around matters of ecosystem governance and the various localised, cumulative dimensions of global environmental degradation. Constructing a body of authorised knowledge and adopting an optimistic stance toward the reduction of uncertainty are problematic at best. Data tend to be unequally distributed, inconsistent, discontinuous, and non-comparable. Universal models of cause-and-effect relations, or key ecosystem processes, are often beyond the current state of the art of environmental science. Boundaries are difficult to specify with precision; feedback linkages are impossibly complex; change, as a rule, is generally non-linear. And in all of this, universal scientific ways of knowing are jarringly different from the various localised, contingent, and culturally bound forms of knowledge that surround local ecosystems. For all of these reasons, it is increasingly recognised that the challenge is to integrate local/contextual and global/systematic ways of knowing (Gilchrist et al. 2005; Hassol 2004) and that the "best available science" may come from many different social sources (Francis et al. 2005). Yet these insights have been painfully slow to penetrate international environmental diplomacy.

Beyond Sovereign Territoriality

Along with authority and knowledge, a third challenge of international institution building is *territoriality*. This is usually rendered as the well-known problem of sovereignty. Nature, it is said, refuses to acknowledge the sanctity of borders, which were rarely drawn with nature in mind. The primary response of international environmental diplomacy to this challenge has been to draw distinctions between "domestic" and "international" environments, with the former assumed to sit still for sovereign governance while the latter transcend that governance in ways that demands inter-sovereign cooperation. By this reasoning, oceans and the earth's climate are properly the stuff of global environmental governance, in ways that local watersheds and airsheds are generally not, at least, not unless harmful effects happen to straddle a sovereign border. This reasoning was enshrined in the famous 'Principle 21' of the 1972 Stockholm Declaration, a product of the UN's first major global environmental conference, which says:

> *States have, in accordance with the Charter of the United Nations and the principles of international law, the sovereign right to exploit their own resources pursuant to their own environmental policies, and the responsibility to ensure that activities within their jurisdiction of control do not cause damage to the environment of other States or of areas beyond the limits of national jurisdiction (United Nations 1973:5).*

When it comes to rivers, lakes, estuaries, and freshwater ecosystems in general, there is an immediately-apparent physical problem with this formulation: water flows, thereby refusing to sit still as domestic nature for sovereign governance. More importantly, in a globalising world economy where water is a scarce and valuable commodity, and in an increasingly up-linked world polity where localised political controversies (say, dam construction) become subjects of global debate, water will not sit still socially any more than it will physically.

Beyond International Diplomacy: Network Mechanisms of Water Governance

Other forms of globalisation in the water sphere force us to ask questions about governance that push beyond the familiar understanding of "international" as the zone of international relations and interstate diplomacy. Water is "governed" across borders not only in the sense of intergovernmental cooperation around physically transnational waters, but also through the social effects of border-crossing networks. I have argued elsewhere (Conca 2006) that we need to see the effects of these networks not simply in terms of their influence on government behavior and intergovernmental arrangements, but also as important sources of institution building in their own right.

Expert Networks: The Rise of IWRM

Expert networks are one important form of this phenomenon. The past few decades have seen the emergence and growing influence of a linked set of transnational expert water networks, exercising power and influence through knowledge-based policy advocacy. Grounded in a technically expert, rationalist understanding of the problem of water, expert networks have put into play the rhetorically powerful concept of integrated water resources management (IWRM). Although there are many different operational definitions of IWRM, most stress a few core themes: the need to manage water with multiple uses (ecological, economic, socio-cultural) in mind, the importance of managing water at watershed scale, the need to integrate management across the historically separate sectors of water use (agricultural, industrial, municipal, and 'in-stream'), and the importance of participatory mechanisms in water management.

As water expert Peter Gleick (1998:9) has pointed out, the emergence of the IWRM paradigm is part and parcel of a broader and ongoing paradigm shift on water use and water resource development:

The twentieth-century water-development paradigm, driven by an ethic of growth, has now stalled as social values and political and economic conditions have changed. More people now place a high value on maintaining the integrity of water resources and the flora, fauna, and human societies that have developed around them. There are growing calls for the costs and benefits of water management and development to be distributed in a more fair and prudent manner and for unmet basic human needs to be addressed. And more and more, efforts are being made to understand and meet the diverse interests and needs of all affected stakeholders.

IWRM has emerged as a potent global conceptual framework for several reasons, including the failure of state-led water developments and UN 'water decade' approaches; the institutional vacuum created by the patchwork of intergovernmental organisations with some sort of water mandate; the growing professionalism of water expertise, replete with global conferences, scholarly journals, and membership societies; and the boost provided by growing concern for "sustainability" as reflected in the 1987 report of the World Commission on Environment and Development and the 1992 Earth Summit (Conca 2006).

By the 1990s, water expertise had arguably eclipsed water diplomacy as a leading source of international institution-building. Expert networks have redefined the conceptual map for thinking about and acting on water in international context. They have also penetrated important water-governance nodes such as the World Bank and created new nodes of their own, such as the Global Water Partnership and the World Water Council. In doing so, IWRM has promoted institutionalised forms of water governance that push against the traditional confines of authority, territory, and knowledge sketched earlier. IWRM embraces a more trans-territorialised understanding of local water as a global problem. Participatory systems are legitimated that begin to devolve authority away from the exclusive purview of government agencies. Watershed-scale knowledge – contingent, location-specific, rooted in experience, dynamic – takes its place beside universal knowledge frameworks. In this sense, the emergence of transnational expert networks can be seen as a manifestation of the Trialogue Model of governance emerging, however partially and unevenly, in some of the international spaces where water-related decision making occurs.

Limits of IWRM Discourse

For all their gains, however, what expert water networks have *not* succeeded in doing is to quell the social controversies that swirl around water. For example, IWRM has very little success in resolving the tension between water as a human right and water as an economic good, with scarcity value. As has been noted by many observers, the seminal 1992 Dublin Principles, which constituted a formalisation of IWRM thought just prior to the 1992 Earth Summit, emphasised both the rights-based and commodity-based approaches (International Conference on Freshwater and the Environment 1992). This controversy has probably been the single most contentious aspect of recent events that bring together water experts, such as the triennial World Water Forum.

The same is true with regard to participatory mechanisms. What exactly does it mean to say that all water users have a stake, and that everyone should consult with everyone else? Some variants of IWRM tend in practice toward the more technocratic end of the spectrum, in which society 'participates' primarily in receiving its marching orders from the state and water experts. Other variants see the participatory dimension of IWRM quite differently, as a way for local stakeholders to guide these institutions in a bottom-up manner.

One consequence of these enduring controversies is that the increasingly well-institutionalised mechanisms of expert water networks – global conferences, international commissions, professional publications, and networked organisations such as the Global Water Partnership – have become the sites where these controversies continue to be aired. In international terms, consensual knowledge-based water governance remains as elusive as law-based governance, even as both have seen substantial institutionalisation.

The Rise of Transnational Advocacy Networks around Water Controversies

This brings us to a second form of transnational network for water governance – cause-oriented advocacy. As law and science have failed to bring closure, protest networks have become an important channel for the globalisation of local water management. The most effective and well-institutionalised protest network on water controversies is the movement against large dams. The construction of large dams has long been a thoroughly transnationalised enterprise, dominated by large international engineering and construction firms and fuelled in most cases by global public and, increasingly, private finance. In recent years, opposition to large dams has scaled up accordingly, as local dam-site opponents have found common cause with international environmental, human-rights, and grassroots-development advocates. The resulting protest network is less a traditional social movement than an expression of the "networks of networks" common to globalisation (Khagram et al. 2002). It involves local social movements, national mobilising organisations in key dam-building countries such as Brazil and India; regional groups such as the European Rivers

Network; and global connective links such as International Rivers Network.

Anti-dam protest has been effective, in part because it has targeted the financial "Achilles' heel" of dam-building: the sensitivity of profits to whether these capital-intensive projects can be brought in on time, at budget, and with a minimum of controversy, particularly when private capital is involved, as is increasingly the case given the neoliberal turn in development financing. A second key to making inroads has been the ability to launch both site-based direct action against dams and technical/institutional critiques that challenge the knowledge-based legitimacy and operating principles of entities such as the World Bank (Moore and Sklar 1998; McCully 1996).

The increasingly effective opposition mounted by anti-dam protest coalitions forced the World Bank and leading dam-building firms to the bargaining table, as they sought some stable rules of the dam-building game in an increasingly uncertain financial climate (World Conservation Union and World Bank 1997). The result of these discussions was the World Commission on Dams (WCD), a mixed-membership body tasked with examining the development contributions and failings of large dams and making recommendations for the future (World Commission on Dams 2000). The WCD was a highly unusual body in two ways. First, rather than papering over deep disputes in favour of a least-common-denominator consensus, it brought together both ardent advocates and equally ardent opponents of large dams. Second, the WCD subordinated inter-sovereign discussions in favour of direct stakeholder dialogue, essentially rendering state-society domestic conflicts over dam construction as a legitimate topic for global debate. The WCD produced several products: a detailed consensus report on the economic, technical, environmental, and social performance of large dams; copious quantities of data, analysis, case studies, and stakeholder testimonials on the development effectiveness of large dams; and an institutionalised initiative taken up by the UN Environment Programme to reinject stakeholder dialogue on dams into national debates.

A second key arena of transnational water protest has involved water privatisation and marketisation. Because water – including bulk water resources, municipal water supplies, sanitation services, and irrigation schemes – has typically been supplied and managed as a public good by the state, it became a ripe target for neoliberal structural adjustment and privatisation schemes during the 1990s (Finger and Allouche 2002). Other motive forces included the perceived need to attract private finance to expand water infrastructure and the perception that "full-cost" pricing for historically subsidised water services would enhance the efficiency of use (Global Water Partnership 2000).

Privatisation initiatives have proved to be highly contentious, leading, in many countries, to protest, political controversy, and even to violent resistance. As with anti-dam protesting, site-specific opposition has melded into a loosely-structured but increasingly institutionalised global protest network. The network links local community opposition, labour unions, grassroots development groups, and social-justice activists; it blends economic critiques, social-justice arguments, and the symbolic power of the idea of water as a human right.

These two expressions of transnationally-networked water protest differ in important ways. Anti-privatisation protests frequently offer a defense of the state as provider of water as a public good, while anti-dam protesters frequently rail against the same state as purveyor of a development model that has steam-rolled over the rights of local communities, ethnic minorities, and indigenous peoples. These differences may help to explain why we have not seen the emergence of a full-blown global water movement that integrates these separate critiques. Nor has either network succeeded in winning the day. The most recent water-sector strategy adopted by the World Bank (2003) articulates a renewed commitment to dam financing and builds centrally on the role of the private sector in water services – ensuring that controversy and protest will continue in both domains.

Although falling short of definitively winning their struggles, these protest networks have succeeded in catalysing processes of institutionalised stakeholder dialogue that push beyond the limits of authority, knowledge, and territoriality – limits that were identified above as barriers to effective intergovernmental cooperation. The authoritative role of non-state actors, and the parallel limits on narrowly state-centred governance, has become increasingly validated. Locally contingent- and socially-grounded knowledge – e.g. about the effects of dams on local communities, the water-provisioning practices of the poor, the rootedness of water in culture – has pushed its way to the bargaining table alongside universalised science frameworks. The validity of treating these "local" controversies, and the state-versus-society conflicts they engender, as transterritorial phenomena that are the proper subject of global debate, has been increasingly affirmed. The processes catalysed by protest networks have also begun to institutionalise themselves: via the World Commission on Dams process in the case of large dams and water infrastructure projects, and via a potentially similar emergent stakeholder dialogue on "private-sector participation" in water services (Conca 2006).

Implications of the Hypotheses

This complex pattern of institutionalisation – halting gains in international water law, the combined growth and fragmentation of transnational expert networks, and the emergence of stakeholder dialogues around contentious socio-ecological water controversies – suggest both validation and refinement of several of the hypotheses presented in Chapter 1. Of particular salience are the hypotheses about the cluster of actors central to the Trialogue Model (hypothesis 2), the different "communities of practice" that these actors reflect (hypothesis 3), the norms and values shaping governance (hypothesis 4), and the role of democracy in governance (hypothesis 7).

The core actor-clusters identified in hypothesis 2 – government, science, and society – are clearly in evidence, albeit to different degrees across the multiple emergent institutional forms identified here. International water diplomacy has very much been the purview of the state. Science has entered the dialogue only to the extent that state bureaucracies have the capacity to bring it to the table, although the growing emphasis on dispute resolution procedures for shared waters may create a second and increasingly important entry point for science-based governance. Social forces are the least effectively represented; most treaty mechanisms create few or no entry points for non-state actors.

Expert water networks reflect a more balanced representation, although the inability to come to consensus on the meaning of stakeholder participation means that the societal sphere is represented more effectively in some domains than in others. The networks springing up around socio-ecological water controversies and the resultant stakeholder dialogues also reflect also represent movement toward the cast of characters of the Trialogue Model. However, there remain enduring questions about who speaks for society in these contexts, with questions raised about whether NGOs involved in the process represent broad public interest or more particularistic forces.

The idea of science, government and society as distinct communities of practice (hypothesis 3) is also apparent, and accounts for many of the difficulties of promoting effective stakeholder dialogue. In transnational context, a critical element to successful dialogue appears to be the emergence of bridging actors: for example, NGOs capable of interfacing with both local community activists and the technical-rational discourse of state bureaucracies and intergovernmental organisations, or interdisciplinarians able to provide a bridge between modern universal science and local ways of knowing.

The evidence for hypothesis 4, that governance occurs across a backdrop of different norms and values at different stages of economic development, is mixed. On the one hand, there is clearly a strong North-South cleavage in global water politics. European states write dramatically different sorts of international water treaties than do, say African states, given the different set of water needs and water uses of industrialised and less-developed societies. The same is true of the networks emerging around transnational water controversies. Many developing countries essentially accused the World Commission on Dams of a failure to recognise their legitimate development needs and of over-emphasizing the environmental and human-rights concerns of Western NGOs. Structural adjustment in the water sector faces similar North-South dichotomies, as for example when sceptics point out that few industrialised countries have fully embraced the privatisation of water utilities being urged on the global South.

However, transnational networks also reveal deep societal cleavages that transcend a stages-of-development model. The struggles of local communities facing displacement by water projects find common cause with northern progressives, even as the beneficiaries of those projects embrace development paradigms that resonate in Washington, Brussels or Tokyo. As suggested above, perhaps the single greatest motive force for the institutionalisation of transnational water governance is that local value-based disputes are being dragged into world politics by transnational coalitions on both sides of the issue. Stages of development are not irrelevant, but neither are they a complete model for understanding normative differences.

Hypothesis 7, that good governance is more likely in the context of democracy, raises complex questions about what we mean by democracy in the transnational sphere. Clearly, the notion that democracy is a purely domestic question cannot be sustained. It is the profound failure of most states to adequately aggregate water-related preferences across all sectors of society that has taken nominally "local" questions of water governance and dragged them into the transnational sphere. Weak actors facing domestic blockage have amplified their voices by going global (Keck and Sikkink 1998). In that sense, one can view the politics of transnational water activism as a form of democratisation, however messy the result.

However, it remains unclear exactly what constitutes democracy in the transnational spaces of world politics. If one's goal is to see performed core functions of democracy that have been notably absent in international water decision making, the growing pluralisation of actors in water-related conversations raises the real hope of greater oversight, monitoring and accountability. The benefits we have seen from the admittedly imperfect stakeholder dialogues emerging in the water sphere would seem to confirm the hypothesis. However, it remains unclear how well stakeholder models can perform the *representative* functions of democracy. Certainly, stakeholder models may offer the possibility of improving democratic discourse by adding unheard voices to the mix. But they also may raise the spectre of a form of "corporatism gone global" in which narrow but well organised interests occupy the key spaces in the global debate, in ways that may actually suppress democratic practice at the national level (Ottaway 2001).

Conclusion

International law has an important role to play in channelling interstate water disputes into peaceful processes of bargaining, arbitration, and conflict resolution. Incorporating emerging legal principles of shared water governance – including peaceful dispute resolution, equitable use of transboundary water resources, prior notification, information sharing, and avoiding significant harm to co-riparians – into robust and dynamic cooperative arrangements among governments in shared river basins is a central challenge of effective freshwater ecosystem governance.

By itself, however, cooperative interstate diplomacy has little hope of either promoting effective ecosystem governance or stabilising socio-ecological controversies around those systems. Rivers are much more than simply border-straddling water-channels, and they engender social relations across boundaries that are difficult to contain through the standard mechanisms of sovereign international relations.

Thus, and particularly in an era of rapid globalisation, transnational social networks will play an increasingly important role in the effort to attain the twin aims of ecosystem governance and conflict management. Expert networks are essential to identify problems, disseminate knowledge, generate creative problem solutions, and validate ecological rationality in resource governance. Also essential, however, are social networks that catalyse discussion of social controversies that cannot be bounded or contained by the routine processes of science and diplomacy.

An important implication of this perspective is the need to nurture the emergence of such network-based institutions. None of the networks described in this chapter was built in the planned, purposeful manner that usually envelops our notions of institutional design, however messy the guts of institution-building actually may be on the inside. They sprang, instead, from the *failure* of existing institutional arrangements to resolve social controversies. The most important innovations to speed along the birth of such new forms are likely to be greater attention to process-oriented mechanisms of dialogue and conflict resolution rather than bargaining processes seeking purely substantive outcomes (Conca 2006).

A second implication is the need to pay much greater attention, not simply to decision-making, but to what Hemmati (2002) has referred to as "decision finding" through dialogue among a wider array of actors than has typically been the case historically. The urgency of many freshwater ecosystem management challenges may tempt managers to seek to limit dialogue in favour of decisive action. Yet, however messy the politics they set in motion may seem, such networks provide a critical societal element in transnational ecosystem debates, dialogues, and governance processes.

References

Beach HL, Hamner J, Hewitt J, Kaufman E, Kurki A, Oppenheimer J, Wolf A (2000) Transboundary freshwater dispute resolution: theory, practice and annotated references. Tokyo and New York, United Nations University Press

Bernauer T (1995) The effect of international environmental institutions: how we might learn more. International Organization 49:351-77

Conca K, Wu F, Mei C (2006) Global regime formation or complex institution building? the principled content of international river agreements. International Studies Quarterly 50:263-285

Conca K, Wu F, Neukirchen J (2003) Is there a global rivers regime? trends in the principled content of international river agreements. Research report. Harrison Program on the Future Global Agenda, University of Maryland. Available online at http://www.bsos.umd.edu/harrison

Conca K (2006) Governing water: contentious transnational politics and global institution building. Cambridge, MA, MIT Press

Dolšak N, Ostrom E (2003) The commons in the new millennium: challenges and adaptation. Cambridge, MA, MIT Press

Finger M, Allouche J (2002) Water privatisation: trans-national corporations and the re-regulation of the water industry. London, Spon Press

Francis T, Whittaker K, Shandas V, Mills AV, Graybill JK (2005) Incorporating science into the environmental policy process: a case study from Washington state. Ecology and Society (10)1:35. Available online at http://www.ecologyandsociety.org/vol10/iss1/art35/

Gilchrist G, Mallory M, Merkel F (2005) Can local ecological knowledge contribute to wildlife management? case studies of migratory birds. Ecology and Society(10)1:20. Available online at http://www.ecologyandsociety.org/vol10/-iss1/art20

Gleick PH (1998) The world's water 1998-99: the biennial report on freshwater resources. Washington, DC, Island Press

Global Water Partnership (2000) Towards water security: a framework for action. Report prepared for presentation at the Second World Water Forum, The Hague, Netherlands, 17-22 March 2000. Stockholm, Global Water Partnership
Haas PM, Keohane RO, Levy MA (eds) (1993) Institutions for the earth: sources of effective international environmental protection. Cambridge, MIT Press
Hassol J (2004) Artic climate impact assessment: impacts of a warming arctic. Cambridge, Cambridge University Press
Hemmati M (2002) Multi-stakeholder processes for governance and sustainability: beyond deadlock and conflict. London, Earthscan
International Conference on Water and the Environment (1992) Dublin statement on water and sustainable development. Available online at http://www.wmo.-ch/web/homs/icwedece.html
Jasanoff S (1998) Contingent knowledge: implications for implementation and compliance. In: Weiss EB, Jacobson HK (eds) Engaging countries: strengthening compliance with international environmental accords. Cambridge, MIT Press
Keck M, Sikkink K (1998) Activists beyond borders: advocacy networks in international politics. Ithaca, NY, Cornell University Press
Khagram S, Riker J, Sikkink K (eds) (2002) Restructuring world politics: transnational social movements, networks, and norms. Minneapolis, University of Minnesota Press
Krasner SD (1982) Structural causes and regime consequences: regimes as intervening variables. In: Krasner SD (ed) International regimes. Ithaca, NY, Cornell University Press
Lipschutz RD, Mayer J (1993) Property rights, constitutive rules, and the renegotiation of resource management regimes. In: Lipschutz RD, Conca K (eds) The state and social power in global environmental politics. New York, Columbia University Press
McCaffrey SC (2001) The law of international watercourses: non-navigational uses. Cambridge, Oxford University Press
McCully P (1996) Silenced rivers: the ecology and politics of large dams. London, Zed Books
Moore D, Sklar L (1998) Reforming the World Bank's lending for water: the process and outcome of developing a water resources management policy. In: Fox JA, Brown LD, The struggle for accountability: the World Bank, NGOs, and grassroots movements. Cambridge, MA, MIT Press
Ottaway M (2001) Corporatism goes global. Global Governance (7)3:265-92, July-September
Ribot J, Larson A (eds) (2004) Decentralization of natural resources: experiences in Africa, Asia and Latin America. London, Routledge
Rittberger V (1993) Editor's introduction. In: Rittberger V (ed) Regime theory and international relations. Oxford, Clarendon Press
United Nations (1973) Report of the United Nations conference on the human environment. United Nations document A/Conf.48/14/Rev 1, 1973

United Nations (1997) Convention on the law of the non-navigational uses of international watercourses. United Nations General Assembly document A/51/869, 11 April 1997

United Nations Committee on Economic, Social and Cultural Rights (2002) Substantive issues arising in the implementation of the international covenant on economic, social and cultural rights: the right to water. General Comment 15. United Nations document E/C.12/2002/11, 26 November 2002

Victor DG, Raustiala K, Skolnikoff EB (eds) (1998) The implementation and effectiveness of international environmental commitments: theory and practice. Cambridge, MA, MIT Press

Weiss EB, Jacobson HK (eds) (1998) Engaging countries: strengthening compliance with international environmental accords. Cambridge, MA, MIT Press

World Bank (2003) Water resources sector strategy: strategic directions for World Bank engagement. Final report. February 2003

World Commission on Dams (2000) Dams and development: a new framework for decision-making. The report of the World Commission on Dams. London, Earthscan

World Conservation Union and World Bank (1997) Large dams: learning from the past looking at the future. Workshop proceedings, Gland Switzerland, 11-12 April 1997. Washington, IUCN

Part 2

Interrogation of the Trialogue Model

From Dialogue to Trialogue: Sustainable Ecosystem Governance and Civil Society

Geoffrey D Gooch

Abstract

This chapter discusses the role of institutions, both formal and informal, in ecosystem governance. The role of different forms of knowledge, and the ways in which these can be combined, are analysed, as is the role of civil society in ecosystem policy processes. Organisational structures, their aims, norms and values, are examined and the problems of cooperation between different institutional cultures are analysed. The chapter presents a number of Trialogues in the context of sustainability, epistemology, and policy processes, and discusses their relevance for ecosystem governance. Finally, a number of recommendations for future research are made.

Keywords: ecosystem governance, civil society, institutions, Trialogue, knowledge, organisational culture, sustainability, epistemology, policy processes

Introduction

Society is faced with a wide variety of rapidly-evolving and intricate policy problems that demand complicated choices between possible solutions. Policy here is defined as political decisions and their implementation in such fields as welfare, social redistribution, the environment, health, education, economics, law etc., all of which involve the interaction of different actors in complicated processes. Unfortunately, important choices in these fields must often be taken under conditions characterised by uncertainty. Predictions of the future are increasingly unsure as societies become more and more complex and interdependent. Also, while science has provided many valuable insights into the workings of ecosystems, the changes resulting from science's findings have also come to affect us all, and we are now 'embarked in the same collective experiments mixing humans and non-humans together' (Latour 2001). Decisions therefore have to be taken in a setting where a lack of knowledge of coming conditions is usual, where large numbers of people will be affected, and it is therefore quite likely that some of the policies formulated under such conditions will prove to be sub-optimal.

When such policies fail to achieve their aims, or where policy-makers are perceived to be inadequate, a lack of public support often develops, and in retrospect, a lack of legitimacy. Research has shown that many societies are in fact faced with problems of decreasing legitimacy, that is, the public now have less faith in government than they did 20-30 years ago. Decisions made by political leaders, managers, and administrators are questioned more and more, and negative reactions to what are considered sub-optimal policies have become more aggressive. How then to make better decisions, or at least to make sure that decisions are accepted by society and seen as legitimate? If the polity (i.e. the political and administrative institutions of government) is unable to solve societal problems by itself, then must government develop into something more than what it currently is? Many contemporary writers feel so, and claim that a way out of the dilemma is to move from government to governance (e.g. Waltz 1999; Stoker 1998; Peters and Pierre 1998). Governance is seen here as including politics and administration, civil society, and economic interests as three different actors with established formal and informal institutions. Cohen and Arato (1994) describe civil society as being between the state and economic interests, and in line with this typology civil society is defined here as a sphere of activity separate from the politico-administrative and business sectors.

Governance and Sustainability

The participation of the private sector and civil society in governance systems (together with the polity), has been seen as a way out of this predicament. Shared decision-making and implementation allows, and may force, formally non-political actors to share responsibility with the polity, and thus possibly increase the perceived legitimacy of decisions and policy. The necessity of increased legitimacy for decisions and policy noted above is especially important in the fields of sustainable ecosystem governance, where the implementation of policies is often dependent on acceptance of the policies by stakeholders and the public. Participation by these groups therefore becomes more and more necessary as problems diversify and become more spatially diffuse.

The problems faced here are complex. The future state of ecosystems will be the result of a combination of societal, economic and ecological influences. Yet the future is uncertain, and our ability to predict it is handicapped by our present state of knowledge, as well as by present values, norms and beliefs. Faced with these dilemmas it is sometimes claimed that *governance*, or even *good governance* can provide possible solutions to problems of ecosystems and sustainability sketched out above. Governance, however, may present advantages, but also creates problems. First, there is the question of definition. As Rhodes (1996:652) pointed out,

...the term 'governance' is popular but imprecise. It has at least six uses, referring to: the minimal state; corporate governance; the new public management; 'good governance'; socio-cybernetic systems; and self-organising networks.

If we content ourselves with the final aspect – self-organising networks, then we can postulate that governance involves the interaction of a number of different actors. These may be individuals or organisations, probably coming from a number of different spheres. The forms of interaction, if we use the self-governing network metaphor, will then take place in informal contexts as opposed to hierarchical organisations. Social network theory (Ward and Williams 1997; Rhodes 1986; Knoke 1990), will lead us to expect that these networks will differ in the number and status of the actors involved, as well as in forms of interaction and the duration of that interaction. Of vital importance in this interaction are the nodal points of the network, actors, organisations or individuals who are able to play a role as communicators and gatekeepers (Richardsson 2000). This leads us to look more closely at the interaction of actors, and at the institutional contexts within which this interaction may take place.

Risk or Uncertainty?

I claimed, in the introduction to this chapter, that decisions must often be taken under conditions of uncertainty, and here it is important to note that there is a difference between risk and uncertainty. Ulrich Beck claims that, during recent decades, human society has experienced a change of policy focus from the distribution of material goods to the distribution of risks (Beck 1986). However, I claim that we may now even be moving away from Beck's *risk society* (ibid.) to a *society of uncertainty*, in which we must attempt to expect the unexpected. The difference between risk and uncertainty is that while risk, at least according to some scientific disciplines, may be mathematically calculated and specified as percentages in numerical terms, uncertainty does not provide this possibility of definition. Uncertainty is simply that – it is quantifiably uncertain. The future has of course always been difficult to calculate, but problems of prediction become even more acute when the issue at stake is sustainability. It will be argued here that sustainability must be built on a combination of social, economic and ecological perspectives, each with their own established forms of epistemology and politics. While it may be possible to glimpse some future developments in each of these fields, albeit with a significant degree of uncertainty, the combination of these three increases the degree of uncertainty. Uncertainty leads to shifts in knowledge bases, and a decrease in the ways that the scientific community can successfully predict changes in nature and society. The move from a modern to post-modern society leads to decreases in levels of confidence in expert opinions, and opens up the debate for other forms of knowledge such as day to day experiences, or *public knowledge*. In extension, uncertainty may also lead to a wider debate about possible futures, futures that are decidedly political in nature yet often hidden behind scientific discourses. As they are political, and affect us all, the choices that have to be made for the future should not only be based on scientific knowledge. Policy choices in sustainable ecosystem management involve trade-offs between alternative uses of scant resources, as well as choices between societal values, norms, and ideologies. The natural sciences – chemistry, physics, biology etc., cannot provide answers to problems that are based on human choices between competing values and norms. In order to manage the use of ecosystems in an uncertain future all representations must be critically analysed, as they involve the exercise of power (Demeritt 1994) and decisions and policy must be based not only on ecological representations, but also on political, social and economic representations of past, present and future conditions.

Power, and different forms of power, lie at the centre of the debate on government, governance, ecosystem management and sustainability, although this is not always apparent, as power can be exercised in various forms. Steven Lukes has developed a characterisation of three forms of power (Lukes 2005), of which the power to determine a discourse, present or future, can be seen as one. The other forms of power are first-level power over others, exerted through control of decision-making procedures. Secondly, Lukes describes the power to determine an agenda, and thus steer the issues that *can* be discussed. Lukes also analyses a form of power that is exerted, consciously or unconsciously, over the very foundations of society and our thoughts. This is the power that may prevent us from thinking freely about ways to achieve sustainability, to govern ourselves, or to formulate a future not simply based on present conditions. In ecosystem governance information is often a form of power, and I will analyse the use of information as a form of power, especially with institutions, later in this chapter.

Do Institutions Matter?

Institutions do matter, and I will try to show why. If so, the move to sustainable ecosystem governance necessitates a greater understanding of the processes of institutions in governance, and involves analyses of the institutions (formal and informal) within which governance can be developed. These may be formal institutions that are created to embody and protect the values of societies, or informal institutions such as liberty, democracy, rights, citizenship, welfare, community and the rule of law. Thoreau, the author of 'Walden' who spent time alone by a pristine lake, held an altogether pessimistic view of institutions, and claimed that *'wherever a man goes, men will pursue him and paw him with their dirty institutions, and, if they can, constrain him to belong to their desperate odd fellow society'* (Thoreau 1854). This point of view is perhaps too negative, as institutions form the basis of much, if not all, of human interaction. We also need to bear in mind the differences between institutional forms, between formal and informal institutions, and between institutional structures. Young (1999) claims that a *'prevalent distinction of institutions is between rules of the game, or settled practices, and the formal organisations who are the players and who have formal hierarchies of decision-making'* and the interaction of these will be looked at later in this chapter. In recent years, concerns have been raised about the effectiveness of our (formal) institutions and the ethics of those who run them. Institutions – including the institutions of government, the market and civil society – have been seen as alienating and overweening, and as failing to live up to the values that they were established to further. Other concerns relate to the reduced power and confidence of institutions to address critical socio-economic issues, especially employment, crime, public safety and the environment. Perhaps these growing levels of doubt can be attributed to an increasingly globalised society and to the apparent weakening of the political and social institutions that underpin liberal democratic values and provide normative support for law. In any case, institutions, in the form of organisational structures or norms and values, are important for sustainable ecosystem governance, as I will attempt to demonstrate. I will also try to show that information, which I claim is a form of power, is treated in different ways in different institutional contexts.

The role of institutions in ecosystem governance is not unproblematic, however, as there is unfortunately no common understanding of the nature and definition of them among scholars. The reason for this is perhaps because there exists here, as in many areas of policy analysis, a lack of comparative studies (Scott 1995). North claims that institutions create society's structural incitement, and that economic achievements are built to a large extent on economic and political institutions (North 1998). He also states that individual's and group's beliefs, which determine their choices, are a result of learning over time, from generation to generation. Members of an institution are also considered to hold common values (Peters 1999), which can be *'webs of interrelated rules and norms'* (Nee 1998:8). Peters and Pierre (1998) also stress the way that informal institutions (norms, values, rules and practices) shape political behaviour, as do many others (Krasner 1983; Krasner 1993). Rowlinson (1997) claims that organisations (formal institutions) are *enclosed* by (informal) institutions and social structures, such as laws and state legal systems, and f*ormal* institutions (or organisations) can be said to be associated with change and action, while *informal* institutions with stability and durability. However, this does not imply that actors within organisations cannot change routines and rules. In some cases they can, and will (Rowlinson 1997:89). However, the question of definition should perhaps not be taken too far, as too rigid definitions may prevent flexibility when analysing the form and effects of institutions (Immergut 1998:13).

The study of institutions takes place in a number of different streams, of which perhaps the most important can be categorised as *sociological institutionalism*, focusing on normative and cultural influences, *rational-choice institutionalism*, that looks for strategic, goal-oriented behaviour, and *historical institutionalism*, that stresses the influence of historical aspects of institutions. Knill (2001), on the other hand, following Mayntz and Scharpf (1975), distinguishes between *institution-based* and *agency-based* approaches. Overviews of, and introductions to, new institutionalist approaches in political science and sociology can be found in March and Olsen (1984), Powell and DiMaggio (1989), Kato (1996) Hall and Taylor (1996), Immergut (1998) and Peters (1999). Anthropologists have also examined institutions, mainly in terms of their internal structures, their cultures of organisation, their roles in wider institutions, their relations to other organs of power and influence, their impact on the communities which they serve, and their roles as producers of ideas and ideologies. Regime analysis has many similarities with institutional analysis, although here the level of analysis is often at a larger geographical scale. Major early publications on regime analysis are Krasner (1975) and Rittberger and Mayer (1993). Overviews can be found in Levy and Young (1994), Levy et al. (1995), Hasenclever et al. (1996) and Hasenclever et al. (1997). Although the field of regime theory has produced many fruitful insights into ecosystem management, it will not be treated in detail here, where institutions in governance are at the centre of interest.

Ecosystem Governance – From Dialogue to Trialogue

Let us now return to the question of ecosystem governance, and attempt to place institutions within this context. Building on the above discussion, governance involves the interaction of the polity, civil society, and the business community. How then can we characterise ecosystem governance? Moving from a dialogue between two political actors, or a political actor and a part of an administration, we can visualise governance in the field of sustainability as the possible interaction between actors from three spheres; the ecological, economic, and societal spheres. Another way of putting this is to say that all three aspects must be taken into account if sustainable ecosystem governance is to succeed. We can also envisage this as an imaginary network-based interaction between three points of a triangle, and interaction would then occur, not as a dialogue between two spheres, but as a *Trialogue* between three spheres (Gooch 2004). These three pillars of the sustainability Trialogue would then be *environment*, *society*, and *economy*.

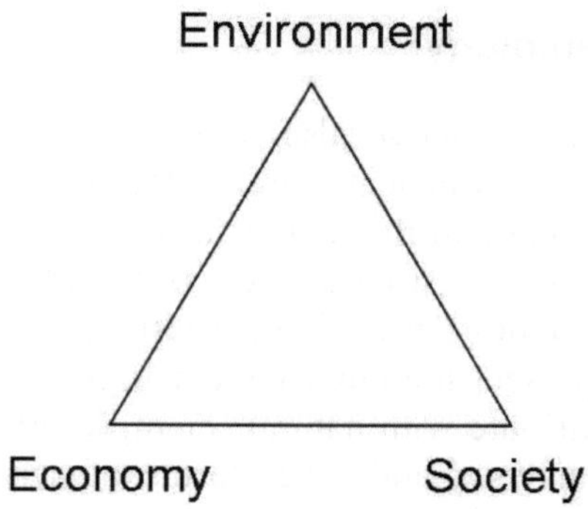

Fig. 7.1. *The trialogue of sustainability (after Gooch 2004)*

The incentive for the interaction of these three spheres in the move towards sustainability may not at first be apparent. For the polity, the advantages are that decisions made in collaboration with the societal and economic spheres are likely to be more firmly anchored in these sectors, and seen by them as more legitimate, and therefore be more likely to succeed in the implementation stages of the sustainability process. The advantages for civil society are that the public, through direct participation or through non-governmental organisations (NGOs) can influence the decision-making process and put pressure on the implementation of those decisions. For the business community, finally, there can be a number of advantages. First, they can influence the decisions made in the political sphere. Secondly, they can gain increased respectability in the eyes of the public, and a goodwill that might be translated into profits. Thirdly, the implementation of many environmental policies is dependent on the active participation of the business community, and involvement in ecosystem governance can provide opportunities to influence the direction of implementation strategies. The emphasis placed on these three spheres may differ. In developing countries it is likely that economy and business will be important aspects during the early stages, and even within Europe there are differences. Countries such as Sweden and Italy are members of the same European Union, yet prioritise different aspects of sustainability, with Sweden emphasising environmental aspects and Italy stressing economic aspects (Gooch et al. 2002).

Ecosystem Governance and Epistemology

Trialogues of Epistemology

The three spheres presented above also imply that successful ecosystem governance will be based on at least three different forms of rationality, that is, three ways of looking at the nature of the world (ontology), and at knowledge (epistemology). First let us look at the role of the experts who provide scientific information on ecological issues. This form of knowledge can be said to represent the environment in the first Trialogue. Experts can of course contribute with natural scientific information, managerial expertise, policy analyses and other forms of academic knowledge. What they have in common is that they all claim to constitute some form of objective knowledge and the special role of experts in ecosystem governance calls for particular attention. Despite what many might claim, expert opinions are far from value-free. Experts occupy their own world of institutions – their own systems of norms and values, or paradigm, to use one of the (many) definitions of the word used by Kuhn (1970). Kuhn writes of *'the entire constellation of beliefs, values, techniques, and so on shared by the members of the given community'* (Kuhn 1970:175) and this definition undoubtedly has some similarities to the definitions of informal institutions presented above. Winch (1958) also stresses the importance of ideas in defining reality and notes that formal institutions must be understood in relation to the rules governing them (informal institutions).

Experts play an important role, due to their control of scientific knowledge, and many environmental issues are defined and communicated by experts (Sundqvist 1991). In the case of ecosystem governance and sustainability, experts are at the centre of the debate. While the view taken here is that ecosystem governance should be defined not only by researchers, but by a *number* of different actors (Wynne 1989) the role of experts and the scientific community is of considerable importance. As Yearley (1995:458) has noted, *'scientific expertise is increasingly at the forefront of environmental policy formulation and of contests over policy.'*

Necessary knowledge for ecosystem governance can, and should, come from a number of different disciplines. Not only must this knowledge be integrated, combining the natural and social sciences (Gooch and Stålnacke 2006), but it is also necessary to improve the ways that this knowledge is presented to actors in ecosystem governance (Gooch 2004; Gooch and Stålnacke 2006). This involves an interaction between scientific knowledge and other forms of knowledge (see below) and the development of methods to improve stakeholder and public participation. In this combination of knowledge forms, natural scientists can contribute with information on fundamental processes (biogeochemical) in ecosystems, while social scientists can contribute with socio-economic information and analyses of policy/governance processes. While natural scientists focus on the physical processes (Mostert 1999) the social sciences, and especially disciplines working with policy and decision-making, contribute to the understanding of the structure, institutions, ideas and strategies of actors and the 'management' of the decision making process (Klijn and Koppenjan 1997). Both aspects are necessary and it is important to remember that ecosystem management should be focussed primarily on the ecosystem and especially on the interaction between actors and ecosystems (Mostert 1999).

However, science is employed not only by the polity, by policy makers and authorities, but also by the other spheres of the sustainability. Trialogue, by civil society and the business community. Environmental groups use scientific knowledge in order to legitimate their claims and utilise scientific concepts and theories in the construction of environmental ideologies (Wildavsky 1991). Mary Douglas (1966:3) also describes how *'the laws of nature are dragged in to sanction the moral code.'* The business community also exploits science and, as Von Wright (1986) points out, science is now influenced by many external and internal factors, factors which may change over time, stating that *'the obedience of science is no longer secured by inquisition, but by ministries of finance'* (ibid. p 123).

The natural science and social science perspectives are, in themselves, the base for ecosystem management, but as has been pointed out they need to be complemented by knowledge of communication processes, as information does not become knowledge automatically. In order to do so information must be communicated and understood (Timmerman and Langaas 2004). Together these three, the natural scientific perspective, the social science perspective and the communication science perspective, make up the first Trialogue of epistemology for ecosystem management.

Is all Knowledge Scientific?

It was noted, above, that both environmental groups and the business community use expert opinions, as do of course the polity. The result is that competing expert opinions often form the basic arguments used in disputes over ecosystem management. Scientific knowledge is however not the only form of knowledge used and it is now time to look at the other forms that occur in ecosystem management and to discuss my second Trialogue of epistemology. Here scientific knowledge will be complemented by other forms, most importantly by political knowledge, by economic knowledge and by local/traditional knowledge.

First let us consider political knowledge. While the polity is in many ways dependent on scientific knowledge, it is also forced to work in an institutional context that has been described as a 'who gets what, when and how' situation. That is, the polity must distribute and redistribute limited resources to the population under its control. It must also compete with other polities for limited resources. For example, a shared watershed or transboundary river can provide a scarce resource, water, for two or more countries, but the resource is always limited. Political knowledge must take into account the benefits or disadvantages of different choices of policy within governance systems. The ideologies behind these choices may differ from pragmatic realism as proposed by the renaissance advisor Machiavelli (Machiavelli 1513) to the theory of justice of John Rawls (Rawls 1971), from competition to cooperation. Most likely, the choice of governance policies will be somewhere in between.

The second form of epistemology in the second Trialogue is that of the business community. Here it is claimed that strict economic rationality rules and that decisions are taken based on the need to generate profits, at least when market economies are concerned. In planned economies the motivations are different. The epistemology of the business community is based on the understanding of what is, and what is not, economically feasible. The polity must also take into consideration this form of knowledge, as political systems are dependent on the taxes that can be levied from economic production in all its forms.

Do political knowledge and business knowledge then constitute different forms of knowledge to scientific knowledge? The polity has an academic discipline that studies it – political science, and business is studied by economists. However, this does not mean that the polity and the business community are ruled by these forms of epistemology. On the contrary, political scientists and economists often complain that their knowledge is used all too seldom. For our purposes here it can suffice to say that the epistemology, and therefore rationality, of the polity and business community are different. They make no claims to being objective in their use of knowledge; on the contrary, they are expected to use information subjectively in order to advance the interests of their population or shareholders.

The third form of epistemology in this second Trialogue is the knowledge held by the public and civil society, the local/traditional forms of knowledge that differ from the other two in that they are not always so clearly expressed. Members of the public and the NGOs that represent them have come to articulate many other, non-scientific arguments in the debates over ecosystem governance. Moral and ethical issues are now an integrated part of many disputes, and the needs of local communities have been given a voice. Many of the developments sketched above point towards an increased need for public participation and the inclusion of local/traditional knowledge. If, as we claim, ecosystem governance is faced with problems of legitimacy, together with most policies, then civil society perspectives need to be integrated into the governance system, both in order to gain legitimacy, and also in order to be exposed to societies prioritising processes, i.e. who gets what, when and how (see above). Ecosystem managers need a greater level of public awareness and public understanding of the problems they face, and they also need active help from the citizenry and other stakeholders in finding acceptable solutions. This help may be provided by local/traditional knowledge. The European Union's Water Framework Directive acknowledges this and states that '*the success of the Directive relies on close cooperation and coherent action at Community, Member State and local level as well as on information, consultation and involvement of the public, including users*' (EU 2000). In many countries local/traditional knowledge can contribute to the success of ecosystem governance, but in order to do so civil society must be able to participate in these governance systems. The ways that they can do so, and the motivations for increased participation, are outlined in the following section.

The concept of the second Trialogue of epistemology can therefore be used to exemplify the vital cooperation between the polity, the business community and civil society. As we have seen, these are also the main spheres in the governance Trialogue. Here we have looked at the role of knowledge as the relationship between 'knowledge' and politics has begun to take a central role in governance. The information produced by the scientific community is not always comprehensible by decision-makers, the public, or the business community. From the scientific point of view, this is often seen as a problem of information preparation. That is, the problem is defined as a case of needing to develop better ways of presenting information. However, this may in many cases not be the main problem. Instead, the central issue is often that decision makers and the public do not always consider the information produced by scientists relevant (Gooch 2004). In order to develop effective governance systems decision-makers and the public (which here includes stakeholders) must play a more active role in the definition of the type and quantity of information necessary to create viable policies, and the scientific community must be prepared to listen to those demands. It is therefore vital that information for governance be produced within both of the epistemological (nature of knowledge) Trialogues, one and two.

Civil Society and Governance

The discussion of the role of local/traditional knowledge above leads us to now focus our attention on the role of civil society in ecosystem governance. Civil society is the third partner in the governance Trialogue, and the focus of increased interest in policy processes. Governance, as defined in this chapter, stipulates the active involvement and participation of civil society and it has also been claimed that the interconnections between ecosystems and human systems need to be studied in more detail (Young and Osherenko 1993). Involvement and participation can of course take place at varying levels, and it is important to identify the best level and form of involvement in each specific case. In some contexts, information exchanges may be sufficient, but in many cases 'information' should not be considered genuine participation, and should be regarded simply as a starting point. It is one of the lowest rungs of the famous 'participation ladder' described by Arnstein (1969). However, which level of participation to choose in a particular case depends on the objectives of the participatory approach and the stage of the policy process. For instance, if managers only want to increase public awareness, a well-formulated campaign at the level of information may be sufficient. However, if managers want not only to raise awareness, but also to take into account the viewpoints and opinions of civil society in the development of a policy plan, the minimum level of involvement should be at the level of 'consultation'. 'Active involvement' in the form of 'partnership', 'delegated power' and/or 'citizen control' will be required if managers want to explore the possibilities to collectively manage problems and to utilise local/traditional knowledge. A major problem of civil involvement facing ecosystem mangers is that the complexity of the systems sometimes makes it difficult for the public and stakeholders to clearly understand the issues at stake. Also, participation takes time, and managers may have to choose between efficiency and democracy. They share this dilemma with most other forms of decision-making but, whereas the public can have strong feelings about levels of taxation or how their children's education should be organised, they may find it more difficult to express an informed opinion on ecosystem governance. A way around this predicament may be deliberative democracy.

The concept of deliberative democracy is becoming increasingly popular, and the question of what it implies to have decisions made in a deliberative democratic fashion, although difficult, can be answered by looking at two related aspects. The first is the cultural, the second the institutional aspect. Most proponents of deliberative democracy consider it important that public decisions are based on reason and dialogue rather than interests, bargaining power, or (purely) on scientific expertise. This may seem somewhat idealistic, yet is in fact based on the concept of practical reasoning. Deliberative democrats reject the idea of a predetermined individual will, and focus instead on the process of the formation of a will, which they assume occurs in deliberative processes. Bohman and Rehg (1997) stress the difference between conflictual interest-based politics and the deliberative and participatory nature of deliberative democracy. One implication of this position is that deliberative democrats tend to be hostile to rational choice or 'utility maximising' frameworks for decision-making such as cost-benefit analysis. In this respect they position themselves against the theoretical assumptions of mainstream economics. While this may appear unnecessarily dogmatic, it should be noted that mainstream economics does in fact have considerable difficulties in formulating theories that can incorporate the specificity's of interaction between different economic systems.

In an institutional sense, deliberative democracy is often associated with direct democracy, self-governance and decision-making at the local level. For those who see deliberative democracy in this way, deliberative democracy can be an alternative to either representative democracy, and/or to decision-making in the market, the courts, and/or centralised bureaucracies. These institutions, particularly representative democracy, are quite often lamented for their lack of possibilities for serious discussion. For instance, deliberative democrats often argue that representative institutions (such as parliaments) are used mainly for 'power trading' and bargaining, and are places where people can only act strategically. Proponents of deliberative democracy claim that the three advantages of their methods over modern representative democracy are: inclusivity (it is not only politicians and technocrats that should decide), deliberation (discussion is vital and not simply power trading), and citizenship (developing opinions and preferences rather than assuming them to be predetermined). This position has been attacked by others, however. Elster, for instance, in Bohman and Rehg (1997) indicates that deliberation should primarily improve political decision-making. Such authors do not necessarily see deliberative democracy as a replacement for representative democracy, but as a supplement. Those who advocate this position stress the importance of rules to structure public debates. Renn and Tyroller (2003) state, for instance, that a deliberative exchange must be based on mutual argument and reflection, and not on the status of the participants. In this context the arguments (and form of knowledge) of civil society would not automatically be given a lower status than, for example, economic arguments. Deliberative democracy may therefore provide a way to integrate the different epistemologies described above. If local/traditional knowledge can be combined with scientific, political and economic knowledge through deliberative processes ecosystem governance should be able to be successfully developed. A variety of techniques are available to facilitate these processes (Gooch 2004). However, the involvement of civil society, as well as the other spheres of the polity and the business community, all occur within institutional contexts, and it is now time to take a closer look at those contexts, and to examine how they may influence the transformation of information into knowledge, and from there the integration of different forms of knowledge into governance systems.

Institutional Frameworks

I have claimed that institutions matter. Now it is necessary to answer the question – why? Let us first look at the two forms of institutions sketched out in the section 'Do institutions matter?' Here a distinction was made between formal and informal institutions. If we are to understand the problems facing ecosystem governance we must take into consideration the interaction of these two forms of institutions. An organisation given the task of managing ecosystems may have a strong or weak formal structure, that is, the organisation may be strictly organised in a hierarchical manner, or it may consist of a loose conglomeration of units with relative autonomy. In a similar way the informal institutions embedded in the organisation or authority may also be characterised as strong or weak, that is, the norms and values that determine the behaviour of the members may strongly influence them or be of lesser importance. These two perspectives are combined in Figure 7.2.

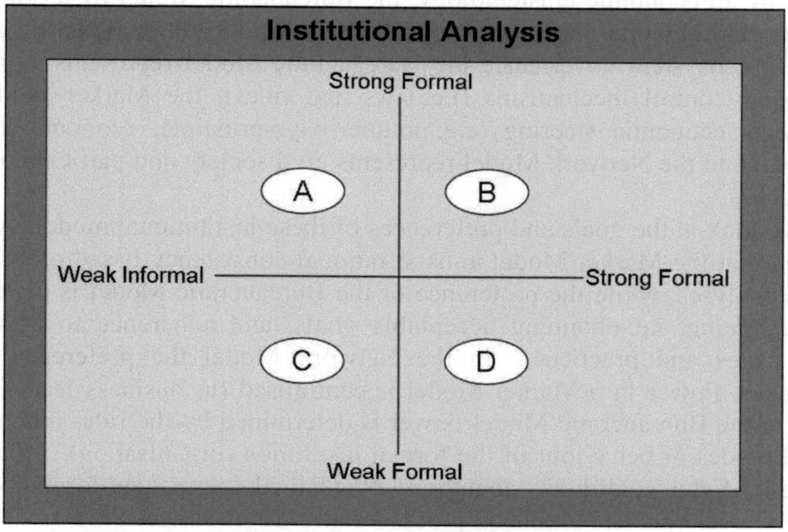

Fig. 7.2. *Formal and informal institutions and their interaction*

The relevance of this approach for ecosystem governance is that it provides us with a tool to analyse the interaction of different organisations and authorities working in the field. Many problems in ecosystem governance arise from the inability of different organisations and authorities to cooperate successfully. If we look at Figure 7.2 we can see how these problems of cooperation might be explained. Let us take the example of an authority with a strong, hierarchical, formal institutional structure (such as a governmental department) and a weak informal structure (of norms and values) that needs to cooperate with an environmental NGO with a weak formal institutional structure but a very strong informal institutional structure. The authority would be placed in the upper left-hand corner of Figure 7.2 as 'A', the NGO in the bottom right-hand corner as 'D'. It is probable that this form of cooperation will encounter problems when the strict formal institutions of 'A' attempts to interact with the weak formal structure of 'D'. Likewise, NGO 'D' will probably be frustrated by what it perceives as a lack of strong informal institutions (norms and values) in authority 'A'.

To return to the concept of Trialogues and their role in ecosystem governance, we can develop this analysis and create a characterisation of institutional forms with the three main types presented in Figure 7.3. The institutional models presented here are the Market Model, the Bureaucratic Model, and the Network Model. The Market Model can be seen to be driven by an economic epistemology, the Bureaucratic Model by a political epistemology and the Network Model by a local/traditional epistemology. In ecosystem governance the Bureaucratic Model represents Command-and-Control mechanisms (i.e. laws and rules), the Market Model represents economic steering (e.g. polluter-pays-principle, economic incentives) and the Network Model represents civil society and participatory approaches.

If we look at the goals and preferences of these institutional models, we can see that the Market Model aims at rational consistency based on economic analyses, while the preference of the Bureaucratic Model is probably *satisficing*, i.e. obtaining acceptable goals, and adherence to established rules and practices. In the Network Model the preference is consensus. Power in a Market Model is centralised (to business leaders), while in the Bureaucratic Model power is determined by the rules and accepted modes of behaviour of the formal institution (organisation). In the Network Model, coalitions – temporary or relatively permanent formations of allies, wield power. Decision processes also differ between the models; and while the orderly, fully logical vision of the Market Model may be a utopia, the procedural mode of the Bureaucratic Model indicates an institutional context where standard operating procedures and rules dominate. The Network Model shows signs of fluctuation and competing interests. Problems arise when organisations and authorities with different rationalities need to cooperate. Conflicts can then arise concerning the institutional preferences that should be adhered to. As noted, the problems of cooperation can be even more acute when organisations based on different institutional norms need to work together.

This leads us to the issue of how information is translated into knowledge in these different institutional contexts. The aim of the Market Model is to utilise information in a rational utilitarian manner. This may be an ideal that is seldom achieved, but the strong informal institutions of this model (create profits, cost-and-benefits) influence the way that Market organisations will attempt to treat information. In the Bureaucratic Model, information is likely to be reduced by rules, that is, information will be formed and filtered by the formal and informal institutions of the organisation, and by cognitive aspects of that organisation. In a strongly politicised organisation information may be used strategically to obtain benefits and advantages by one coalition over another. In both cases information will be used selectively, and the production of only one set of informative factors may lead to the information being rejected or distorted. In the Network Model the translation of information into knowledge will probably be heavily influenced by the informal institutions of the organisation. These may determine the acceptance of new information in situations where it is constantly compared with local/traditional forms of knowledge. These different uses of information and knowledge constitute potential problems for ecosystem governance, as will be shown below.

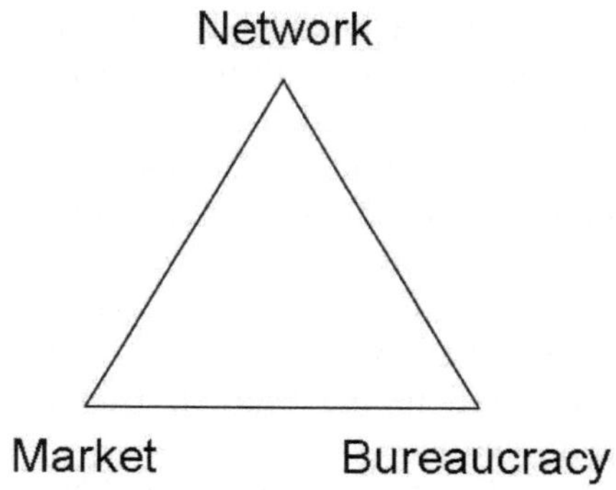

Fig. 7.3. *The trialogue of institutional organisation*

Ecosystems Governance and the Policy Process

Governance is, to a large degree, a matter of policy, public or private. Public policy can be defined as a course of action, usually pursued by a government or other part of the polity. The origin of the term can be traced to the Greek *politeia* (citizenship), from *polités* (citizen) and from *polis* (city). Public policy may be concerned with the allocation of goods, predominantly under conditions of scarcity, and/or with the allocation of values. Public policy may be concerned with public services (welfare state, education, public health, pensions, housing, etc.), regulating the economy (subsidies, tax concessions, public ownership etc.), limiting the impact of activities (social control, environmental protection, consumer protection, limiting monopolies and cartels, etc.), or regulating various other aspects of society. Policy may be distributive, regulatory, or constituent.

A traditional way of seeing the policy process is as a series of stages, starting with forms of input, continuing over the policy process, and ending up with some form of output (Sabatier 1999). The input stage, according to this model, includes information, demand, support, perceptions etc. The policy stage consists of the aspects described above, and the output stage of application, enforcement, interpretation, evaluation modification etc. However, this 'stages' model has been criticised for failing to recognise the existence of, for example, different forms of input in all stages of the policy process, including during implementation, and it has been pointed out that the policy process does not lend itself easily to a division into clear and separate stages (Sabatier 1999). Perhaps it is therefore more correct to see policy processes as resulting from the push and pull of a considerable number of actors and influences. Among these can be named the distribution of power, economic conditions, the organisation of interests, institutional arrangements, political leadership and parties, social demography, cultural attitudes, levels of technology, past and present policies, and bureaucracies etc. These can be simplified into a Trialogue of policy-making, consisting of economic, political, and social forces, a Trialogue that can be compared to the Market, Bureaucratic and Network models in the preceding section.

Policy is also about choices between tools, and here the decision-maker or manager is faced with a wide choice of options, ranging from legal rules and sanctions, tax incentives, procurement, demand management, information, rewards, public-private partnerships, subsidies, regulation, monitoring and investigation etc. These too can be simplified into a Trialogue of policy tools – economic, command and control, and social, sometimes known as the 'stick' (command and control, regulation), the 'carrot' (economic), and the 'love affair' (creating loyalties). Another way of characterising these policy alternatives that was noted above is as hierarchical/bureaucratic methods (command and control), market inspired methods (economic) and community/network methods (loyalty). A major challenge for the move to sustainable ecosystems is to identify the most efficient mix of policies for each problem and context. In a historic perspective it can be claimed that hierarchical/bureaucratic, command and control regulation dominated the preferences of policy makers and managers during the 1970s, and still does in many parts of the world and in many situations. However, during the 1980s many policy makers adopted the ideology of the market, and sought to introduce economic tools such as the polluter pays principle. This change of focus was as much the result of experiences with the problems encountered in command and control procedures, especially when attempting to deal with diffuse problems such as water pollution from agriculture, as it was a result of a belief that the market was a better model to work towards. Combinations of command-and-control mechanisms and market tools continue to be applied but, during the last decade or so, an awareness has developed that neither of these two groups of policy tools, either individually or combined, can solve the problems associated with ecosystem management. The inclusion of the third set of tools in the policy tool Trialogue has begun to be discussed in greater detail, and public participation, information, and the creation of voluntary incentives have risen high on the agenda in many national and international organisations.

The Way Ahead for Ecosystem Governance?

This chapter has identified some of the problems facing ecosystem governance and has examined some of the central aspects of the issues at stake. How, then, to move ahead? Further analyses of ecosystem governance should include comparisons between different systems, different spatial levels, and different geographical contexts. The comparative perspective is important for a number of reasons. First, comparative analysis is an area of study that can provide valuable insights into related problems existing in diverse societies. Secondly, comparative studies are at present an underdeveloped and underrepresented field of research that has the potential to contribute to our understanding of governance processes.

The level of analysis is also important. Ecosystem governance takes place at many spatial and administrative levels, and we should look not only at international and national arenas but also at the ways that policy is locally formed and influenced. It is especially important to look at the ways that institutional contexts influence ecosystem governance, and to analyse the ways that different forms of knowledge are utilised. Unfortunately, while these institutional procedures, and the ways that information are used, are crucial, it is exactly here where we are handicapped by a lack of knowledge. Much work remains to be done on the post-decisional stages of the policy process (Jordan 1997). We need, therefore, to look closely at institutional interactions, at organisational cultures, and at the transformation of information into knowledge. We need to study in more detail the move from government to governance. Analyses of the role of civil society and of deliberative forms of influence can contribute to this move, as they indicate a possible, but not necessary, direction in which ecosystem management could attempt to move.

The transformation from *government* to *governance* in ecosystem management is therefore a major challenge for the international community, involving the development of innovative forms of communication, cooperation, and interactive Trialogues between actors, forms of knowledge and organisations. As *governance* also occurs at different administrative and geographical levels in systems known as *multi-level governance systems*, we also need to examine the interaction between these levels in more detail. In August 2004 the Stockholm Water Week Symposium raised the issues of governance and Trialogues (Gooch 2004). We now need to take our studies further along this road, and to develop the necessary tools for improved ecosystem governance. This is a major task for the future.

References

Arnstein S (1969) A ladder of citizen participation. Journal of the American Institute of Planners **35**:216-224

Beck U (1986) Risk society: towards a new modernity. Sage

Bohman J, Rehg W (1997) Introduction. Deliberative democracy. Essays on reason and politics. Bohman J, Rehg W. Cambridge MA, MIT Press

Cohen J, Arato A (1994) Civil society and political theory. Cambridge, MA, MIT Press

Demeritt D (1994) Ecology, objectivity and critique in writings on nature and human societies. Journal of Historical Geography (20)1:22-37

Douglas M (1966) Purity and danger. London, Routledge and Keagan Paul

EU (2000) Community legislation: directive 2000/60/EC of the European Parliament and of the Council of 23 October 2000 establishing a framework for Community action in the field of water policy. 2001

Gooch G (2004) The communication of scientific information in institutional contexts: the specific case of transboundary water management in Europe. Environmental information in European transboundary water management. Laangnäs S and Timmerman J. Netherlands, IWA

Gooch GD (2004) Improving governance through deliberative democracy – initiating informed public participation in water governance policy processes. 14th Stockholm Water Symposium 2004, Stockholm

Gooch GD, Höglund P, et al. (2002) Review of existing structures, models, and practices for transboundary water management. 2nd International conference on sustainable management of transboundary water in Europe, Miedzyzdroje, Poland

Gooch GD, Stålnacke P (eds) (2006) Integrated transboundary water management in theory and practice: experiences from the new EU eastern borders. London, IWA Publishing

Hall PA, Taylor RCR (1996) Political science and the three new institutionalisms. Political Studies **44**:952-973

Hasenclever A, Mayer P, et al. (1997) Theories of international regimes. Cambridge, Cambridge University Press

Hasenclever P, Mayer P, et al. (1996) Interests, power, knowledge: the study of international regimes. Mershon International Studies Review **40**:177-228
Immergut EM (1998) The theoretical core of the new institutionalism. Politics and Society (26)1:5-34
Jordan AG (1997) 'Overcoming the divide' between comparative politics and international relations approaches to the EC: what role for 'post-decisional politics'? West European Politics (20)4:43-70
Kato J (1996) Institutions and rationality in politics: three varieties of neo-institutionalists. British Joumal of Political Science **26**:553-82
Klijn EH, Koppenjan JFM (1997) Beleidsnetwerken als theoretische benadering: een tussenbalans. Beleidsnetwerken **2**:143-167
Knill C (2001) The Europeanisation of national administrations. Cambridge, Cambridge University Press
Knoke D (1990) Political networks: the structural perspective. Cambridge, Cambridge University Press
Krasner S (ed) (1975) International regimes. Ithaca, NY, Cornell University Press
Krasner S (1983) Structural causes and regime consequenses: regimes as intervening variables. International regimes. Krasner S. Ithaka, Cornell University Press, pp 1-22
Krasner SD (1993) Sovereignty, regimes, and human rights. Regime theory and international relations. Rittberger V, Mayer P. Oxford, Clarendon Press, pp 139-167
Kuhn TS (1970) The structure of scientific revolutions. Chicago, Unversity of Chicago Press
Latour B (2001) What rules of methods for the new scientific experiments? Accessed 2006, Keynote speech, 13th Darmstadt Colloquium
Levy MA, Young OR (1994) The effectiveness of international regimes. Annual convention of the International Studies Association, Washington DC
Levy MA, Young OR, et al. (1995) The study of international regimes. European Journal of International Relations (1)3:267-330
Lukes S (2005) Power: a radical view. Houndmills, Palgrave Macmillan
Machiavelli N (1513) The prince. Oxford, Oxford University Press
March J, Olsen JP (1984) The new institutionalism: organizational factors in political life. American Political Science Review **78**:734-749
Mayntz R, Scharpf FW (1975) Policy-making in the German federal bureaucracy. Amsterdam, Elsevier
Mostert E (1999) Perspectives on river basin management. Physics and chemistry of the earth part B: hydrology, oceans and atmosphere **24**:563-569
Nee V (1998) Sources of the new institutionalism. The new institutionalism in sociology. Brinton MC and Nee V. New York, Russel Sage Foundation
North DC (1998) Economic performance through time. The new institutionalism in sociology. Brinton MC and Nee V. New York, Russel Sage Foundation
OUP (1989) Oxford English Dictionary. Oxford, Oxford University Press
Peters BG (1999) Institutional theory in political science: the 'new institutionalism'. London, Pinter

Peters GB, Pierre J (1998) Governance without government? rethinking public administration. Journal of Public Administration Research and Theory (8)2:223-244

Powell and DiMaggio (1989) Rediscovering institutions. New York, Free Press

Rawls J (1971) A theory of justice. Harvard, Harvard University Press

Renn O, Tyroller A (2003) Citizens' juries within the river dialogue project. recommendations for the practice. Stuttgart, CTA

Rhodes RAW (1986) European policy-making, implementation and sub central governments: a survey. Maastricht, European Institute of Public Administration

Rhodes RAW (1996) The new governance: governing without government. Political Studies (44)3:652-667

Richardsson J (2000) Government, interest groups and policy change. Political Studies **48**:1006-1025

Rittberger V, Mayer P (eds) (1993) Regime theory and international relations. Oxford, Clarendon Press

Rowlinson M (1997) Organisations and institutions. London, Macmillan Press

Sabatier P (ed) (1999) Theories of the policy process. Boulder, CO, Westview Press

Scott R (1995) Institutions and organisations. London, Sage

Stoker G (1998) Governance as theory: five propositions. International Social Science Journal (50)1:17-28

Sundqvist G (1991) Vetenskapen och miljöproblem. Sociologiska institutionen. Göteborg, Göteborgs Universitet, p 240

Thoreau HD (1854) Walden.

Timmerman JG, Langaas S (eds) (2004) Environmental information in European transboundary water management. IWA Publishing

Waltz KN (1999) Globalization and governance. PS: Political Science and Politics (32)4:693-700

Ward S, Williams R (1997) From hierarchy to networks? sub-central government and EU urban environment policy. Journal of Common Market Studies (35)3:439-464

Wildavsky A (1991) If claims of harm from technology are mostly false, or unproven, what does that tell us about science? health, lifestyle and environment: countering the panic. New York, The Social Affairs Unit/Manhattan Institute

Winch P (1958) The idea of social science. Routledge

Von Wright GH (1986) Vetenskap och förnuftet. Stockholm, Fakta Förlag

Wynne B (1989) Sheepfarming after Chernobyl: a case study in communicating scientific information. Environment (31)2:10-15, 33-45

Yearley S (1995) The environmental challange to science studies. Handbook of science and technology studies. Jasanoff S. Thousand Oaks, California, Sage Publications, pp 457-479

Young OR (1999) Dimensions of global environmental change science plan. Bonn, International Human Dimensions Programme on Global Environmental Change

Young OR and Osherenko G (1993) Testing theories of regime formation: findings from a large collaborative research project. Regime theory and international relations. Rittberger V, Mayer P. Oxford, Clarendon Press, pp 223-251

Evolution of International Norms and Values for Transboundary Groundwater Governance

Raya Marina Stephan

Abstract

Given the fact that the largest number of cases of shared water occur in aquifers, it is surprising that so little is written about this. The Trialogue Model of Governance assumes interfaces between three actor-clusters; Governance, Science and Society. International law plays a significant role in facilitating the quality of these interfaces by codifying norms and values to the extent that these can become building blocks of future cooperation. Within the IWRM paradigm, international law has the capacity to facilitate linkages at different levels. The aim of this chapter is to discuss the evolution from the traditional approach to groundwater in international law to the latest trends both at the International Law Commission and in environmental treaties. International law is not only about regulating international relations because core norms and values have the potential to be incorporated into national law.

Keywords: groundwater governance, international law, International Law Commission, aquifer, transboundary groundwater, United Nations (UN), Southern African Development Community (SADC), Global Water Partnership (GWP), UNESCO

Introduction

The Global Water Partnership Framework for Action at the Second World Water Forum (The Hague 2000) stated that the water crisis that humankind is facing today is mainly a crisis of water governance (Global Water Partnership 2003), i.e. "the ways in which individuals and societies have assigned value to, made decisions about, and managed the water resources available to them" (UNDP 2004). The ministerial declaration of the Second World Water Forum (2000) identified *governing water wisely* as one of the key challenges for the future. Since then, water governance has assumed increasing importance in the international debate on water, and the second UN World Water Development Report (WWAP 2006) presented at the 2006 Fourth World Water Forum in Mexico focuses on the importance of governance in managing the world's water resources. However, until recently, this debate concerned surface water more than it did groundwater.

Groundwater represents as much as 97% of the earth's freshwater fraction in liquid form, i.e. excluding that locked in polar ice caps (Foster 1999). In arid and semi-arid regions this is often the only source of water. Most of this groundwater occurs in transboundary aquifers. Despite an increased dependency on it, leading to over-exploitation, depletion and pollution (WWAP 2003, Yamada Add1 2003), governance of groundwater at the international level has received less attention than has surface water. Until recently, international norms and values for groundwater were almost non-existent. As a result, new questions are arising in the international arena, concerning the use, protection and ownership of transboundary groundwater resources, which should be addressed by international law in order to ensure its proper management and good governance.

International law has typically considered groundwater only as a subsidiary to surface water. The UN Convention on the Law of Non-Navigational Uses of International Watercourses (1997) represents the latest authority in international water law. It includes groundwater in its coverage, but in a very limited way. The Convention was drafted by lawyers at the International Law Commission and adopted at the UN General Assembly by diplomats who had no specific knowledge of hydrology in general, or hydrogeology in particular. The limited approach of the UN Watercourse Convention was followed by other conventions and treaties, and is similar to the view that the doctrine had adopted.

However, a new evolution seems to be occurring, tending to give transboundary groundwater the degree of attention it needs. The International Law Commission has added to its programme of work the topic of "Shared Natural Resources", which includes transboundary groundwater. The Special Rapporteur on the topic has already submitted three reports (International Law Commission 2003, 2004 and 2005). For the preparation of his reports and the draft articles, the Special Rapporteur has turned to the scientific community and has received "valuable assistance from experts under the auspices of UNESCO" (Yamada 2004). The codification process is considering the scientific knowledge and information for elaborating the rules. In addition, an environmental trend and evolution, under the influence of scientific findings, is taking place in the international arena that considers groundwater among other natural resources, and intends to give it the adequate protection it deserves.

The aim of this chapter is to discuss the evolution from the traditional approach to groundwater in international law, to the latest trends both at the International Law Commission and in environmental treaties. This evolution shows the increased awareness of the need to understand groundwater, to integrate the scientific knowledge about it, to set up adequate rules for its sustainable management and to improve its governance.

Through the study of the evolution of international norms and values on groundwater, this chapter highlights the role of law in the overall debate on ecosystem governance and, more particularly, in the Trialogue Model.

The Traditional Approach to Groundwater in International Law

In International Law the traditional approach to groundwater is represented by the UN Convention on the Law of Non-Navigational Uses of International Watercourses, hereinafter the UN Watercourse Convention, and inter-State agreements on water (UN Doc. A/RES/51/229, available at www.un.org/law/cod/watere.htm). The doctrine, represented mainly by the work of the International Law Association, has remained close to this approach, even if more innovative.

These instruments, with different legal forces, have considered groundwater in an incomplete manner, both as concerns its definition and the principles to be applied to it.

The Definition of Groundwater

Following the recommendation of its General Assembly in 1970, in 1971 the International Law Commission (ILC) initiated the study of the law on the non-navigational uses of international watercourses with a view to its progressive development and codification. It presented the final version of the draft articles to the General Assembly in 1994.

Up until 1991, the ILC focused its studies on surface water. The question of groundwater was introduced only in 1991, when the Special Rapporteur Stephen McCaffrey presented a detailed study on the subject (McCaffrey 1991). The Commission debated his proposals and finally agreed to include groundwater related to surface water in the draft Convention. The members of the ILC considered that in cases where surface and groundwaters formed a unitary system, human intervention at one point in such a system might have effects elsewhere within the same system. What the members of the Commission called "confined groundwater," i.e. groundwater that was unrelated to surface water, remained excluded from the scope.

In 1992 Robert Rosenstock, who succeeded Stephen McCaffrey as Special Rapporteur, suggested including *all* groundwaters in the scope of the Convention (Rosenstock 1993), but his proposal was not accepted. The members of the ILC felt that including groundwater in the scope of the Convention came at a time when they had already drafted many principles and rules which related only to surface water. Thus, the ILC members felt uneasy about applying these principles and rules to something they had not investigated before formulating them.

The Unitary Whole and the Common Terminus

The UN Watercourse Convention applies to international watercourses (Article 1 of the Scope). In Article 2, paragraph a, a watercourse is defined as *"a system of surface waters and groundwaters constituting, by virtue of their physical relationship, a unitary whole and normally flowing into a common terminus."* An international watercourse is "a watercourse, parts of which are situated in different States" (Article 2, paragraph b).

Regarding groundwater, the Watercourse Convention is therefore limited in its scope. It only considers groundwater when it is related to surface water, **and** flowing to a common terminus. Groundwater unrelated to surface water is excluded. This leaves out important systems such as the Nubian Sandstone Aquifer System (Chad, Egypt, Libya and Sudan). On the other hand, groundwater and surface water, even when they are related, do not necessarily "share" a common terminus. In reality, surface water and groundwater rarely flow to a common terminus.

Following the UN Watercourse Convention, some interstate agreements on water have adopted the same definition for groundwater. The Revised Protocol on Shared Watercourses in the Southern African Development Community (SADC) (Windhoek, 7 August 2000)[1] refers directly to the UN Watercourse Convention in its preamble and represents a first application of the Convention, although it has not yet entered into force (Sohnle 2001). The Revised Protocol adopts exactly the same definition for a watercourse (Article 1), except that it specifies "common terminus" as "a sea, lake or aquifer." In the application of the Revised Protocol, the Republic of Mozambique, the Republic of South Africa and the Kingdom of

---

[1] The Protocol entered into force 22 September 2003. The Parties and/or signatories are: Angola, Botswana, Congo, Lesotho, Malawi, Mauritius, Mozambique, Namibia, Seychelles, South Africa, Swaziland, United Republic of Tanzania, Zambia, Zimbabwe.

Swaziland have adopted an Interim Agreement for Co-operation on the Protection and Sustainable Utilisation of the Water Resources of the Incomati and Maputo Watercourses (Johannesburg, 29 August 2002). The Interim Agreement refers in its preamble to the Revised Protocol and to the UN Watercourse Convention. The Interim Agreement adopts the same definition for a watercourse as that in the Revised Protocol (Article 1). In Europe, the Framework Agreement on the Sava River Basin (Kranjska Gora, 3 December 2002) defines the Sava River Basin as *"the geographical area extended over the territories of the Parties, determined by the watershed limits of the Sava River and its tributaries, which comprises surface and groundwaters, flowing into a common terminus"* (Article 1, paragraph 2). The Framework Agreement refers to the Convention on Cooperation for the Protection and Sustainable Use of the River Danube of 1994, and does not mention the UN Watercourse Convention.

Groundwater as a Subsidiary to Surface Water

Interstate treaties and agreements concluded on transboundary waters concern in a very large majority international rivers and rarely address transboundary aquifers. One exception is the Agreement on the Franco-Swiss Genevese Aquifer (1978) which is the only example of a treaty dealing ***exclusively*** with the management of a transboundary aquifer. Most treaties and agreements on transboundary waters view groundwater only in so far as it is related to the surface water body of concern, as illustrated in Table 8.1.

Table 8.1. Selection of treaties on transboundary waters, and the surface and groundwater bodies they cover

Agreement	Surface water body	Groundwater body
Convention on Cooperation for the Protection and Sustainable Use of the River Danube (Sofia, 29 June 1994)	The Danube River	Groundwater in the catchment area of the river (Article 2§1)
Convention on the Protection of the Rhine (Berne, 12 April 1999)	The Rhine	Groundwater interacting with the Rhine (Article 2§a)

Agreement	Surface water body	Groundwater body
Agreement between the Federal Republic of Nigeria and the Republic of Niger concerning the equitable sharing in the Development, Conservation and Use of Their Common Water Resources (Maiduguri, 18 July 1990)	The Maggia/Lamido River basin The Gada/Goulbi og Maradi River Basin The Tagwai/El Fadama River basin The lower section of the Komadougou-Yobe River basin (Article 1§2)	Groundwater contributing to the flow of surface waters (Article 1§3)
Convention on the Sustainable Development of Lake Tanganyka (Dar es Salaam, 12 June 2003)	Lake Tanganika (Article 3)	"groundwaters that flow into the Lake" (Article 1)
The Protocol for Sustainable Development of Lake Victoria basin (Arusha, 29 November 2003)	Lake Victoria	"underground waters flowing into Lake Victoria" (Article 1)

The Consideration of the Doctrine: A Step Forward

Groundwater was considered by the Helsinki Rules, adopted in 1966 by the International Law Association (ILA). The ILA is an association of experts and has as its main objectives the study, clarification and development of both public and private international law. The resulting rules, such as the Helsinki Rules, are non-binding. Contrary to the UN Watercourse Convention, the Helsinki Rules have adopted the basin approach, and have considered an international drainage basin as *"a geographical area extending over two or more States determined by the watershed limits of the system of waters, including surface and underground waters, flowing into a common terminus"* (Article II). In 1986, at the Seoul Conference, the ILA adopted a set of rules consisting of four articles specifically designed to govern groundwaters that intersect international boundaries regardless of whether or not they are associated with a system of surface waters. The Seoul Rules introduced the term *aquifer* and defined it in a note under Article 1. According to this definition an aquifer "comprehends all underground water-bearing strata capable of yielding water on a practicable basis, whether these are in other instruments of contexts called by another name such as *"groundwater reservoir"*, *"groundwater catchment area"*, etc., including the waters in fissured or fractured rock formations and the structures containing deep, so-called *"fossil wa-*

ters"." By including the term 'aquifer', the ILA considers more than just the groundwater itself, it also considers its container. Here the ILA meets the concerns of the hydrogeologic community, and brings a scientific component into the legal tool. Furthermore, the ILA acknowledged the gap in the Helsinki Rules regarding groundwaters and opened the rules to all sorts of aquifers as long as they are part of an international drainage basin. An aquifer intersected by an international boundary, or hydraulically connected to a surface water body intersected by such a boundary, was considered as part of an international drainage basin and thus fell under the Rules.

Another important contribution by the doctrine is the "*Model Agreement Concerning the Use of Transboundary Waters*" drafted at Bellagio (Italy). This Model Treaty also refers to 'aquifer', which it defines as "*a subsurface water-bearing geologic formation from which significant quantities of water may be extracted*" (Article I). The Model Treaty applies exclusively to transboundary groundwaters, defined as "the waters in transboundary aquifers."

Applicable Principles: The Extension from Surface to Groundwater

When considering groundwater, the above mentioned instruments, with the exception of the Bellagio Draft Treaty, have simply extended the international water law rules as conceived and applied to surface water.

Basic Principles: From Surface Waters to Connected Groundwaters

There are two key principles in international water law, developed for surface water resources (Buirette 1991; Caflisch 1997; Sohnle 2002):

1. Equitable and reasonable use, and
2. The obligation not to cause significant harm.

Considering the existing international law custom for water resources, the ILA had already included in the Helsinki Rules the equitable and reasonable use principle (Article IV), and specified relevant factors to be considered for the application of the principle in each particular case such as the geography of the basin, the economic and social needs of each basin State and the availability of other resources (Article V).

After the development of the Helsinki Rules, the UN Watercourse Convention codified 'equitable and reasonable use' in Articles 5 and 6, as well as the 'obligation not to cause significant harm' in Article 7. Long debates and discussions took place at the ILC and at the 6th Committee of the UN General Assembly (the Legal Committee) to decide which rule would have supremacy over the other (Sohnle 2002). The formulation finally adopted in the UN Watercourse Convention gives precedence to the equitable and reasonable use over the no harm rule (Caflisch 1997, McCaffrey 1998). The equitable and reasonable use principle becomes the core principle of international water law. In the Gabcikovo-Nagymaros case of 21 September 1997, the International Court of Justice refers twice to the equitable and reasonable use principle (paragraphs 85 and 147) and does not mention the obligation not to cause harm.

Another important principle in international water law is the obligation to cooperate. This principle is a general principle of international law. It is drafted in Article 8 of the UN Watercourse Convention, and one of its important applications in this Convention and in international water law in general is the regular exchange of data and information (Article 9 of the UN Watercourse Convention).

Even if the UN Watercourse Convention is not yet in force, these principles can be considered to be part of international water law as customary rules. They appear in the agreements mentioned above, and in other agreements. In that sense, they apply to transboundary groundwaters, at least when they are related to surface waters.

The Extension to all Groundwaters

Robert Rosenstock, the ILC Special Rapporteur on the law of the non-navigational uses of international watercourses had suggested the inclusion of "unrelated confined groundwaters" in the draft articles (paragraph 1.1.1). He considered that the principles and norms already drafted for surface waters and related groundwaters were applicable to unrelated groundwaters, with a few minor changes (Rosenstock 1994).

When completing their work on the law of non-navigational uses of international watercourses, the members of the ILC recognised that, by including groundwater in the scope of the draft articles only when it is related to surface water, they excluded other types of groundwater. Thus, in 1994 they adopted a "*Resolution on confined transboundary groundwater*". It is worth mentioning that the ILC does not use here the word "confined" in its hydrogeological meaning. Instead, it uses it to designate an aquifer unrelated to surface water, whereas, for hydrogeologists, a confined aquifer is an "*aquifer overlain and underlain by an impervious or almost impervious formation*" (UNESCO/WMO 1992). This misuse of scientific terms by the legal community has caused a misunderstanding between lawyers and hydrogeologists, or more generally between lawyers and water experts, for years.

In this Resolution the ILC commends States to be guided by the principles contained in the draft articles (the Convention), "***where appropriate***", in regulating transboundary groundwater (paragraph 1), and to consider entering into agreements with the other State or States in which the "*confined*" transboundary groundwater is located (paragraph 2). In sum, with this Resolution the ILC suggested transposing the principles it had drafted to all types of groundwater. However, the ILC also recognised in this Resolution that these principles may not be appropriate and it acknowledged "*the need for continuing efforts to elaborate rules pertaining to confined transboundary groundwater.*" In 1998, the Planning Group of the ILC identified shared natural resources (i.e. confined groundwater and single geological structures of oil and gas) as one of the topics for inclusion in the ILC's long-term programme of work.

The ILA had adopted the same attitude with the Seoul Rules of 1986. By adapting the definition of an international drainage basin to include aquifers in specific cases as detailed above, the ILA has extended the application of the Helsinki Rules to aquifers.

The Need for Specific Rules

The Resolution of the ILC and the ILA rules are non-binding instruments and are only authoritative. Even if their recommendations are followed, i.e. their principles and rules apply to all groundwaters, it is not certain that these rules are really fit and adequate. It is not even certain that these principles are fit enough for groundwaters that seem to be already covered by existing rules. It is worth saying again that these rules were originally drafted for surface waters, and then the groundwater component was added, but only in situations where it has a connection with the surface water. This is true not only for the UN Watercourse Convention, but also for the existing State practice expressed in the treaties on transboundary waters. However, in the case of the Seoul Rules, the ILA had already acknowledged one of the specificities of groundwater as compared to surface water. The Rules had provided a specific provision for the protection of groundwater in Article III, strengthening the Helsinki Rules on Pollution (Chapter 3). By including an article on groundwater protection in the Seoul Rules, which consist of only four articles, the ILA gives special weight to these measures and highlights the importance of protection and pollution prevention in the case of groundwater as compared to surface water. Article III, paragraph 1, recommends that "*Special consideration...be given to the long-term effects of the pollution of groundwater.*" In paragraph 2, Article III insists on the "*exchange of relevant available information and data*" for the purpose of preserving the groundwaters.

Paragraph 3 relates to the obligation to cooperate between "*basin States*" for the purpose of collecting additional information and data on the international aquifer.

In the case of the core principles of international water law, it is clear that an adaptation is necessary before they can be applied to groundwater. For instance, the principle of equitable and reasonable use is not independent, and relies on a list of factors for its implementation. The principle appears as a balance of the needs of the States sharing the waters, and factors are involved in determining the equitable distribution of the waters. These factors can be adapted to each specific case. The UN Watercourse Convention has set up a non-limitative list of factors in Article 6. However, no factor amongst those listed is specific to groundwater. Of course, any hydrogeological factor could fall under "*other factors of natural character*" (Mechlem 2003) and can always be considered whenever necessary. But the rule of principle and equitable use is already considered vague and difficult to implement (Stitt 2004). The inclusion in the list of factors specific to groundwaters is advisable in an international instrument ruling these waters. For instance, in the Agreement between the Federal Republic of Nigeria and the Republic of Niger concerning the Equitable Sharing in the Development, Conservation and Use of their Common Water Resources (1990), in the factors to be considered by the Parties "*in determining the equitable share*" to which each one is entitled (Article 5) a reference to hydrogeology is made, without more details being given.

Considering the vulnerability of groundwater, the obligation not to cause significant harm appears especially important. A more precise idea of what is "*significant harm*" would be appreciable regarding groundwaters, but this has been a long debate in international law.

Finally, the duty to exchange data and information on a regular basis under the general obligation to cooperate, is fundamental in case of transboundary aquifers. Scientific knowledge on a transboundary aquifer may not yet be highly developed, and data are often not available and/or not accurate. These difficulties make it especially important for States to share data on international aquifers.

The traditional approach to groundwater in international law has shown its limits. Emerging rules in environmental treaties, and the work currently under preparation at the ILC, may fill the gap.

The Emerging Trend: Scientific and Environmental Inputs to International Groundwater Law

The Topic at the International Law Commission

In 2002 the International Law Commission decided to include in its programme of work an item entitled "*Shared Natural Resources*", under which the nominated Special Rapporteur on the topic decided to deal with "*confined*" transboundary groundwaters, oil and natural gas. He also decided to adopt a step-by-step approach to the study of the topic and to start with groundwaters (Yamada 2003).

From "Confined Transboundary Groundwaters" to Transboundary Aquifers

The First Approach

In his first report on outlines (UN Doc. A/CN.4/533), the Special Rapporteur addressed the background of the topic at the International Law Commission with a brief summary of the consideration of the law of non-navigational uses and the proposals by the different Special Rapporteurs on the topic to include groundwaters in the scope of the draft articles. He then concluded that the scope of groundwater which he would address in his studies covered "*water bodies that are shared by more than two States but are not covered by Article 2(a) of the Convention on the Law of the Non-navigational Uses of International Watercourses*". He would continue to use the terminology "confined transboundary groundwaters" for purposes of convenience until a precise definition could be formulated "*on the basis of a correct understanding of...hydrogeological characteristics.*" The Special Rapporteur thus acknowledged, since his first report, the need for the ILC, whose members are lawyers and diplomats, to take into account the scientific knowledge and approach to groundwaters before formulating any definition. He referred to the Internationally Shared Aquifer Resources Management (ISARM) project of the International Hydrological Programme of UNESCO (Puri 2001). Under this project the following five focus areas were identified: scientific, environmental, socio-economic, legal and institutional. Within the framework of ISARM, UNESCO-IHP has set up an ad-hoc group of groundwater experts, lawyers and hydrogeologists to assist the Special Rapporteur from a scientific and technical point of view.

The Introduction to Aquifers: Considering Science

In his second report (UN Doc. A/CN.4/539), the Special Rapporteur proposes using the term "*aquifers*", which is "*a scientific and more precise term*", as he acknowledged in his oral introduction of the second report at the ILC (Report 2004). As "*the concept of aquifer consists of both the rock formation which stores waters and waters in such a formation*", the Special Rapporteur considered that "*it suffices to say...aquifer(s)" as opposed to the case of the UN Watercourse convention where reference is made to both the "international watercourses" and "their waters*". In the report the Special Rapporteur specifies that he chose this terminology "*after consultation with hydrogeologists" (Yamada 2004), underlining his intention to adapt the law on transboundary aquifers to the science of groundwater. The following definition was adopted: ""Aquifer" means a permeable water-bearing rock formation capable of yielding exploitable quantities of water*" (draft Article 2(a)). This definition comes from the UNESCO/WMO International Glossary of Hydrology, prepared by an international panel of water experts. However, the adoption of this definition for an aquifer met with questions and criticism.

The questioning came from the members of the ILC during the debates on the second report. They first wondered about the concept of "*exploitability*" whether it referred to the evolution of technology which makes more water accessible, or if it referred to quantities of water and to commercial viability. Other criticism of this definition came from groundwater experts, who felt that the term "*exploitability*" might be understood as excluding aquifers that are not currently exploitable, for economic or technological reasons, but which might become so in the future (Eckstein 2005).

The reaction of the members of the ILC to this scientific definition of an aquifer makes it clear that the terminology needs to be adapted. If used amongst water experts, from whatever background, the term "*exploitability*" does not present a problem, but this is not the case in other scientific communities dealing with water issues.

Another limit of the definition of aquifer, recognised by the Special Rapporteur himself, is that it does not address recharge and discharge zones (Yamada 2004). For hydrogeologists, recharge and discharge areas are integral parts of an aquifer. The Special Rapporteur expressed, in his oral introduction to the report, his intention to formulate draft articles to regulate these areas "*for proper management of aquifers.*" This exclusion also met with some criticism, as recharge and discharge zones are crucial to the protection of an aquifer (Eckstein 2005).

Finally, the Special Rapporteur expressed his decision not to use the term "*confined*" (Yamada 2004), recognising its meaning from a hydrogeological point of view. In his oral introduction he stressed that "*it is bet-*

ter not to use the term "confined", in order to avoid confusion between lawyers and groundwater experts, as the latter will be involved in the implementation of the proposed convention."

An Adapted Definition and New Concepts

In his third report (UN Doc. A/CN.4/551), presented at the 57th session of the ILC in 2005, Ambassador Yamada, reformulated the definition of an aquifer to meet the concerns expressed. The new definition is as follows: " *"Aquifer" means a permeable [water-bearing] geological formation underlain by a less permeable layer and the water contained in the saturated zone of the formation.*" The term "exploitable", which had created controversy, was thus avoided.

Furthermore, the third report innovates by introducing a definition of a "*recharging aquifer*" and of a "*non-recharging aquifer*". The need for introducing this distinction was questioned during the debates.

The adoption of the "*aquifer*" as the unit for the rules to be drafted in the process at the ILC is a first achievement towards integrating science into law and policy. The term "*aquifer*" transposes the unity of natural processes across political borders (Puri 2005).

What Legal Principles for Transboundary Aquifers are Contained in the Draft Articles

In the second report the Special Rapporteur presented seven draft articles; four of these relating to principles applicable to transboundary aquifers. These principles are:

- The obligation not to cause harm;
- The general obligation to cooperate; and
- The regular exchange of data and information.

These principles are drafted following closely the UN Watercourse Convention, with a small adaptation to the case of aquifers.

Two of the principles identified (in the paragraph above) are applicable to transboundary groundwaters were drafted. The equitable and reasonable use was not yet introduced, as the Special Rapporteur was not sure it could be transposed to aquifers. He expressed the need to conduct further research (Yamada 2004).

A complete set of draft articles was proposed in the third report. The drafted principles were the following:

- The equitable and reasonable use (Articles 5 and 6);
- The obligation not to cause harm (Article 7);

- The general obligation to cooperate (Article 8); and
- The regular exchange of data and information (Article 9)[2].

These are followed by articles on the monitoring, the protection, preservation and management, the activities affecting other States and miscellaneous provisions, which include an article on the scientific and technical assistance to developing States, emergency situations, on the protection in times of armed conflict and data and information vital to national defence and security.

As the Special Rapporteur has continued to receive scientific assistance from groundwater experts, due consideration is given to the specific characteristics of aquifers, and the principles are adapted consequently. For example, in draft Article 5 on the equitable and reasonable use principle a distinction was made between recharging and non-recharging aquifers. In the list of factors related to this principle, "*the natural condition of the aquifer*" was added. A provision on monitoring was introduced. This provision, and the provision on the regular exchange of data and information, constitutes the basis for a proper management of a transboundary aquifer. A special provision is introduced for the protection of recharge and discharge zones (draft Article 13). And, finally, draft Article 18, on the scientific and technical assistance to developing States, considers the difficulties for such States in acquiring the necessary knowledge on their groundwater resources, whether because of lack of adequate training or because of the cost of the equipment for monitoring and collecting data.

Groundwater in Environmental Instruments

More and more treaties, conventions and agreements are concluded with an environmental purpose and concern, and consider groundwaters, whether domestic or transboundary. Environmental principles are also introduced in inter-state water treaties.

---

2 Two agreements on transboundary aquifers have been recently signed, and their main objective is the regular exchange of data and information. These agreements are on the:
(1) Nubian Sandstone Aquifer System (2000): Chad, Egypt, Libya and Sudan; and
(2) Northwestern Sahara Aquifer System (2002): Algeria, Libya and Tunisia.
These two aquifer systems are non-recharging systems with no relation to surface water.

Incorporating Environmental Principles into International Groundwater Law

The introduction of environmental considerations and principles into the provisions of water treaties and conventions is recent. It has been a principal driver for several water agreements and treaties since the 1990s (Lautze et al. 2005). This is the case for the UN Economic Commission for Europe Convention on the Protection and Use of Transboundary Watercourses and International Lakes (Helsinki, 17 March 1992), acting as a framework convention for transboundary waters in Europe. Its example is followed by the Convention on Cooperation for the Protection and Sustainable Use of the River Basin (Sofia, 29 June 1994) and the Convention on the Protection of the Rhine (Berne, 12 April 1999), which both refer, in their preambles, to the UN ECE Water Convention. The environmental consideration can also be found in the Convention on the sustainable development of Lake Tanganyka (Dar es Salaam, 12 June 2003) and the Protocol for Sustainable Development of Lake Victoria Basin (Arusha, 29 November 2003). However, in these conventions, groundwater is considered only when related to a surface water body.

The emerging environmental principles from such treaties and conventions are the following:

- The precautionary principle;
- The 'polluter pays' principle; and
- The sustainable development/management [principle].

It is worth noting here that, in the draft articles of the law on transboundary aquifers proposed in Ambassador Yamada's third report, environmental considerations were introduced, such as the precautionary approach or the protection of ecosystem. The concept of sustainable management is also present in draft Article 5 on the application of the equitable and reasonable use to recharging transboundary aquifers (Article 5, paragraph 2(a)).

The special vulnerability of aquifers to depletion or to pollution is always a subject of concern to groundwater experts. The environmental principles can offer adequate tools for aquifer protection.

From Transboundary Groundwater to All Groundwaters

Some international environmental instruments with a strong emphasis on the protection of natural resources have jumped a step forward and regulate natural resources, even when they are domestic, within only one national State. This is the case for the Protocol on Water and Health adopted under the UN ECE Water Convention (London, 17 June 1999), which entered into force in August 2005; and also of the Framework Convention on the Protection and Sustainable Development of the Carpathians (Kiev, 22 May 2003), and the African Convention on the Conservation of Nature and Natural Resources (Maputo, 11 July 2003). In these treaties groundwater is considered as a natural resource to be protected and conserved.

Principles to be applied are the same as above, but the notable issue is that they apply to all groundwaters, and not only those that are related to a surface waters.

The International Law Association has also adopted this trend. In August 2004 the Helsinki and Seoul Rules were replaced by the Berlin Rules. The rules apply to all waters, whether domestic or part of an international basin (Article 1), and extend principles of international law to domestic waters. Regarding aquifers, the Rules consider all aquifers, and promote the application of the precautionary management, the duty to acquire information, and the sustainability and the protection of aquifers. In transboundary situations, the Rules enhance the obligation to cooperate and the obligation not to cause significant harm.

Conclusion

While transboundary aquifers have long remained ignored or inadequately covered by international water law, the tendency seems now to be reversing, and a clear change occurring. The growing pressure on water resources in general, and on groundwater resources in particular, especially where surface waters have already been exhausted or are non-existent, such as in arid regions, has slowly led to an awareness at the international level of the need for reaching cooperation in the management of transboundary aquifers, with adequate and adapted rules. When it adopted its "*Resolution on confined transboundary groundwaters*" in 1994, the ILC already acknowledged the necessity to complete the work it has started with the law on the non-navigational uses of international watercourses by developing rules to address what it then called "*transboundary confined groundwaters*". And, in 2002, when it included in its programme of work the topic of "*Shared Natural Resources*", its first intention was to limit itself to groundwaters not already covered by the UN Watercourse Convention (Yamada 2003). It was only after consultations with hydrogeologists that the Special Rapporteur decided to cover all transboundary aquifers (Yamada 2004). The consideration of scientific evidence and reality has therefore influenced the process of codification at the ILC and changed the scope of the topic. Since the beginning of his work, the Special Rapporteur has received support and appreciation not only from the ILC members, but also from States' delegates at the 6th Committee (legal committee) of the UN General Assembly for the consultation of groundwater experts (ILC 2005). At the last meeting of the 6th Committee (New York, October-November 2005), the delegations welcomed the ILC's current work on "*transboundary aquifers*", noting that groundwater was often a shared resource, pointing to the need for international regulation on the basis of international law (Topical summary of the 6th Committee Debate on Shared Natural Resources, 2005, on file with the author). With the codification process at the ILC, the science of groundwater has slowly made its way to the State delegations' sitting at the 6th Committee, in other words to their governments.

To summarise, the development of rules and principles of international water law applicable to transboundary aquifers is evolving, not only under the consideration of the scientific understanding and knowledge on aquifers, but also under the global environmental trend. These trends are not completely independent, as the environmental influence in law also comes from the scientific observation of the degradation of natural resources, landscape, biodiversity and ecosystems. Hence, more generally, the international legal rule has started a slow evolution of connection with science. In the case of groundwater, this step is seen as crucial by the hydrogeological community, who have been appealing for international rules to be applied to the management of transboundary aquifers (Margat 1992).

On the other hand, law is the tool for a government to implement its policy, whether on the domestic or the international level. Therefore, when integrating science, law acts as an interface between government and science. On the other hand, law is also an interface between government and society, as it applies the policy to the society, but also it translates the evolution of the society into legal rules. Today society claims and awaits the sustainable management of natural resources, and of water in particular.

The recent evolution of international norms and values applicable to transboundary groundwater seems to fit, the Trialogue Model, at least in part. It is connecting successfully between science and governments or science and policy makers, and the results are rules, or draft rules, considering the scientific characteristics of groundwater. These rules are to be applied to society, but are also the expression of some of its concerns about freshwater. In water management issues, civil society is usually involved through a participatory mechanism and institutions. In the case of transboundary aquifers, the draft articles presented in the Special Rapporteur's third report encourage the establishment of joint commissions in order to facilitate cooperation between the concerned States (draft Article 8). This is the open door for the involvement of civil society, through a mechanism to be decided between the States, or at the domestic level of each State. In order to play its role in the management of transboundary aquifers adequately, society needs to understand properly the functioning of aquifers, and therefore needs both education and science. When considering the role of stakeholders in the management of transboundary aquifers, and giving them adequate knowledge, the Trialogue Model linking science, government and society will be achieved.

References

Berlin Rules (2004) Reprinted in: ILA, Report of the 71st conference held in Berlin. Available online at http://www.ila-hq.org/html/layout_committee.htm, under "List of committees", "Water resources law" (Accessed 30 September 2005)

Buirette P (1991) Genèse d'un droit fluvial international général. Revue Générale de Droit International Public, 1991, p 5-70

Eckstein G (2005) Protecting a hidden treasure: the UN International Law Commission and the international law of transboundary ground water resources. Sustainable Development Law and Policy (V)1

Foster S (1999) Essential concepts for groundwater regulators. In: Salman SM (ed) Groundwater legal and policy perspectives. World Bank, Proceedings of a World Bank seminar, Washington DC

Global Water Partnership (2003) Effective water governance, learning from the dialogues. Available online at http://www.gwpforum.org/gwp/library/-Effective%20Water%20Governance.pdf

International Law Commission (2005) Report of its fifty-seventh session

Lautze J, Giordano M, Borghese M (2005) Driving forces behind African transboundary water law: internal, external, and implications. Paper presented at: International Workshop on African water laws: plural legislative frameworks for rural water management in Africa, 26-28 January 2005. Johannesburg, South Africa

Margat J (1992) Problèmes spécifiques aux nappes souterraines transfrontières. In :Atelier sur la "Gestion Environmentale des Bassins Internationaux". UNEP International Office of Water

McCaffrey S (1991) Seventh report. Yearbook of the International Law Commission 1991, vol II (Part One), document A/CN.4/436

McCaffrey S (1998) The UN Convention on the olaw of the non-navigational uses of international watercourses: prospects and pitfalls. In: Salman MA and De Chazournes LB (eds) International watercourses enhancing cooperation and managing conflict. World Bank Technical Paper 414

Mechlem K (2003) International groundwater law: towards closing the gaps? Yearbook of International Environmental Law (14)

Puri S (ed) (2001) Internationally shared aquifer resources management, their significance and sustainable management, a framework document, UNESCO-IHP, Series on Groundwater 1. Available online at http://unesdoc.unesco.org/-images/0012/001243/124386e.pdf
Puri S (2005) Challenges arising from the politics of water – with some examples from Central Asia. Paper submitted at the Workshop on Transboundary water governance as a manifestation of a trialogue
Rogers P and Hall AW (2003) Effective water governance. Global water partnership. Technical Background Papers 7. Available online at http://www.gwp-forum.org/gwp/library/TEC%207.pdf
International Law Commission (2004) Report of the International Law Commission, 56th session, 2004, A/59/10, paragraph 88, p 126
Rosenstock R (1993) First report. Yearbook of the International Law Commission vol II (Part One), document A/CN.4/451
Rosenstock R (1994) Second report. Yearbook of the International Law Commission vol II (Part One), document A/CN.4/462
Seoul Rules (1986) Reprinted in: ILA, Report of the 62nd conference held at Seoul
Sohnle J (2001) Les développements conventionnels récents en droit international des ressources en eau douce sous l'emprise des Conventions-cadre. L'observatoire des Nations-Unies°11
Sohnle J (2002) Le droit international des ressources en eau douce: solidarité contre souveraineté
Stitt T (2005) Evaluating the preliminary draft articles on transboundary groundwaters. Presented by Special Rapporteur Chusei Yamada at the 56th session of the International Law Commission in Geneva, May 2004. Georgetown International Environmental Law Review (17)
United Nations Development Programme (UNDP) (2004) Water governance for poverty reduction. Key issues and the UNDP response to the millenium development goals. Available online at http://www.undp.org/water/docs.html
United Nations Educational Scientific and Cultural Organisation (UNESCO) and World Meteorological Organisation (WMO) (1992) International glossary of Hydrology. 2nd rev edn
World Water Assessment Program (2003) Water for people, water for life. United Nations World Water Development Report
World Water Assessment Program (2006) Water, a shared responsibility. Second United Nations World Water Development Report UN-WATER/WWAP/-2006/3, UNESCO
Yamada C (2003) Special Rapporteur. Shared natural resources: first report on outlines. UN A/CN.4/533
Yamada C (2003) Special Rapporteur. Shared natural resources: first report on outlines. UN A/CN.4/533/Add.1
Yamada C (2004) Special Rapporteur. Second report on shared natural resources: transboundary groundwaters. UN A/CN.4/539
Yamada C (2005) Special Rapporteur. Third report on shared natural resources: transboundary groundwaters. UN A/CN.4/551

Dynamics of Transboundary Groundwater Management: Lessons from North America

Michael E Campana, Alyssa M Neir and Geoffrey T Klise

Abstract

Transboundary groundwater management in the North American countries of Canada, the United States of America, and Mexico is truly dynamic. Institutions such as the International Boundary and Water Commission (US-Mexico) and the International Joint Commission (US-Canada) were originally established to consider surface water. However, they have been adapted to consider groundwater, and the North American Free Trade Agreement, implemented in 1994, may prove to be applicable to groundwater, although in some cases may eventually prove inimical to the interests of border regions as the three countries attempt to manage their transboundary groundwater resources. These institutions, coupled with the ad hoc approach of individual stakeholder groups, illustrate that transboundary groundwater management is functioning quite well in North America. Eight case studies, involving both water quality and quantity, illustrate our premise. Seven of the studies describe very specific issues; the final one involves the groundwater resources of the Great Lakes basin of the US and Canada, and provides a brief discussion of some of the issues that might arise in this region.

Keywords: transboundary groundwater, Canada, United States, Mexico, international law, water use, aquifer, institutions

Introduction

Our working hypothesis is that the degree to which governance is successful depends upon several essential elements. The first three of these, described as a Trialogue, are effective scientific, governmental, and societal processes. But successful governance also requires not only the aforementioned elements but also effective interfaces among the three elements.

We will explore the postulated hypothesis in the context of transboundary groundwater management in North America – Canada, the United States (US), and Mexico (Figure 9.1). Insofar as this chapter is concerned, "transboundary" is synonymous with "transnational"; we will not explore transboundary groundwater issues within a single nation *unless* it has bearing on transnational issues.

Although all three countries are developed, Canada and the US are more so than Mexico. Similarly, although Mexico is a democracy, it is not quite as mature as its two North American partners. If we assume that good governance and its essential elements are necessary for effective transboundary groundwater management, then we would expect that North America would offer excellent examples of said management. The International Joint Commission (IJC) and International Boundary and Water Commission (IBWC) are two international institutions that provide the means for the countries to communicate and resolve problems based on both scientific findings and political compromises. The societal process is also a factor in the international arena, typically initiating the political process that leads to the involvement of the IJC or the IBWC, or creating a cooperative, transboundary interest group that addresses the issue before it reaches higher levels of government. These actions illustrate the presence of a government-society interface – societal processes can lead to involvement by the government. As we shall see, however, transboundary groundwater management in the three countries is more of an ad hoc proposition due to the presence of, and relationship among, the government, scientific, and societal processes.

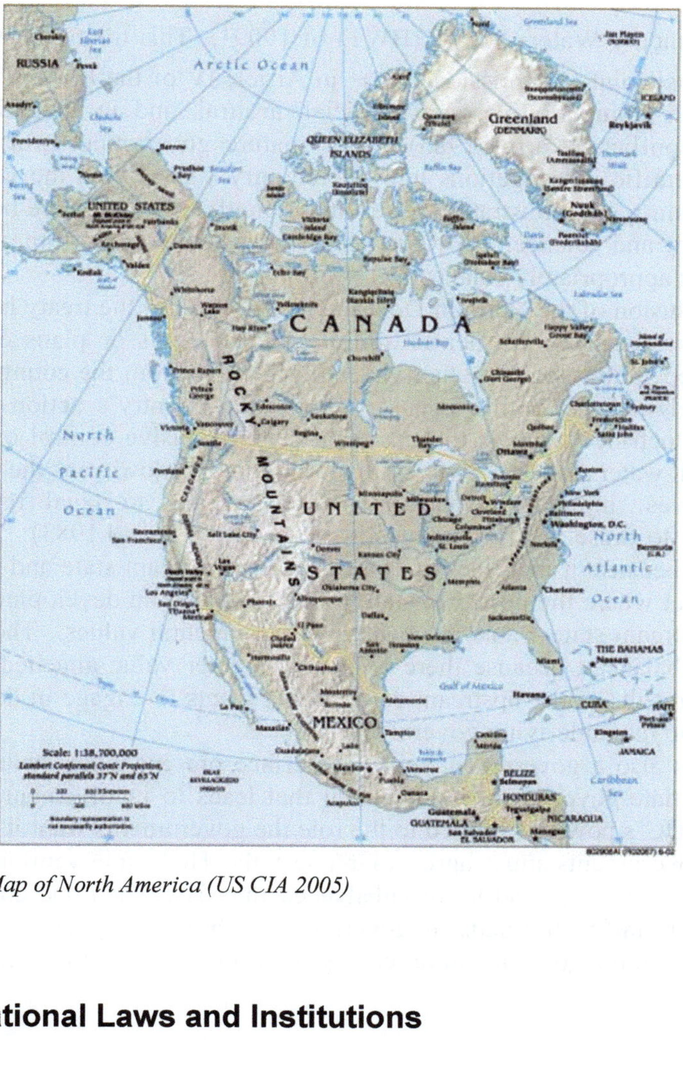

Fig. 9.1. *Map of North America (US CIA 2005)*

International Laws and Institutions

The Boundary Waters Treaty of 1909 and the International Joint Commission

Transboundary water conflicts between the US and Canada can occur all along the 8,000 km border (Carroll 1986) (Figure 9.2). Rules concerning transboundary waters were created almost 100 years ago with the signing of the Boundary Waters Treaty (BWT) of 1909[1]. This treaty created the International Joint Commission (IJC)[2] in Article IV of the BWT which is involved in administrative, quasi-judicial, arbitral, and investigative aspects of conflicts (Carroll 1983). This mature government process includes scientific investigations into one country's impact on the water in another country and illustrates the results of an effective interface between government and science. However, this government-science interface is not always appropriately balanced.

The inclusion of the Harmon Doctrine in Article II of the treaty has specific implications for the permissibility of the actions or plans of each country depending upon whether water flows to, or from the country, and can outweigh the scientific considerations of a country's actions. The Harmon Doctrine "...gives the upstream state exclusive control over the use of all waters on its own side of the line...[h]owever, the article [also]...gives...injured downstream interests rights to legal remedies equivalent to those in effect domestically (p 43)" (Carroll 1983). This is important because it increases the power of the upstream state and can incite conflict where the upstream state wants to engage in development and the downstream state wants to preserve environmental values. The interface is unbalanced because there is always greater value attached to the scientific results of the upstream country that wants to engage in development due to the underlying government process.

There is also a government-society interface present when stakeholder groups initiate government involvement that leads to involvement by the IJC. The IJC's power is limited to the role the governments want it to play – both governments must agree to request the IJC's intervention (Carroll 1983). This can lead to an unbalanced interface when one country's stakeholders fail to persuade its government to bring the issue to the IJC for resolution; the government processes are more powerful than the societal processes.

---

[1] A copy of the treaty can be found on the International Joint Commission's website http://www.ijc.org/rel/agree/water.html. For a description of the waters covered by the Boundary Waters Treaty, refer to Carroll (1983:42). A summary of the principles outlined by the treaty is provided by Le Marquand (1986:233).

[2] IJC (1997) provides a good introduction into the IJC's role in US-Canada water issues.

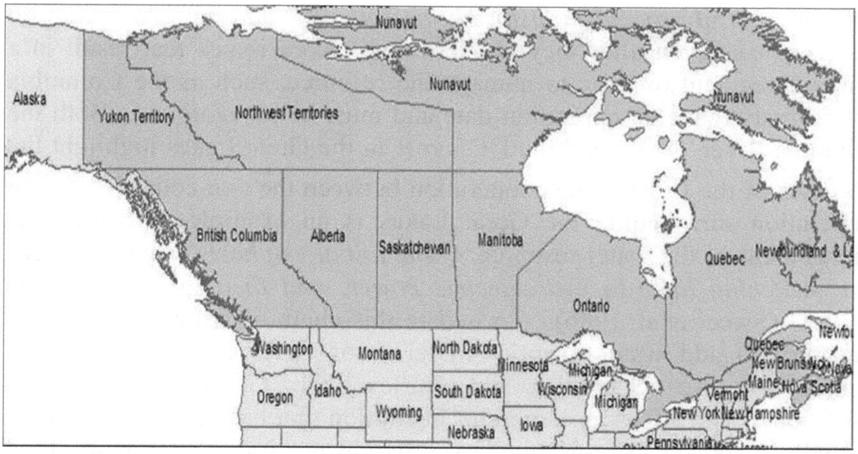

Fig. 9.2. *Map of US-Canada border region. The United States is in green and Canada is in blue; The Great Lakes are between the province of Ontario and the USA. (created by authors using ArcGIS)*

The absence of groundwater from the treaty and the IJC's jurisdiction is a real issue. However, as Everts (1991) explains, there is a way that the IJC can include groundwater in its deliberations in certain situations:

...the IJC has no legal mandate to investigate and make recommendations on groundwater issues which are isolated from possible impacts on surface waters; unless both countries stipulate that the IJC may do so. However, the legal loophole utilized by the IJC to bypass the legal mandate issue in conducting groundwater impact studies in the Flathead River Basin could be the possible adverse effects that polluted groundwater might have on surface waters. By linking polluted groundwater as a possible cause to surface water degradation across the border, the IJC would not overstep its legal mandate under the BWT [Boundary Waters Treaty], but would be legally pressed to investigate the source of the degradation [Emphasis in Original] (p 73-74).

This means that both countries must cooperate in order for groundwater issues to be addressed. This has been done on a smaller scale, province to state, for the Abbotsford-Sumas Aquifer underlying Washington, US, and British Columbia, Canada.

Past water issues between Canada and the US have set precedents for resolving future disagreements[3]. Schwartz (2000) discusses how current

---

[3] Carroll (1983) provides an excellent description of the major issues of concern up to 1983. It covers the High Ross Dam-Skagit River, Champlain-Richelieu,

issues are "*firmly rooted in historical precedents*," and how the issues do not disappear, even after they are resolved. Even issues that result in a signed agreement on how to manage the resource, such as the Columbia River Treaty, have an expiration date and must be renegotiated. Both the Columbia River Treaty and the IJC's role in the Great Lakes highlight the successes of the IJC and the cooperation between the two countries.[4] The cooperation surrounding the Great Lakes is an example of continuous management of the water resource. "*The initial emphasis was on navigation, switching later to hydroelectric power, and then to water quality (p 204)*" (Sweel et al. 1986). To update this chain of issues, it would be necessary to add water quantity in terms of water exports as well as groundwater use (Schindler and Hurley 2004; Galloway and Pentland 2005). The IJC is an important institution that has the potential to be officially extended to transboundary groundwater issues that arise along the US-Canada border. The International Boundary and Water Commission plays a similar role along the US-Mexico border.

International Boundary and Water Commission (US-Mexico)

The US and Mexico share an international border of 3,110 km, with river boundaries making up around 66% of that border (Figure 9.3). The Rio Grande/Rio Bravo borders the US state of Texas and the Mexican states of Chihuahua, Coahuila, Nuevo Leon, and Tamaulipas for 2,020 km. The Colorado River separates Arizona, US, and Sonora, Mexico for 27 km of the international border. The international border between both countries was first established by the Treaty of Guadalupe Hidalgo in 1848, just prior to the end of the Mexican War.

The first border water issues dealt with the location of the international boundary. The Convention of 12 November 1884 was adopted to help deal with the ever-changing international boundary as a result of meandering by the Rio Grande and Colorado River (IBWC 1884). Five years later, the International Boundary Commission was created in 1889 (changed to the International Boundary and Water Commission (IBWC) in 1944) to deal specifically with boundary and water issues (IBWC 2005). The

Chicago Diversion, Cabin Creek, Garrison Diversion, Poplar River, Columbia River, and St Mary and Milk Rivers issues.

4 IJC (1997) includes succinct summaries about the issues and the IJC's role in the solution for the High Ross Dam-Skagit River, the Garrison Diversion, the Columbia River, the St Croix River, the Flathead River (Cabin Creek), and the Milk and St Mary Rivers.

IBWC is also a mature governmental process that incorporates both scientific and societal processes into its decisions and actions.

Water use for irrigation was important to both countries and controversies surfaced in the late 1800s and early 1900s about the equitable distribution of water. The Convention of 21 May 1906 was the first treaty regarding water quantity and stated that the US must deliver 74 million cubic meters (MCM) per year to Mexico via the Rio Grande (IBWC 1906). The 1944 Treaty was more comprehensive and laid out specific actions that each country must take to reduce water sharing conflicts (IBWC 1944).

Water quality issues were addressed through the passage of "Minutes" or legally binding agreements between both countries. Specifically, water quality minutes addressed salinity from irrigation return flows and wastewater treatment plants on both sides of the border (IBWC 2005a, 2005b). Mexico's farmers (stakeholders) were being affected by the poor water quality and the subsequent societal pressures led to scientific investigations and a permanent decision by the IBWC. In the steps to resolving the issue, there is again a slight imbalance within the three interfaces because the three processes are used at different points in time and do not carry equal weight. However, as with the IJC, all three are intertwined and the use of at least government and scientific processes is necessary for an issue to be resolved.

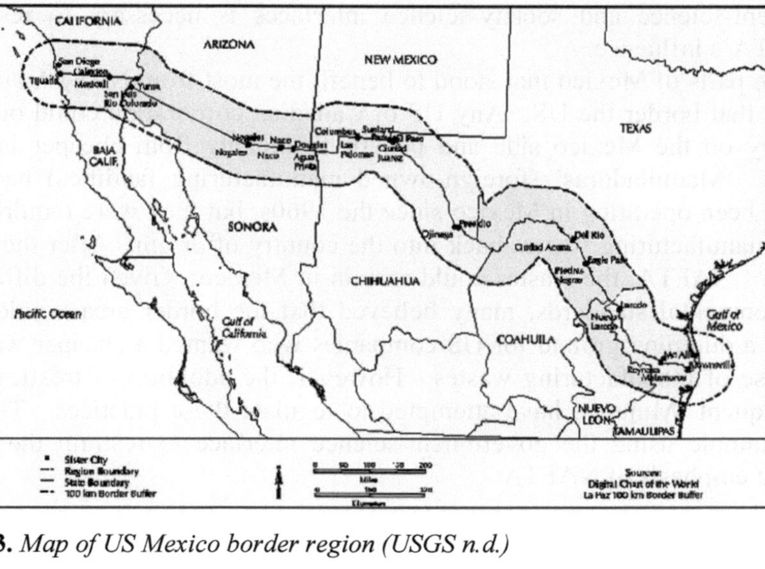

Fig. 9.3. *Map of US Mexico border region (USGS n.d.)*

North American Free Trade Agreement (NAFTA)

In 1994, the North American Free Trade Agreement (NAFTA) was adopted by Canada, Mexico, and the US as a way of cooperating on trade issues. This agreement among the three countries essentially removed tariffs to facilitate increased trading which would lead to greater economic opportunities for all countries involved.

The successful passage of the 1992 constitutional amendment that allowed for increased water privatisation paved the way for Mexican participation in NAFTA, as it was necessary for the government to let the private sector have some sort of autonomy to stimulate investment (Castro 2004). However, the citizenry is cautious about these moves because of real and potential abuse by private companies. If the government decides it would rather let the market operate in the realm of water supply, it must actively enforce its own regulations to the benefit of its citizenry and the environment. If the government looks the other way and allows abuses, such as those alleged by urban water customers, then it might cause conflict if it does not act to regulate over-pumping of border aquifers and discharge of pollutants into streams and aquifers. The balance among government, societal, and scientific processes is vital to the success of applying NAFTA to groundwater resources. Due to NAFTA's economic emphasis, it would be easy to make decisions which do not include the scientific process and only address some of society's concerns. The presence of balanced government-science and society-science interfaces is necessary to restrain NAFTA's influence.

The parts of Mexico that stood to benefit the most from NAFTA are the states that border the US. Any US or Canadian corporation could open a factory on the Mexico side and benefit financially from cheaper labour costs. 'Maquiladoras' (foreign-owned manufacturing facilities) had already been operating in Mexico since the 1960s, but they were required to take manufacturing wastes back into the country of origin. After the passage of NAFTA, the wastes could remain in Mexico. Given the differing environmental standards, many believed that the border area would become a dumping ground for US companies who wanted a cheaper way to dispose of manufacturing wastes. However, the adoption of treaties and subsequent 'Minutes' have attempted to regulate these practices. This is an example using the government-science interface to restrain the economic emphasis of NAFTA.

International Transboundary Groundwater

Abbotsford-Sumas Aquifer

The Abbotsford-Sumas Aquifer (Figure 9.4), in the Fraser River Basin, underlies British Columbia (Canada) and Washington State (US); its water flows southward from Canada to the US. The aquifer is unconfined and provides water for over 115,000 people (Mitchell et al. 2003; Cox and Liebscher 1999). The current concern is the high concentration of nitrate in the aquifer from agricultural practices in both British Columbia and Washington (Washington State Department of Ecology 2003; Mitchell et al. 2003). The presence of the Abbotsford-Sumas Aquifer International Task Force demonstrates the presence of cooperation (A-S Task Force, n.d.). This task force is the product of the 1992 Environmental Cooperation Agreement between the province and state and was created specifically to address transboundary problems concerning the aquifer (A-S Task Force, n.d.). The agreement covers the broad area of "groundwater protection" which can be expanded to include future issues. There is a Trialogue present in this groundwater basin that uses all six of the elements and is illustrated by the example of the proposal to create Aldergrove Lake Regional Park.

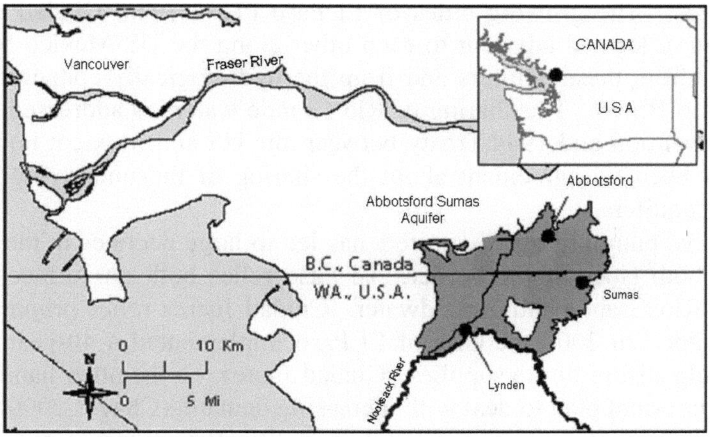

Fig. 9.4. *Map of Abbotsford-Sumas Aquifer (from Mitchell et al. 2005). Reprinted with permission*

A recent example of small-scale cooperation surrounding the protection of the aquifer is British Columbia's proposal to reclaim a gravel pit and transform it into Aldergrove Lake Regional Park, which would use biosolids and biosolids compost to re-vegetate the area (Van Ham et al. 2000). The public on both sides of the border was concerned about the effects that biosolids would have on the aquifer's water quality in general, and specifically for regions of the aquifer that people rely on for their drinking water (Van Ham et al. 2000). In order to allay people's fears, open meetings were held and stakeholders (elected officials, Abbotsford-Sumas Aquifer International Task Force, residents within a one-kilometer radius of the park, and local and US interest groups) were informed about the project (Van Ham et al. 2000). The project, which was shown potentially to improve the aquifer's water quality, was approved and demonstrates how open cooperation from the beginning of a project led to success. This open cooperation included government (the task force), societal (the initial concern and subsequent meetings), and scientific (the study on the impact of biosolids on the aquifer's water quality) processes. In this case, the interfaces in the Trialogue were balanced – everyone was satisfied with the outcome and no element was used in isolation or at the expense of another element.

Hueco Bolson and Mesilla Bolson Aquifers

The Mesilla and Hueco Bolsons are transboundary aquifers that span the US states of New Mexico and Texas and the Mexican state of Chihuahua (Figure 9.5). The growing cities of El Paso (Texas) and Ciudad Juarez (Chihuahua), located adjacent to each other along the US-Mexico border, use water from these aquifers and from the hydrologically-connected Rio Grande/Rio Bravo. The sharing of Rio Grande waters is addressed in the 1906 Convention and 1944 Treaty between the US and Mexico; however, there has been no agreement about the sharing of the underlying transboundary aquifers.

Extensive pumping of the aquifers has led to large declines in the water table on both sides of the border. El Paso relies both on surface water from the Rio Grande and groundwater. Ciudad Juarez relies primarily on groundwater. In 2000 the City of El Paso implemented a 40-year water plan to help ensure future supplies. Ciudad Juarez, on the other hand, does not have a formal plan to deal with increasing demand (Chaves 2000).

Agricultural water use of the Rio Grande/Rio Bravo has degraded surface and groundwater quality by increasing salinity. The saline waters seep into the ground, recharging both aquifers and increasing groundwater salinity. Conversely, saline water from the shallow aquifer recharges the Rio Grande via irrigation drains (Walton and Ohlmacher n.d.). Due to these water quality and potential water quantity issues, a bill has been introduced in the US Senate that will appropriate money to study the transboundary aquifers (Senate Bill 214 2005), which, if passed, may help prevent conflict and lead to future water sharing agreements.

The precursor to this bill was a joint effort by the US and Mexico to create a groundwater database for the El Paso-Ciudad Juarez area. The title of the work is 'Transboundary Aquifers and Binational Groundwater Database.' In 1998, studies completed on both sides of the border were brought into this database by the IBWC to help understand existing data gaps and make recommendations for future studies (IBWC 1998). This study is an example of cooperation of both countries by using government and scientific processes via the IBWC. Understanding the physical properties of the aquifers may minimise future conflict on both sides of the border since both cities will continue to grow and rely on the same source of water.

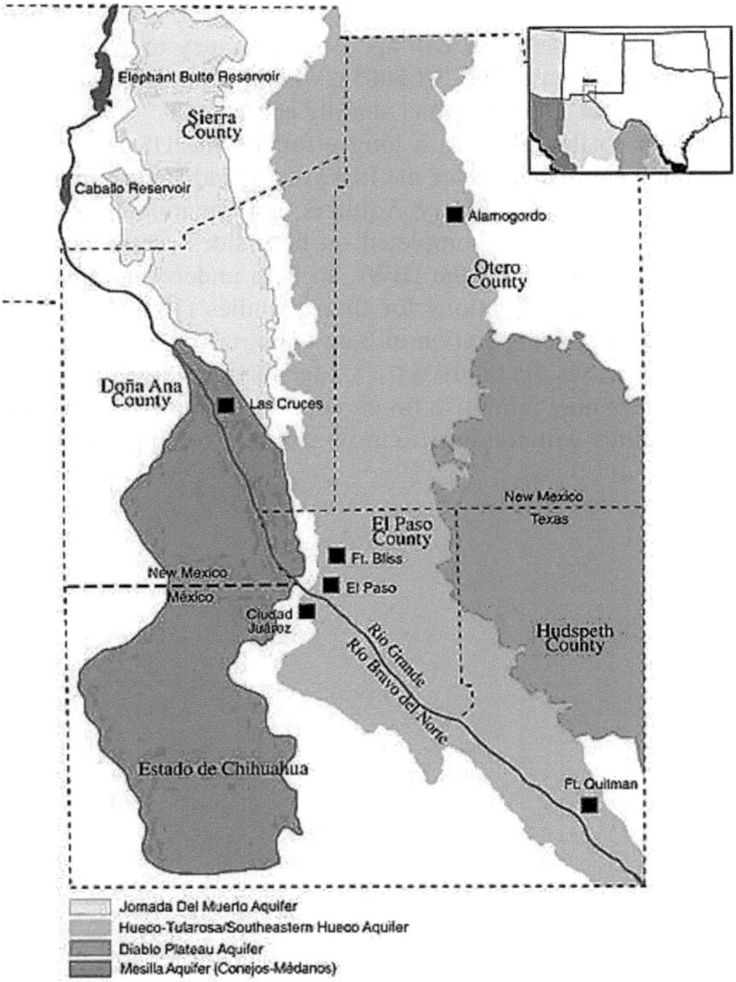

Fig. 9.5. *Map of the Hueco Bolson and Mesilla Bolson Aquifers (McHugh 2005). Reprinted with permission*

Nuclear Waste Sites

The issue of nuclear waste storage facility siting in the US has created a great deal of tension between the US. and Mexico. In 1991, the US state of Texas determined that a low-level nuclear waste repository would be built near the town of Sierra Blanca (Figure 9.6). This town is located approximately 25 km north of the US-Mexico border. The repository would store waste from Texas and, through an interstate agreement, low-level wastes from both Maine and Vermont (Boren 1997). Both US and Mexican citizens strongly opposed this repository because it was to be located close to the international border in one of the most seismically active areas in Texas, right above an aquifer that discharges to the Rio Grande/Rio Bravo (Boren 1997).

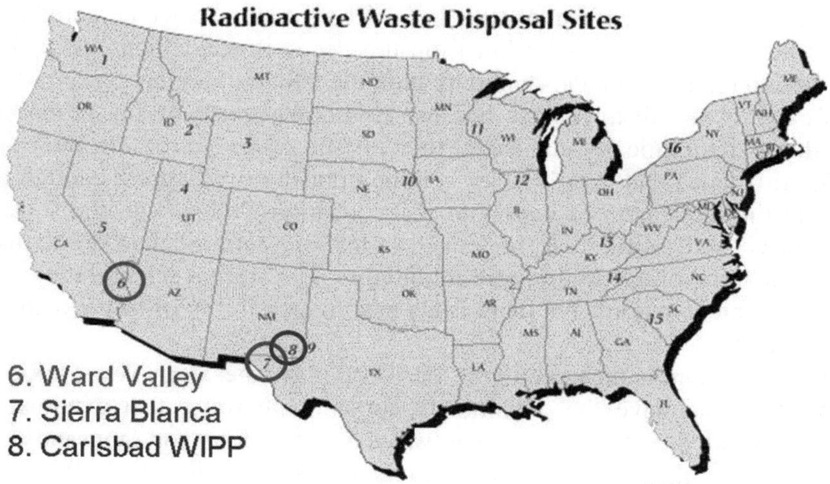

Fig. 9.6. *Map of proposed and used nuclear waste sites (Woodard 1998). Reprinted with permission*

Those within Mexico who opposed the repository used the 1983 La Paz agreement as an argument against it. They interpreted the treaty as banning the siting of new pollution generating facilities. The US Environmental Protection Agency (EPA) interpreted the agreement as requiring 'consultation and notification' (Boren 1997). In the end, the facility was not built in Sierra Blanca, due to a lack of research into the geologic hazard and a lack of planning to understand the socioeconomic impacts to the 'surrounding community'.

Around the same time, a nuclear waste repository was proposed in Ward Valley, California (Figure 9.6). This site would have taken nuclear waste and placed it in dirt trenches above an aquifer that feeds the Colorado River. The proposed location of the repository was on US Bureau of Land Management (BLM) land, adjacent to Indian tribal land. The potential for impacts to the Colorado River prompted Mexicans to join the tribes in opposing the repository (Greenaction 1999).

Out of the opposition came a group called the "Binational Coalition against Radioactive and Toxic Waste Dumps" that pressured lawmakers to amend the La Paz Treaty explicitly to ban siting of nuclear waste facilities near the border. Also mentioned was that, if the Sierra Blanca site was to be approved, the tribal and environmental groups would sue under the NAFTA environmental side agreement (Borderlines Updater 1998). These actions did not happen, but brought attention to the volatile issue of siting nuclear waste facilities near the international border.

One site that was on the radar, and which did eventually open, was the Waste Isolation Pilot Plant (WIPP) in Carlsbad, New Mexico (Figure 9.6). This site is located about 1,5 km from the Pecos River, which eventually drains into the Rio Grande, shared by both the US and Mexico.

Mexico is at a disadvantage because it is downstream of any proposed site that is hydraulically connected to a surface water source. In the case of Sierra Blanca and Ward Valley, public opposition was great enough to prevent the movement of wastes to these locations. However, the southwestern US will continue to be looked at for long-term nuclear waste disposal due to favorable hydrologic conditions. The current science supports the government actions, but there is always the possibility of future technical difficulties.

A case in point is the opening of the WIPP in southeastern New Mexico. This site could have the potential to contaminate water that eventually reaches Mexico. Because nuclear waste is hazardous for thousands of years before it decays to innocuous levels, the potential negative effects of a failed repository make Mexico vulnerable to nuclear waste policy decisions made by the United States. However, the influence of the public opposition groups in the government decisions points to a very powerful societal process in decisions regarding nuclear waste disposal, no matter how the government entities interpret the agreements.

Hermosillo Basin

The Hermosillo Aquifer is located in the state of Sonora, Mexico (Figure 9.7). This aquifer does not straddle the US and Mexico border. However, its use for agricultural production as a result of NAFTA has caused local conflict due to competing demands. This region typically grew crops for local consumption; however, after the removal of trade barriers, many higher-valued fruits and vegetables replaced these traditional crops and are primarily shipped to the US for consumption. The change in what was produced led to the consolidation of many farms in the region with larger farms controlling most of the acreage. The resulting shift in agricultural production has placed a strain on the coastal aquifer with sea-water intrusion threatening many wellfields (Rodriguez 2002; Steinich et al. 1998).

At the same time, the municipal government has decided to expand its industrial sector and needs water to do so. The government proposed pumping salt-water from coastal wells and desalting the water; however, this has created tension with the growers who hold the current monopoly over the coastal aquifer.

This example shows the relationship among government, science, and society and how the decision will affect the different areas. This tension, as a partial result of trade between the US and Mexico, leaves the aquifer vulnerable to over-exploitation and is a direct result of economic growth in the Hermosillo Valley. Any expansion can further reduce groundwater quality by drawing in more sea water. The government will have to decide if the value of new industry outweighs existing agricultural exports, and it may be that a switch to a different industry has a positive effect on the aquifer but might bring in less money to the region. The government knows the scientific and potential societal impacts of its decisions and must determine how to manage the aquifer in a "sustainable" manner so that the entire region does not suffer.

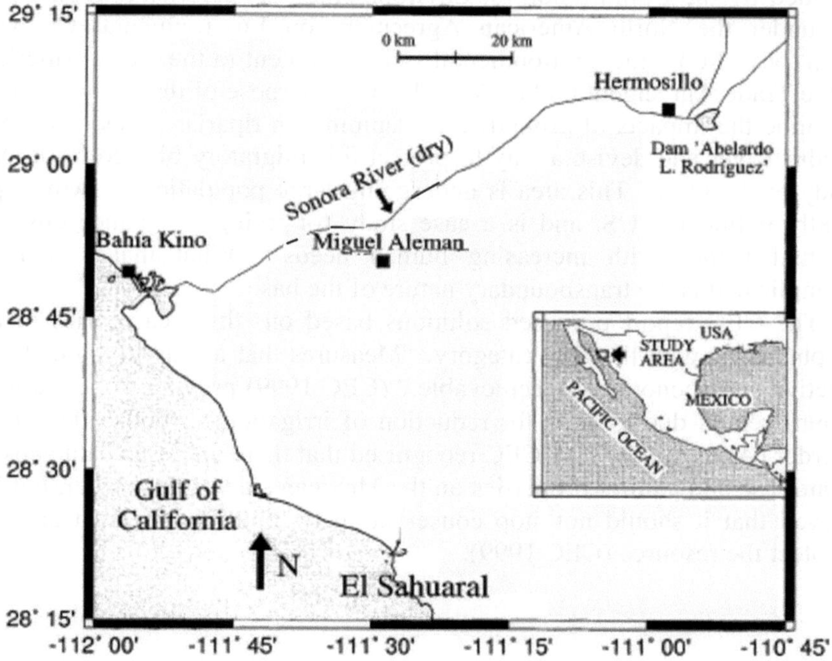

Fig. 9.7. *Map of the Hermosillo basin (Steinich et al. 1998). Reprinted with permission*

San Pedro River Basin

The Upper San Pedro basin is the location of a unique desert ecosystem that has international importance (Figure 9.8). The watershed originates in Sonora, Mexico, and water flows north across the international border into the US state of Arizona; 1,900 m^2 of the basin are located in Mexico and 4,500 m^2 are located within the US (Arias 2000). Groundwater in the basin flows from Mexico to the US (Arias 2000). This watershed has a large number of migratory birds that use the riparian area of the San Pedro River before continuing their journey. Due to the importance of this ecosystem, a portion of the riparian area in the US has been given special status as the San Pedro Riparian National Conservation Area (SPRNCA). Current water use in the basin occurring on both sides of the border is for irrigation, mining, municipalities and domestic purposes. These uses are primarily satisfied by groundwater pumping, which exceeds recharge by an estimated 6-12 million cubic meters (MCM) per year (Varady et al. 2001).

This transboundary basin is the subject of a multi-national study spearheaded by the Commission on Environmental Cooperation (CEC), which is under the North American Agreement on Environmental Cooperation (NAAEC) – the environmental side-agreement to the North American Free Trade Agreement (NAFTA). The main purpose of this study is to determine the impacts of groundwater pumping on riparian areas of the San Pedro River and devise a way to protect this migratory bird corridor (Varady et al. 2001). This area is undergoing large population growth, especially within the US, and is a case study for trying to balance environmental values with increasing human needs. What makes it more complicated is the transboundary nature of the basin.

The CEC report proposed solutions based on 'three categories of acceptability,' with the first category, "Measures that are hydrologically effective and economically achievable," (CEC 1999) proving to be the most controversial due to calls for reduction of irrigation on both sides of the border (Arias 2000). The CEC recognised that there are large data gaps on water use and aquifer properties on the Mexican side of the border, but believed that it should not stop conservation by both countries in order to protect the resource (CEC 1999).

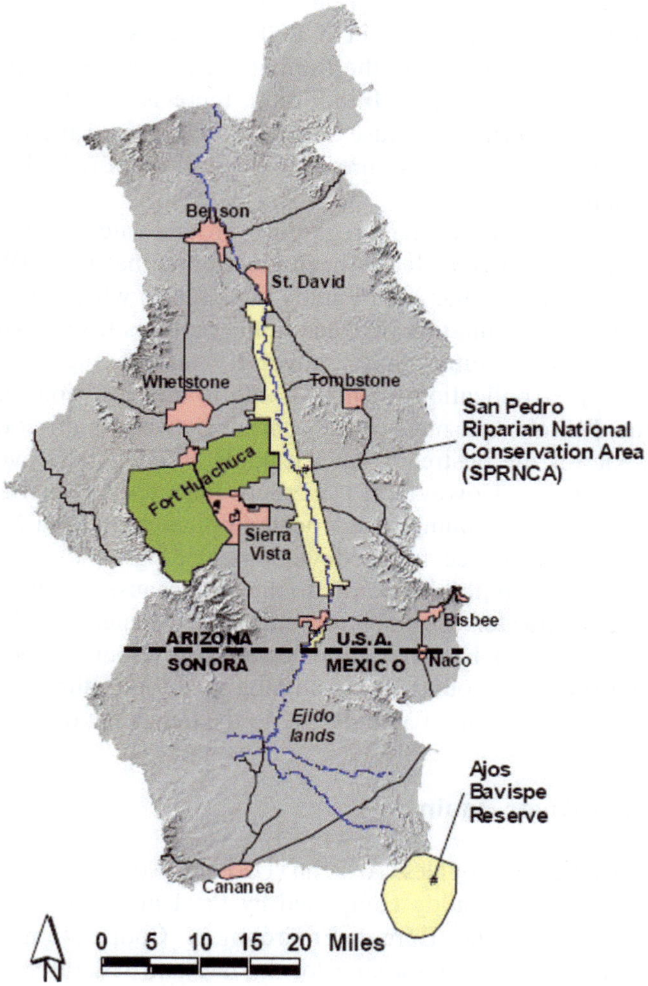

Fig. 9.8. *Map of the San Pedro basin (Harris 2001)*

The initiation of this CEC study created a great deal of conflict among groups in the US, as they felt that growth should not be restricted and irrigation should not be curtailed. Some of the largest US water users besides irrigation include municipalities and a military base. On the Mexican side, the largest water user is a copper mine and there is not as great an interest in protecting habitat for migratory birds as there is in the US (CEC 1999).

No major decisions have been implemented as a result of this study, but it has helped stakeholders on both sides of the border to understand the tradeoffs of protecting one 'use' of water for another, and has brought about bi-national communication and cooperation between local agencies and advocacy groups (Varady et al. 2001). This study demonstrates the government-science interface: the findings of the scientific study have led to international cooperation. In addition, there is also the presence of a science-society interface because science has informed the citizens of the situation. The basis for the future decision will illuminate the balance between the three elements.

In order for any agreement regarding groundwater to take place, the IBWC must be involved. It has been mentioned that the IBWC has been unwilling to apply Minute 242 (Varady et al. 2001) which, when written in 1973, called for consultation between both countries if groundwater development in one country may 'adversely affect the other country' (IBWC 1973). Application of this Minute to any decision-making in the Upper San Pedro basin may cause conflict between both countries. Because Mexico is the upstream state, it holds an advantage over the US in terms of water use. However, if US groups that support riparian habitat by reducing groundwater pumping advocate implementation of Minute 242, it may significantly reduce future development of groundwater in Mexico. Minute 242 would apply differently to the US because it is the downstream state and groundwater use does not currently threaten Mexican supplies. However, if Mexico becomes more concerned about protecting riparian habitat and sees US groundwater pumping as a future threat, then it could argue for the application of Minute 242 to the detriment of the US.

Santa Cruz River Basin

The border cities of Nogales, Arizona (US) and Nogales, Sonora (Mexico) were the subject of a study completed by the Binational Technical Committee, headed by the Border Environment Cooperation Commission (BECC). This group, comprising local, state and federal water agencies, developed a plan to mitigate wastewater runoff originating in Mexico and flowing into the US (BECC 2004). The purpose of this project is to fix existing leaky wastewater pipes on the Mexican side, which will in turn improve water quality in the Nogales Wash that flows into the US. These actions used both government and scientific processes.

The project is jointly funded by the US Environmental Protection Agency (EPA) and the Mexican government, and is an example of cooperation between both countries in response to deteriorating water quality on both sides of the border, and public health issues that arose due to the presence of untreated wastewater. This area was also the subject of the first binational groundwater quality monitoring project between the two countries and set the stage fixing the wastewater leaks in Nogales, Sonora (Castaneda 1998).

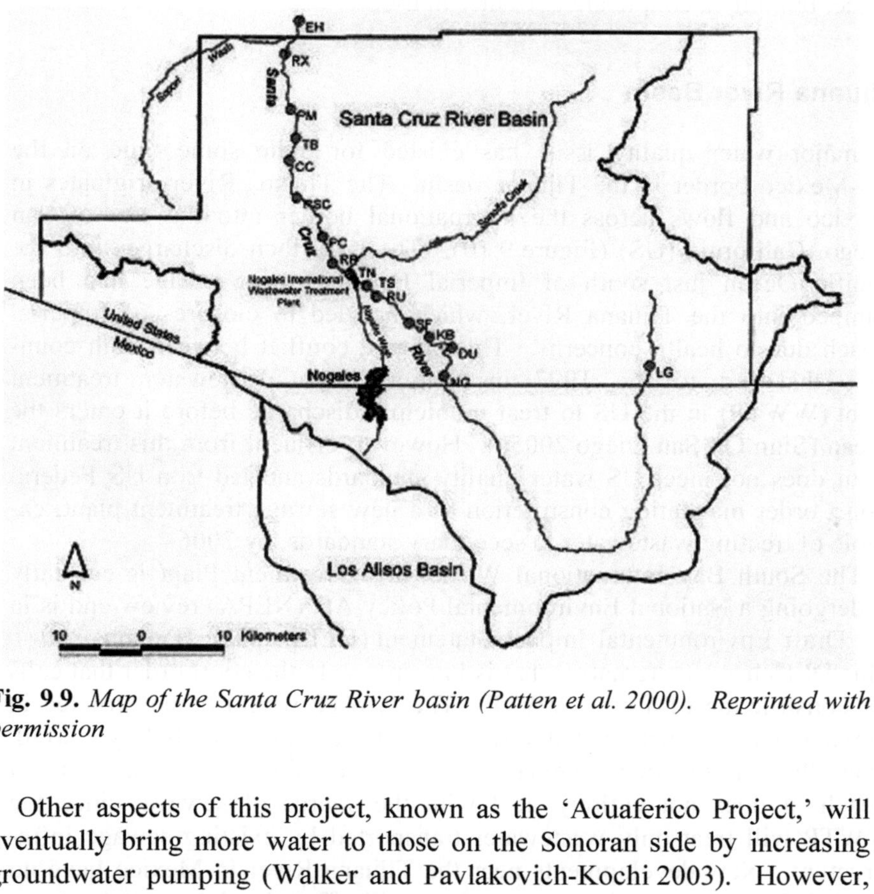

Fig. 9.9. *Map of the Santa Cruz River basin (Patten et al. 2000). Reprinted with permission*

Other aspects of this project, known as the 'Acuaferico Project,' will eventually bring more water to those on the Sonoran side by increasing groundwater pumping (Walker and Pavlakovich-Kochi 2003). However, some in Arizona are worried that growing water use on the Sonoran side will lower water tables on the Arizona side, increasing pumping costs (Walker and Pavlakovich-Kochi 2003). The direction of water flow in the transboundary aquifer is from south to north, with those in Sonora having the ability to use the water first before it flows across the international border (Figure 9.9). This renders those in Nogales, Arizona, vulnerable to the increasing population in Nogales, Sonora, especially since there is no agreement on the apportionment of groundwater. While government processes exist, an agreement would create more specific government processes in managing the transboundary groundwater in the Santa Cruz River basin. In addition, concerns by citizens on groundwater use may result in future societal processes playing a role in the management and instigating scientific studies on the impact of groundwater use by each country on the water table levels. This basin has all the factors required to create a Trialogue, but only a couple are currently in use.

Tijuana River Basin

A major water quality issue has existed for quite some time on the US-Mexico border in the Tijuana basin. The Tijuana River originates in Mexico and flows across the international border into the city of San Diego, California (US) (Figure 9.10). The river then discharges into the Pacific Ocean just south of Imperial Beach. Raw sewage had been dumped into the Tijuana River, which has led to closures of Imperial Beach due to health concerns. This created conflict between both countries that led to the 1997 installation of a wastewater treatment plant (WWTP) in the US to treat municipal discharge before it enters the ocean (Sign On San Diego 2005a). However, effluent from this treatment plant does not meet US water quality standards and led to a US Federal Court order mandating construction of a new sewage treatment plant, capable of treating wastewater to secondary standards, by 2006.

The South Bay International Wastewater Treatment Plant is currently undergoing a National Environmental Policy Act (NEPA) review and is in the Draft Environmental Impact Statement (DEIS) phase (Parsons 2004). This DEIS has an alternative that is in response to the court order that calls for construction of a new WWTP. The WWTP would be constructed in Mexico, with effluent piped directly to the existing South Bay outfall. Even though this may alleviate the water quality issue near Imperial Beach, it does not entirely deal with the source of sewage, since the WWTP will treat only wastewater transported by existing sewage infrastructure. New developments near the Tijuana River in Mexico have no sewer infrastructure and raw sewage can still affect the river and cause water quality concerns near Imperial Beach (Sign on San Diego 2005b).

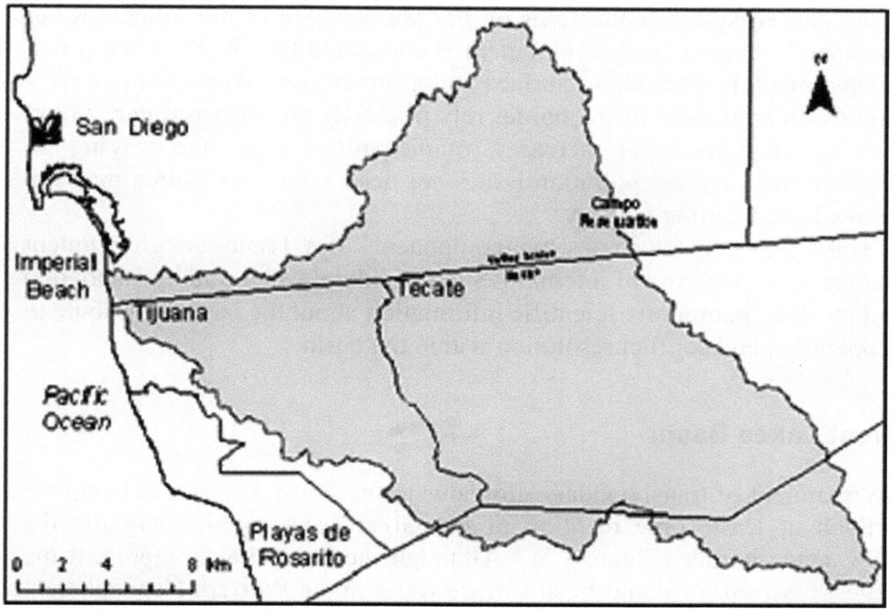

Fig. 9.10. *Map of the Tijuana River basin (SDSU 2004). Reprinted with permission*

One consequence of pollution in the Tijuana River is its impact on the underlying aquifer. Currently, Tijuana only uses about 5% of available groundwater supplies, and the quality is poor due to surface pollution and salt-water intrusion. At best, the aquifer could be used to augment existing surface water supplies or used to store treated wastewater for future use (US EPA 2005). On the US side, San Diego has plans to develop approximately 3,1 MCM of water from the San Diego Formation in the lower Tijuana River Valley (SDCWA 1997).

To help assist in the planning process, San Diego State University published an atlas of the Tijuana River watershed to allow decision-makers access to the same information (Sign On San Diego 2005b). The atlas was created through collaboration with universities and agencies on both sides of the border and will help those involved with the international watershed that deal with environmental issues ranging from water quality to ecosystems (SDSU 2005).

Currently, cooperation in the Tijuana River basin has led to efforts that have helped clean up surface water in the Tijuana River in Mexico and near Imperial Beach in the United States. With the completion of the new WWTP, water quality near Imperial Beach will continue to improve, as sewage will be treated to secondary standards. One problem that remains is the lack of sewer connections on the Mexico side of the Tijuana River watershed. Unless new development is connected to a WWTP, there will be continued degradation of surface and groundwater. At present, municipalities on both sides of the border rely primarily on surface water. However as the population increases, municipalities may start relying on groundwater. Unless pollution issues are dealt with, this source may not be available to either country.

This case study illustrates two components of a Trialogue: government and science. The role of internal US laws (federal water quality standards) and an atlas that reports scientific information about the basin contribute to cooperation and conflict resolution within the basin.

Great Lakes Basin

No treatment of transboundary groundwater in North America is complete without at least some mention of the Great Lakes, which straddle the US-Canada border (Figure 9.2). Although the Great Lakes represent the largest reservoir of liquid fresh surface water in the Western Hemisphere – almost 23,000 km^3 (Galloway and Pentland 2005) – little attention is paid to the groundwater resources of the region, whose volume is approximately equal to that of Lake Michigan, 4168 km^3 (Grannemann et al. 2000). Despite the large amount of groundwater in storage, groundwater provides only about 5% of the total water use in the basin and relatively little is known of the quality and quantity of groundwater in the Great Lakes region (IJC 2000). However, evidence indicates that groundwater is an important component of the water balance of the Great Lakes, either directly as seepage into the lakes or indirectly as baseflow of streams which discharge into the lakes (Holtschlag and Nicholas 1998; Grannemann et al. 2000). Baseflow contributions to streams entering the lakes range from a low of under 20% on the Canadian side to about 42% to both Lakes Ontario and Huron. Low contributions in Canada are the result of less permeable groundwater reservoirs (Holtschlag and Nicholas 1998). Lake level changes can effect changes in the groundwater flow into/out of the lakes.

The Boundary Waters Treaty is silent on the issue of groundwater, although there is a way the IJC can consider groundwater (Everts 1991). Certainly, the groundwater resources of the Great Lakes basin will come under increasing scrutiny as the competition for water becomes more intense among the basin riparians – two Canadian provinces and eight US states. Climate change may also affect water availability. In the case of water transfers outside the basin, all the aforementioned provinces and states must concur. Not only will there be quantity issues, but also water quality and ecosystem health issues. Galloway and Pentland (2005) suggested that, by 2050, a variety of issues – climate change, unfettered diversions, overuse and pollution – could mount to the point that the social and economic fabric of the region would be adversely affected. The problems may be daunting, but the potential for solving them in the context of the Trialogue exists. Indeed, Great Lakes transboundary groundwater issues may be the Trialogue's ultimate test.

Summary

The majority of the examples presented demonstrate cooperation between two countries – they are voluntarily collaborating and using the institutions available to them in those regions as well as creating new institutions to deal with specific problems and to work together more effectively. These institutions tend to incorporate a Trialogue in identifying and resolving transboundary groundwater issues. A task force was created for the Abbotsford-Sumas Aquifer to address water quality issues impacting both Canada and the US. Additionally, Mexico and the US are funding a study to address water quality problems in the Santa Cruz River basin on the US-Mexico border. The use of a Trialogue creates relationship between the three elements; however, that interface relationship is not always balanced.

The IJC and the IBWC are both the products of bilateral treaties between the US and Canada, and the US and Mexico, respectively, but have different roles and powers over transboundary water resources. The IJC has limited power because it cannot become involved in disputes until both countries refer the matter to the Commission. Even then, the IJC acts as an information-gathering body and its decision on the matter, while carrying weight, is not binding, but is respected. The IJC's jurisdiction is also limited to surface water, but has bypassed this limitation in specific cases where the surface water problem was directly related to groundwater resources.

The IBWC, in contrast, is not just a mediator, but also an active participant in the apportionment and utilisation of the transboundary water resources. The actions of both Commissions require consent by the two sovereign governments which constrain their effectiveness as an institutional entity. In addition, IBWC's decisions are not binding, which means that both countries (US and Mexico) need to be willing to abide by their treaty obligations – this is true for the IJC as well. The IBWC's role in constructing water storage and conveyance systems can be seen as a symbol of cooperation between the two countries because they are two riparian countries that completely share the use of the river.

These two mature institutions are based on a Trialogue. The societal process instigates the government process, which uses scientific processes to resolve an issue. However, the interfaces between these three elements are not always balanced. The fact that the three elements are used in a process does not mean that they carry equal weight. Societal processes may initiate concern and involvement by the government and/or scientific community but may not play as large a role in the outcome as the science or government processes. NAFTA, by way of the CEC, also plays a role in US-Mexico transboundary issues and allows for the resolution of disputes on a more local level, and has the potential to play a bigger role in Canada-US transboundary issues.

Some of the more prominent issues have to do with the sharing of groundwater in transboundary aquifers, and creating more specific government processes for each basin. There have been no direct conflicts between these two countries, as the water levels in transboundary aquifers have not lowered to a point to where one country's use is threatened. Minute 272 was set in place to deal with groundwater, but it has not been utilized in sharing the resource; another option is the Bellagio Draft Treaty (Hayton and Utton 1989), which was written to help promote dialogue over transboundary aquifers. Just recently, the scientific element has been included with the completion of studies on these transboundary aquifers, which should help decision-makers with future apportionment issues and define the relationship between science and government. Examples include the Upper San Pedro River in Arizona and Sonora and the aquifers underlying the cities of El Paso and Ciudad Juarez. What has been consistent in these projects is that the US portion of the aquifer has been studied extensively, but the Mexican portion is relatively unknown. When the study phase of the aquifers is complete, decision-makers should look at ways to apportion water 'sustainably' from these transboundary aquifers.

One of the most complicated issues that can impact the quality and use of transboundary waters is the implementation of NAFTA. The three conflict and cooperation examples that have NAFTA implications have added another layer to the agreements already set forth by US and Mexico through the IBWC. NAFTA has created tension over the Hermosillo aquifer in Mexico. Although it is not a transboundary aquifer, many of the agricultural products grown with water from the Hermosillo aquifer are in high demand due to the easing of trade restrictions. This increased demand has created internal conflict over the rights and use of the Hermosillo aquifer. In this case, the people must decide between the benefits of exporting high-valued agricultural products over the use of water for other local purposes, such as commercial and industrial development. However, NAFTA's power has the potential to create an unbalanced government-society interface and severely limit the scientific process. The interface has the potential to be unbalanced because the societal processes are constrained by NAFTA, which recognises only economic uses of water and disregards environmental and/or ecological uses of water. This focus on economics also has the potential to limit the scientific process by only acknowledging the results of science when an economic value can be attached to the results that outweighs other economic considerations.

The outliers are the cases of the Sierra Blanca and Ward Valley nuclear waste facilities. The siting of nuclear waste is a water quality issue that has the ability to strain relations between both countries if not handled carefully. It is difficult enough to site a facility within the US, due to the unknown future impacts to water quality in both streams and aquifers. Even more difficult is siting a facility near the US-Mexican border, because the largest surface water bodies in the southwestern US flow into Mexico. A few of these controversial projects that could have impacts to US and Mexican water quality were scrapped due to public opposition on both sides of the border. It remains to be seen how these projects will strain binational relations in the future; however because of the highly volatile attitudes towards the storing of nuclear waste, there will undoubtedly be conflict over a facility that threatens to impact water that flows into Mexico. These are cases where societal processes (i.e. public opposition) play a powerful role in the outcome.

The structure of the IJC and IBWC is such that any action taken is a reflection of the perceived or actual harms to each country due to their relative position – upstream or downstream. In general, Canada and the US are typically proactive when it comes to water quality issues, especially when they are the downstream country, as in the case of the Abbotsford-Sumas aquifer. Examples along the southern border include the Tijuana watershed dispute over dumping of raw sewage in the Tijuana River, where the US was the downstream recipient and the Mexicali sewage flowing into the US via the New River. In these situations, the US was reactive in approaching the IBWC for solutions in these types of water quality. The US has become more proactive initiating the involvement of NADBank (North American Development Bank) for projects within Mexico that have the potential to affect downstream communities in the US.

Groundwater has been dealt with on a local level, and has been party to cooperative scientific efforts to delineate the physical and chemical properties and understand human impacts to the resource. The nature of groundwater is such that human impacts have recently been the focus of attention, and the problems that have surfaced do not have a 'quick fix'. Their solutions, therefore, have been undertaken through local concerns and, in the case of transboundary groundwater resources, this has led to task forces and international exchange of data to prevent future crises associated with the quantity and quality of the resource. These actions have acknowledged that successful cooperation requires using a Trialogue (government, scientific, and societal processes) that recognises the importance of making decisions that incorporate and weigh multiple factors and concerns.

Conclusion

The adaptability of institutions such as the IBWC and the IJC and their ability to resolve bilateral disputes and promote cooperation between the countries is demonstrated in the examples of conflict and cooperation. Cooperation is demonstrated by the voluntary use of the institutional entities available to each country such that effective management of transboundary groundwater resources is accomplished. Institutions like the IJC and IBWC, while not specifically established to consider groundwater, have managed to function properly whenever groundwater is an issue, thus effecting transboundary groundwater management, if on an ad hoc basis. There is no predetermined process that clearly defines the role of government, science, and society in transboundary groundwater management; however, at least two of the three Trialogue elements are usually used in making decisions and agreements. NAFTA has also shown that it, too, can treat groundwater, although in the US, NAFTA's approach to groundwater as an economic good may jeopardise the use of scientific processes.

The majority of the examples from North America demonstrate that there is a Trialogue working to manage transboundary groundwater resources. The cooperation surrounding the Abbotsford-Sumas Aquifer demonstrates the strongest presence of the Trialogue. However, Trialogues are also in use in the Hueco Bolson-Mesilla Bolson aquifers, nuclear waste sites, Hermosillo basin, San Pedro River basin, and Santa Cruz River basin. The Tijuana River basin example shows very strong government and scientific processes at work; however, at this time there seems to be very little societal input, aside from the exercise of society's desires through laws. It is evident, however, that more attention needs to be paid to groundwater and its unique characteristics, and not to try to "fit" groundwater into the existing surface water compacts and agreements.

References

Abbotsford-Sumas Aquifer International Task Force (A-S Task Force) (n.d.) Available online at http://wlapwww.gov.bc.ca/wat/aquifers/absumas.html [Accessed 16 February 2005]

Arias HM (2000) International groundwaters: the upper San Pedro river basin case. Natural Resources Journal 40:199-221

Border Environment Cooperation Commission (BECC) (2004) Wastewater collection system rehabilitation for Nogales, Sonora. Available online at http://www.cocef.org/aproyectos/ExcomNogalesSonWW2004_07ing.pdf [Accessed 3 April 2005]

Borderlines Updater (1998) Sierra Blanca protests sweep both sides of the border. Available online at http://www.americaspolicy.org/updater/1998/oct20-sierra_body.html [Accessed 20 March 2005]

Boren R (1997) Waste on the way? Borderlines 37(5), no 7. Available online at http://www.americaspolicy.org/borderlines/1997/bl37/bl37wast_body.html [Accessed 20 March 2005]

Carroll JE (1983) Environmental diplomacy: an examination and a prospective of Canadian-US transboundary environmental relations. University of Michigan Press, Ann Arbor, Michigan

Carroll JE (1986) Water resources management as an issue in environmental diplomacy. Natural Resources Journal (26)2:206-220

Castaneda M (1998) Binational water quality monitoring activities along the Arizona-Sonora Border region. In: Proceedings of the NWQMC National Monitoring Conference. Monitoring: critical foundations to protect our waters. 7-9 July 1998. Reno, NV. Available online at http://www.nwqmc.org/98proceedings/Papers/34-CAST.html [Accessed 3 April 2005]

Castro JE (2004) Urban water and the politics of citizenship: the case of the Mexico City metropolitan area (1980s-1990s). Environment and Planning (36)2:327-346

Chaves OE (2000) Mining of internationally shared aquifers: the El Paso-Juarez case. Natural Resources Journal (40)2:237-251

Commission for Environmental Cooperation (CEC) (1999) Ribbon of life: an agenda for preserving transboundary migratory bird habitat on the upper San Pedro river. Available online at http://www.cec.org/files/pdf/sp-engl_EN.pdf [Accessed 31 March 2005]

Cox SE, Liebscher H (1999) Ground-water quality data from the Abbotsford-Sumas aquifer of southwestern British Columbia and northwestern Washington State, February 1997. US Geological Survey Open-File Report 99-244. Available online at http://pubs.er.usgs.gov/pubs/ofr/ofr99244

Everts TM (1991) Canadian/American transboundary groundwater issues: the forgotten resource. Thesis/Dissertation. University of Oregon

Galloway G, Pentland R (2005) Securing the future of groundwater resources in the Great Lakes basin. Groundwater (43)5:737-743

Grannemann N, Hunt R, Nicholas J, Reilly T, Winter T (2000) The importance of groundwater in the Great Lakes region. US Geological Survey Water Resources Investigation Report 00-4008. Lansing, Michigan, USGS

Greenaction (1999) Mexico groups join ward valley fight. Available online at http://www.greenaction.org/wardvalley/pr011999.html [Accessed 20 March 2005]

Harris E, Huntley C, Mangle W, Rana N (1998) Transboundary collaboration in ecosystem management: integrating lessons from experience. Available online at http://www.snre.umich.edu/ecomgt/pubs/transboundary/San%20Pedro.pdf [Accessed 17 October 2005]

Hayton RD, Utton AE (1989) Transboundary groundwater: the Bellagio draft treaty. Natural Resources Journal (29)3:668-722

Holtschlag D, Nicholas J (1998) Indirect groundwater discharge to the Great Lakes. US Geological Survey Open-File Report 99-529. Lansing, Michigan, USGS

International Boundary and Water Commission (IBWC) (1884) Convention between the US and Mexico. Available online at http://www.ibwc.state.gov/-Files/TREATY_OF_1884.pdf [Accessed 24 February 2005]

International Boundary and Water Commission (IBWC) (1906) Convention between the US and Mexico. Available online at http://www.ibwc.state.gov/-Files/1906Conv.pdf [Accessed 24 February 2005]

International Boundary and Water Commission (IBWC) (1944) Treaty between the US and Mexico. Available online at http://www.ibwc.state.gov/html/-treaties.html [Accessed 10 March 2005]

International Boundary and Water Commission (IBWC) (1973) Minute 242. Available online at http://www.ibwc.state.gov/Files/Minutes/Min242.pdf [Accessed 31 March 2005]

International Boundary and Water Commission (IBWC) (1998) Transboundary and binational groundwater database, for the city of El Paso/Ciudad Juarez area. Available online at http://www.ibwc.state.gov/html/body_binational_waters.htm#Figures [Accessed 3 April 2005]

International Boundary and Water Commission (IBWC) (2005) History. Available online at http://www.ibwc.state.gov/html/about_us.html [Accessed 24 February 2005]

International Boundary and Water Commission (IBWC) (2005a) Minutes 1-179. Available online at http://www.ibwc.state.gov/html/index.htm [Accessed 24 February 2005].

International Boundary and Water Commission (IBWC) (2005b) Minutes 180-311. Available online at http://www.ibwc.state.gov/html/body_-minutes.htm [Accessed 24 February 2005]

International Joint Commission (1997) The IJC and the 21st century: response of the IJC to a request by the governments of Canada and the United States for proposals on how to best assist them to meet the environmental challenges of the 21st century. Available online at http://www.ijc.org/php/publications/-html/21ste.htm [Accessed 8 February 2005]

International Joint Commission (2000) Protection of the waters of the Great Lakes. Ottawa, Canada, International Joint Commission

LeMarquand DG (1986) Preconditions to cooperation in Canada-United States boundary waters. Natural Resources Journal (26)2:221-242

McHugh R (2005) Cross border management tool could set trend: Paso del Norte water task force fills void. Available online at http://www.sharedwater.org/-News/2005Aug/crossborder_jointmanagement_to.htm [Accessed 17 October 2005]

Mitchell RJ, Babcock RS, Gelinas S, Nanus L, Stasney DE (2003) Nitrate distributions and source identification in Abbotsford Sumas aquifer, northwestern Washington state. Journal of Environmental Quality (32)3:789-800. Available online at http://www.jeq.scijournals.org/cgi/reprint/32/3/789?ijkey=8dcf-0bf0a6aa37eb74455b5f7545df5c1bfbb14c

Mitchell RJ, Babcock RS, Hirsch H, McKee L, Matthews RA, Vandersypen J (2005) Water quality: Abbotsford-Sumas final report. Available online at http://www.kula.geol.wwu.edu/rjmitch/Report_2005.pdf [Accessed 17 October 2005]

Parsons (2004) Draft supplemental environmental impact statement clean water act compliance at the South Bay international wastewater treatment plant. Prepared for IBWS and US EPA. Available online at http://www.ibwc.-state.gov/Files/SEIS/Draft_SEIS.pdf [Accessed 26 March 2005]

Patten DT, Stromberg JC, Sommerfeld MR (2000) Water and riparian resources of the Santa Cruz river basin: best management practices for water and resource quality. Available online at http://www.scerp.org/projects/WQ93_12.html-#fig1 [Accessed 17 October 2005]

Rodriguez JMM (2002) Aquifers and free trade; a Hermosillo coast cse study. Red Fronteriza de Salud y Ambiente, AC and Texas Center for Policy Studies. Available online at http://www.texascenter.org/publications/sonora-english.pdf [Accessed 22 March 2005]

San Diego County Water Authority (SDCWA) (1997) Groundwater report executive summary. Available online at http://www.sdcwa.org/manage/-groundwater-exec-summary.html [Accessed 26 March 2005]

San Diego State University (SDSU) (2004) Tijuana river watershed. Available online at http://trw.sdsu.edu/English/WshdOverw/wshOverwFrame.htm [Accessed 17 October 2005]

San Diego State University (SDSU) (2005) Institute for regional studies of the Californias publications. Available online at http://www.rohan.-sdsu.edu/~irsc/newpub.htm [Accessed 26 March 2005]

Schindler DW, Hurley AM (2004) Rising tensions: Canada/US cross border water issues in the 21st century. Available online at http://www.powi.ca/-uvic_finalnotes_remarks.pdf [Accessed 16 February 2005]

Schwartz AM (2000) The Canada-US environmental relationship at the turn of the century. American Review of Canadian Studies (30)2:207-226

Senate Bill 214 (HR 469) (2005) 109th Congress. United States-Mexico transboundary aquifer assessment act. Available online at http://thomas.loc.gov/-cgi-bin/bdquery/z?d109:s.00214 [Accessed 31 March 2005]

Sweel W, Derrick R, Utton AE (1986) 'Getting to yes' in United States Canadian water disputes. Natural Resources Journal (26)2:201-205

Sign on San Diego (2005a) Public-private funding is seen for sewage treatment plant. Available online at http://www.signonsandiego.com/news/metro/-20050319-9999-6m19ibwc.html [Accessed 26 March 2005]

Sign on San Diego (2005b) Atlas records Tijuana river watershed. Available online at http://www.signonsandiego.com/uniontrib/20050228/news_1m-28atlas.html [Accessed 26 March 2005]

Steinich B, Escolero O, Marín LE (1998) Salt-water intrusion and nitrate contamination in the valley of Hermosillo and El Sahuaral coastal aquifers, Sonora, Mexico. Hydrogeology Journal (6)4:518-526

US Central Intelligence Agency (US CIA) (2005) The world fact book: North America. Available online at http://www.cia.gov/cia/publications/factbook/-reference_maps/north_america.html [Accessed 17 October 2005]

US Environmental Protection Agency (2005) Mexicali, Baja California: wastewater and treatment project for the Mexicali II service area. Available online at http://www.epa.gov/region09/border/mexicali/ [Accessed 3 April 2005]

US Geological Survey (USGS) (n.d.) US-Mexico border regions. Available online at http://international.usgs.gov/images/projects/prjusmexborderpic2.gif [Accessed 17 October 2005]

Van Ham M, Lee K, McLean B (2000) Pit to park: gravel mine reclamation using biosolids. Available online at http://www.gvrd.bc.ca/nutrifor/pdfs/Pitto-Park.pdf [Accessed 17 October 2005]

Varady RG, Browning-Aiken A, Moote MA (2001) Watershed councils along the US-Mexico border: the San Pedro Basin 2. In: Globalization and water resources management: the changing value of water. AWRA/WLRI-University of Dundee International Specialty Conference, 6-8 August 2001. Available online at http://www.awra.org/proceedings/dundee01/Documents/Vara-dyandMoote.pdf [Accessed 29 March 2005]

Walker MP, Pavlakovich-Kochi V (2003) The state of the Arizona border region: shared pollution, shared solutions. In: "Border institute V" the state of the border and the health of its citizens: multinational indicators of progress, 1990-2020. 28-30 April 2003. Rio Rico, AZ. Available online at http://www.scerp.org/bi/BIV/AZSon.pdf [Accessed 3 April 2005]

Walton J, Ohlmacher G (n.d.) Surface and groundwater interactions: El Paso Ciudad Juarez region. In: The US-Mexican border environment: water issues along the US-Mexican border. Paul Westerhoff (ed) Southwest Center for Environmental Research and Policy (SCERP), San Diego, CA. Available online at http://www.scerp.org/pubs/monographs.htm [Accessed 31 March 2005]

Washington State Department of Ecology (2003) Transboundary water: feasibility of conducting negotiations with other states and Canada on water bodies shared with Washington. Available online at http://www.ecy.wa.gov/pubs/-0311009.pdf [Accessed 7 February 2005]

Woodard T (1998) The grassroots confront the nuclear waste nightmare. Available online at http://www.groundworknews.org/nuke/waste/nuke-waste-sites.gif [Accessed 17 October 2005]

Organisational Culture as a Function of Adaptability and Responsiveness in Public Service Agencies

Nyambe Nyambe, Charles Breen and Robert Fincham

Abstract

Public service agencies are increasingly expected to adapt effectively and respond to dynamic and discontinuous changes in their external environments. Organisational culture is often presented as a strong determinant of both adaptation and responsiveness. In this paper, we provide a descriptive analysis of this claim. We base the paper on the view that agencies involved in managing the use of ecosystems have strong organisational cultures partly because of their origins, which are rooted in certain ethical and moral precepts. We suggest examining assumptions – the core of organisational culture – as a critical starting point in seeking adaptation, responsiveness and managing for change in such agencies. Explicit analysis of assumptions is critical for securing support for, and reducing prospects of resistance to change. The examination of assumptions also has an important role in harmonising the goals of society, science and government.

Keywords: organisational culture, public service agencies, adaptation, responsiveness, institutions, managing for change, Trialogue, Southern African Development Community (SADC)

Introduction

Public service agencies are increasingly finding themselves in a period of hitherto unforeseen turbulence and far-reaching change, with equally far-reaching implications (Walker et al. 2002). Various aspects of the work of these agencies – whether funding, policies or stakeholder engagement, are open to the changes under way in the external environment at local, national and global levels (WCPA 2003). Effective responsiveness and adaptation to shifting public values, expectations and other environmental developments may be their only route to ensuring survival and relevance in the future.

Global trends indicate growing dissatisfaction with many public service agencies in terms of their responsiveness and adaptation capabilities. Agencies managing the use of ecosystems, or otherwise known as resource management agencies, are not an exception (Lubchenco 1999). Thus, this chapter is premised on the view that adaptation and responsiveness in public service agencies are better assured by managing for change. However, managing for change can be an elusive aspiration and activity. One of the reasons for this situation is the prevailing organisational culture which often determines the extent to which managing for change is practiced.

The question we pose is whether the agencies responsible for managing the use of ecosystems are prepared for the challenges imposed by a turbulent external environment. In other words, do such agencies have organisational cultures that enable them to be adaptive and responsive to change? This focus on organisational culture is important if we are to reconceptualise public service agencies in ways that will create prospects for them to be more practically and analytically attuned to issues of responsiveness and adaptation.

The Trialogue Model

A worthy starting point is an interpretation of the Trialogue Model on ecosystems governance (Turton et al. 2005) in the context of the central themes of this chapter: responsiveness, adaptation and organisational culture. The model underscores the relationships among the three actor clusters: society, government and science. It suggests that, under democratic conditions, society, government and science processes are highly correlated to one another, with each being dependent on the others in meeting its roles and responsibilities. As societies aspire to become more democratic, careful consideration of the relationship among science, government and society processes will increasingly become paramount.

The Trialogue Model encourages the appreciation of the fact that the changes underway in society are bringing science, government and society processes into a new reality – one of unavoidable convergence arising from the nature of problems facing humanity. The three actor clusters must be managed differently than previously in order to contribute collectively to resolving problems facing humanity. Hence, convergence in this context refers to the increasing realisation of the need to harmonise the goals and responsibilities of the three actor clusters. The harmonisation of these processes underpins the concept of good governance – the stepping-stone to democracy (Turton et al. 2005).

Harmonisation means that society ought to have a say in what science does and how government is run; government must ensure that there are direction and proper mechanisms for conducting scientific work; and science ought to produce new and appropriate knowledge and skills consistent with society and government needs, values and aspirations.

And yet, this envisaged harmonisation cannot happen, let alone be sustained, without institutionalising collaborative behaviour among the three actor clusters. Institutionalisation of collaboration requires accepting that the status quo is not an option. But change is inconceivable without a deep understanding of organisational culture as a strong determinant of behaviour, with considerable implications for adaptation and responsiveness.

Public Service Agencies: An Overview

In simple terms, public service agencies represent societal and governmental responses to felt needs, immediate or otherwise. Typically, they are an expression of the principles by which society wishes to guide and regulate the management of an identified issue, e.g. the use of ecosystems. The work of public service agencies is proscribed by policy and legislation, and reflects in most cases, the public meaning in the form of norms, beliefs and values at a particular point in time (Hoekstra 1999). In the context of fostering adaptation, responsiveness and appropriate organisational cultures, three aspects are emphasised here: public service agencies as integral parts of their external environments; why public service agencies ought to be seen as organisations and the influence of bureaucracy.

Public Service Agencies and the External Environment

Public service agencies, regardless of the sector in which they operate, are continuously interacting with their external environments – the social, political, economic and broader societal processes. The emphasis on the external environment is essential because "*every organisation (i.e. public service agency) is a subsystem of a "wider social system (environment), which is the source of 'meaning', legitimation, or higher-level support which makes the implementation of the organisation's goals possible*" (Parsons 1960: 63-64). It is therefore important to locate public service agencies and the work they do in the broader society, appreciating the various external pressures coming to bear on their management, planning and strategy processes.

Societies and their governments provide support to, and invest in public service agencies (the practitioners of science) predicated on social contracts expressed in policy. These contracts are underpinned by the expectation that public service agencies will deliver services and products (e.g. knowledge, economic opportunities, technologies and skills) that reflect the felt needs of, and are useful to society (Lubchenco 1999). As such, public service agencies are constantly under intense scrutiny for their adaptation and responsiveness capabilities from various groups – citizens, professional bodies, decision-makers, academics and many others.

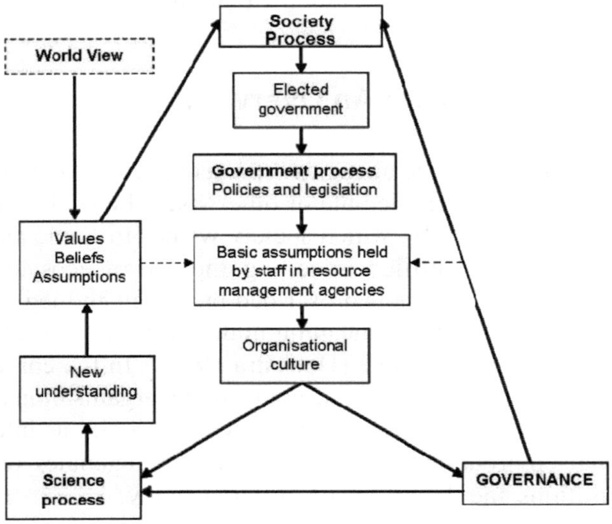

Fig. 10.1. *A conceptual framework depicting linkages of the elements of the Trialogue Model and their roles and responsibilities in relation to worldviews and governance*

In the context of the Trialogue Model, the work of public service agencies can be traced back to worldviews that shape the assumptions, beliefs and values that underpin science, society and government processes (Figure 10.1). Further, society processes influence many aspects including the type of government and those elected into office. Government process in a democracy, typically in the form of policy and legislation, is a reflection of societal values, beliefs and assumptions. It follows that the basic assumptions (deeply-held views about how things ought to be) held by staff in agencies involved in managing the use of ecosystems are a reflection of the interpretation of society, science and government processes.

Evolving values, beliefs and assumptions or meanings require public service agencies to adapt and respond so as to meet the emerging felt needs in society. Understanding public service agencies including those that manage use of ecosystems as 'organisations' is a prerequisite to appreciating the significance of organisational culture in relation to adaptability and responsiveness.

Public Service Agencies as Organisations

To the extent that they pursue certain defined goals, public service agencies can be regarded as organisations because they are social units created to meet defined goals (Etzioni 1961). The goals of an organisation usually reflect some form of interpretation by its members of the mission – "*an organisation's primary task, or reason to be, a set of beliefs about its core competencies and basic functions in society*" (Schein 1992:53). Accepting the view that public service agencies are organisations demands two further acknowledgements.

The first is that "*every organisation, whether a business or not, has a theory of business*" (Drucker 1995:20). The foundations for a theory of business are the assumptions about the environment or context in which that organisation operates, their mission and core-competencies (Drucker 1995). The second acknowledgement relates to the importance of institutions in human affairs. So variously defined as to make it virtually impossible to agree on a single definition (Ostrom 1986), broader conceptualisation of institutions includes both formal and informal dimensions and underscores sociological aspects. According to Bellah et al. (1991), an institution is a pattern of expected action of individuals or groups enforced by social sanctions, both positive and negative. Institutions provide mechanisms for societies and governments to confront challenges facing their societies. They also illuminate the underlying thinking and fundamental values that influence action and behaviour in organisational settings (Cortner and Moote 1994).

Formal and Informal Institutions

Formal institutions, in the context of managing the use of ecosystems include entities such as resource management agencies. Informal institutions, on the other hand, denote the customs and practices of such agencies. Informal institutions, while not the formally accepted way of thinking and behaving, may indeed be the 'rules in use'. They can significantly influence the thinking and behaviour of actors in an agency. In both formal and informal institutions, customs and practices become entrenched through the process of institutionalisation, defined as "*...something that happens to an organisation over time, reflecting the organisation's own distinctive history, the people who have been in it, the groups it embodies and the vested interests they have created, and the way in which it has adapted to the environment*" (Selznick 1957:16-17).

Institutions "*consist of cognitive, normative and regulative structures and activities that provide stability and meaning to social behaviour*" (Scott 1995:33). Accordingly, it can be postulated that the influence of institutions would be particularly strong in organisations whose theory of business and consequently mission, goals, products and or services, are tightly bound to certain values and are grounded in particular ethical and moral positions, e.g. utilitarianism and protectionism. Examples of such agencies would include those responsible for managing protected areas, wildlife, forests, rivers and fisheries. A common feature of agencies is bureaucracy, which is often believed to be deeply rooted (Rogers et al. 2000).

Bureaucracy and its Implications in Agencies

Ecosystem management agencies largely reflect the classical bureaucracy – a type of organisation that has characteristics such as hierarchical structures, compartmentalised departments, a downward flow of policy and information, detailed rule and clear lines between the organisation and its environment (Weber 1893). In general, policy is set at the uppermost level of the organisation. Entrenched bureaucracies are a common feature in agencies, but they exist at a huge cost to managing for change.

Resource management agencies are notorious for their resistance to change or lack of adaptation and responsiveness (Walker et al. 2002; Rogers et al. 2000, Holling and Meffe 1996). Most agencies exhibit "*recalcitrance or inertia ... and the almost pathological inability to renew or restructure*" (Gunderson et al. 1995:495). An ingrained inability to adapt or respond, with systemic and structural roots prevents change, as there is no toleration for innovation or other behavioural variance. Adaptation and responsiveness both require new ways of thinking, functioning and structuring, but all these are stifled in most bureaucracies (Rogers et al. 2000). The lack of adaptation and responsiveness has also been attributed to institutions – both formal and informal, that characterise agencies. Regarding formal institutions, Gunderson et al. (1995:496) encapsulate the view shared by many others, "*Most institutions are established to carry out some set policies, or mission, and spend most of their time and energy becoming more efficient in the implementation of these policies. That narrowing of attention causes strategic analysis to wither. The resulting myopia inevitably leads to crises in resource management that seem to occur when expectations are not met or shift with changes in underlying myths."*

Concerns about informal institutions in respect of change have been expressed in the sense of organisational practices and customs which, although unofficial, are widespread throughout the organisation and create mindsets and beliefs that are unadaptive (Stoll-Kleeman 2001). A fundamental aspect about informal institutions is their subtlety arising from the fact that they are entrenched in their thinking and actions which, in turn, manifest themselves in how we are taught to think and speak of resources as being divided into clear-cut types such as water, timber, fisheries and wildlife (Cortner et al. 1996). Such has been the perception of agencies, suggesting that their resistance to change is built-in their institutions – both formal and informal, and therefore, hard to deal with (Schlager and Ostrom 1992).

But we are in an era of great and rapid change in which organisations, including those responsible for managing the use of ecosystems, face new challenges and are required to be adaptive and responsive (WCPA 2003). A blend of changes in the legislation, and seemingly unstoppable forces of liberalisation and democratisation, variations in social structures and attitudes, technological innovations and the orientations of political and administrative structures are collectively challenging the traditional approach to managing the use of ecosystems. For example, public service agencies serve constituencies, citizens and governments with far different needs and expectations than they originally meant to meet. Adaptation and responsiveness across spatial and temporal scales are therefore critical to dealing with changes in demands for use of ecosystems and changes in ecosystems as they respond to use and environmental change.

Adaptation, Responsiveness and Organisational Culture

To *adapt* means making something suitable for a new use or purpose, or becoming adjusted to new conditions, while to *respond* is to do something in reply or as a reaction (Concise Oxford Dictionary 2004). Applied to the world of organisations, adaptation implies being prospective and purposeful in identifying, proposing and implementing changes. In this particular context, adaptation refers to changes based on an organisation's abilities to learn from its interaction with the environment with a view to sustaining relevance and survival into the future. Responsiveness, on the other hand, is guided by a short-term view – 'stopping the fire', with a distinct inclination towards immediacy. Collectively, adaptation and responsiveness can help organisations identify opportunities and risks emerging in a changing environment, and in the process addressing both short term and long terms challenges. Neither is a replacement for the other, both are necessary in dealing with the different pressures facing organisations. Their realisation, or otherwise, is a function of many factors, not the least being organisational culture.

If institutions are expressions of collective experience (Cortner et al. 1996), then organisational culture is one of the transmitters of institutions (Scott 1987). Organisational culture stems from continued association and interaction or, put simply, social relations within group or organisational settings. There can be no culture without a group and conversely, no group without a culture. There are two common challenges that all groups face and must deal with (Schein 1992). First is the challenge of survival, growth and adaptation in the environment. Second is that of integration which permits functioning and adapting for survival. Both challenges must be dealt with successfully for a group to accomplish its goals and thrive. Organisational culture partly evolves from the process of dealing with these challenges, hence its definition as: "*A pattern of basic assumptions – invented, discovered, or developed by a given group as it learns to cope with its problems of external adaptation and internal integration – that has worked well enough to be considered valuable and, therefore, to be taught to new members as the correct way to perceive, think and feel in relation to those problems*" (Schein 1992:12).

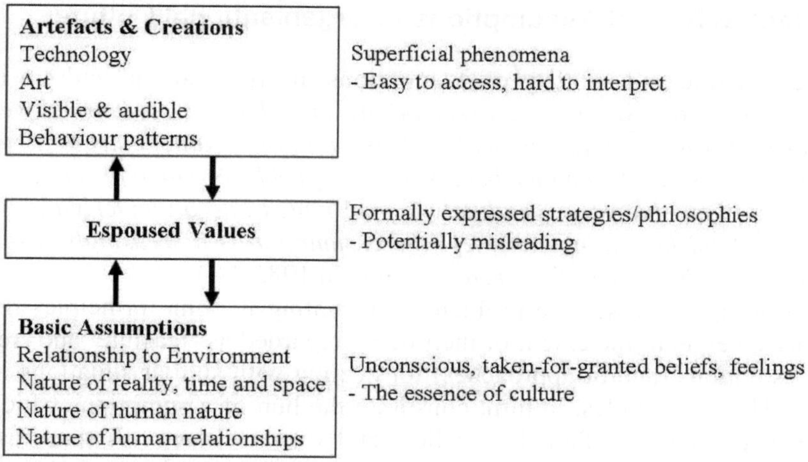

Fig. 10.2. *Levels of culture. After Schein (1992)*

Schein divides organisational culture into three levels (Figure 10.2): **artefacts** – the physical manifestations and behaviours, such as dress, which can easily be discerned yet are hard to understand, **espoused values** – social principles, philosophies, goals and standards seen as worthwhile; and **basic assumptions** – these represent the essence of a culture, normally existing as core beliefs which are taken for granted, about what is correct and real. Basic assumptions are hard to discern, mainly because they exist at an unconscious level. They form around deeper dimensions of human existence such as the nature of human relationships and activity, reality, time and truth, and therefore, provide a key to understanding why things happen the way they do.

Organisational culture influences perceptions of the fundamental mission, the environment and even decisions considered relevant for surviving environmental turbulence (Schein 1992; Senge 1990). It is also credited with the ability to influence behaviour and reduce uncertainty (Hampden-Turner 1990; 1994), enhance organisational learning (Argyris 1993; Argyris and Schöen 1978; Senge 1990), provide credence to decisions and resultant actions (Pettigrew 1985), make sense of rituals, myths, ceremonies and norms (Bolman and Deal 1991) and leverage economic performance (Kotter and Heskett 1992; Peters and Waterman 1982). All these happen against the backdrop of assumptions as the core of a culture.

The Central Role of Assumptions in Organisational Culture

The central role and subtlety of assumptions in organisational culture is well captured by Schein: *"...assumptions are themselves learned responses that originated as espoused values. But, as a value leads to a behaviour, and as that behaviour begins to solve problems which prompted it in the first place, the value gradually is transmuted into an underlying assumption about how things are. As the assumption is increasingly taken for granted, it drops out of awareness*" (Schein 1985:3-4).

As a group addresses the problems confronting it, some principles repeatedly emerge, to the extent of their being regarded as 'feasible' and 'realistic' solutions (and/or approaches) for dealing with similar situations in future. These principles, in turn, constitute the hub of assumptions which influence perceptions, thoughts, beliefs and overt actions. In time, assumptions become more or less like 'templates' for any future decisions and problem-solving. In this way, culture and its underlying assumptions can function as a cognitive defence mechanism for individuals and the group (Schein 1992). Assumptions can produce positive and negative influences.

On the positive side, they can enhance group identity, foster stability through minimising uncertainty and anxiety while at the same time make things 'manageable' – a shorthand description of 'coping with complexity' (Kotter 1996). Negatively, assumptions can be problematic, for example in dealing with change, because of their tendency to embed themselves in the corporate and individual subconscious, thereby creating stability against change, even when change is necessary. In this way, assumptions have the capacity to limit an organisation's adaptation and responsiveness to changes in the environment (Schein 1992). Sustained responsiveness and adaptation cannot occur without understanding, exposing and, where necessary, altering an organisational culture's underlying assumptions. But altering assumptions is not easy, because of their latent nature – a situation usually compounded by their history and dynamic reinforcing processes (Senge et al. 1999).

In the case of resource management agencies, difficulties in responsiveness and adaptation are worsened by the fact that both organisational culture and assumptions have historically not been given sufficient attention. More specifically, organisational culture commonly appears as an implied subject, and therefore, lies in the shadow of other subject matter such as the need to use social sciences in conservation (McCool 2003; Nyphus 2002; Patterson and Williams 1998; Harberlain 1988), resistance to change (Gunderson et al. 1995; Rogers et al. 2000; Caldwell 1990); promotion of adaptive management (Salafsky et al. 2001; Lee 1999, McLain and Lee 1996; Rogers et al. 2000), protected area evaluation (Conley and Moote 2003; Presber-James 1997), protected area funding (Alexander 1999), collaboration in protected area management (Stoll-Kleeman 2001), transformation or reform of protected area institutions (Brinkerhoff 1999; Bensted-Smith and Cobb 1995); promoting ecosystem approach to natural resources management (Cortner et al. 1996); global change (McNeely and Schutyser 2003; WCPA 2003) and collaboration in resource management programme evaluation and monitoring (Folke et al. 2002; Bellamy et al. 1999).

In addition to being shadowed by other subject matter, assumptions and culture have also been addressed in ways that focus more on the internal dynamics of agencies. Typically, approaches to managing organisational culture are inward looking, focusing on establishing a culture within the agency and paying little, if any, attention to managing the relationships between the cultures of involved constituencies so as to achieve co-evolution of compatible cultures.

Discussion: Founding Values and Strength of Organisational Culture

What can be discerned from the evolution of organisational cultures, in the context of public service agencies involved in resource management? While the concept of organisational culture and its implications are generally applicable to most organisations, it takes a different dimension when specifically applied to resource management agencies. Two major reasons are responsible for this situation: the ethical/moral origins of such agencies and the resultant strong organisational cultures.

Regarding the ethical/moral origins of resource management agencies, the focus ought to be placed on the pervasive and lasting influence of founding values in such agencies. The founding values were, and sometimes continue to be, anchored in certain world views consistent with ethical positions and moral slants of a particular period. In fact, the original values of most resource management agencies were informed by the philosophy of preservation. Such deep-rooted values are partly responsible for the strong organisational cultures that characterise most resource management agencies, which therefore are a good example of organisations with strong cultures (Nyambe 2005).

A strong culture, defined as "*a set of norms and values that are widely shared and strongly held throughout the organisation*" (O'Reilly and Chatman 1996:166), facilitates organisational cohesion. The benefits of a strong organisational culture are believed to derive from three effects of widely common and strongly held norms and values: increased dexterity and direction within the organisation, enhanced goal alignment between the organisation and its employees, and boosted employee effort (Kotter and Heskett 1992; Gordon and DiTomaso 1992; Burt et al. 1994).

There is, however, a down-side associated with strong organisational cultures: entrenchment of certain values due to predisposition to particular world views. Such cultures characterised by insularity, inadequate responsiveness, inadequate communication with stakeholders, and paternalistic management and leadership styles have historically made agencies aloof to environmental developments. This situation potentially leads to an inevitable trade-off for agencies with respect to their responsiveness and adaptive ability in the face of environmental change. Thus their adaptive and responsive capabilities become compromised. Organisational cultures of public service agencies are prone to be environmentally 'unadaptive' and therefore 'limiting', and this causes them to perceive and experience change as a disruptive force. The strength of an organisational culture becomes a 'shadow' opportunity cost: *"Cultural assumptions are the product of past successes. As a result, they are increasingly taken for granted and operate as silent filters of what is perceived and thought. If the organisation's environment changes and new responses are required, the danger is that the changes will not be noticed or, even if noticed, that the organisation will not be able to adapt because of embedded routines based on past success. Culture constrains strategy by limiting what the CEO and other senior managers can think about and what they perceive in the first place"* (Schein 1992:382).

Implications

In considering the implications of this matter, a point to emphasise is the fact that resource management agencies are not immune from the negative effects of a traditional organisational culture, i.e. one that is unadaptive, unresponsive and therefore, resistant to change. Such a culture positions any organisation, public or otherwise, to experience change as a disruptive force. But resource management agencies need not experience change only as a disruptive force.

Change can be experienced as a useful force if the prevailing organisational culture is adaptive and responsive. The key element in an agency's ability to be responsive and adaptive is the paradigm – the pattern of underlying assumptions that members of an organisation intuitively operate with (Schein 1992). It is these underlying assumptions and values that predispose an agency either to learn, to be open and to adapt, or else to be stagnant, insular and dogmatic in dealing with change. One key implication can be drawn: there is a need to make assumptions explicit for communication purposes with both internal and external stakeholders.

Making Assumptions Explicit

Left unattended and allowed to operate at the subconscious level, assumptions can facilitate the status quo and subvert novelty (Schein 1992). Prolonged inattention to assumptions can be costly, especially if there is significant turbulence in the environment. Not addressing assumptions explicitly can easily lead to a strong culture which, among other things, predisposes an agency to rigidity, insularity and other characteristics that can lead an agency to experience change as a disruptive force. The virtuous promises of a strong culture can turn into vicious threats for an agency if the associated assumptions remain implicit.

The inherent challenge, therefore, is to make the assumptions behind our ideas explicit, so that people can start discussing the potentially conflictive and 'undiscussable' subjects without invoking defensiveness and appreciating deeper causes of problems and their interdependencies (Senge 1990; Senge et al. 1999). Real change depends on making assumptions explicit, understanding their influence, evolution and how they can be brought into the organisational and individual conscious.

Central to making assumptions explicit is the nature and process of dialogue. Paternalism, command and control, hierarchy and top-down communications systems have historically prevented dialogue in organisations (Senge 1990) including public service agencies. Promoting dialogue around assumptions raises a fundamental need for a shared commitment to open, engaged and extensive communication. Dialogue can enable agencies track changes over time in assumptions so that they may accord with the changes in both the internal and external environments. Dialogue can also be useful as a basis for questioning the appropriateness of prevalent assumptions in relation to changes in the broader society. If there is a process for dialogue specific to assumptions, and if it precedes strategy processes, incorporating the visioning exercises and the setting of the mission, it is likely to lead to a shared understanding of the underlying pattern of assumptions (Nyambe 2005). If understood and appreciated, such a pattern of underlying assumptions would lead to support for the mission, vision and strategy processes. Importantly, should a need to make changes arise, the prospects for opposition are reduced. This suggests two emphases in approaching responsiveness and adaptation.

First is the need to address the assumptions held by individuals, both within and outside the agency. At this level, assumptions are reflected in individual attitudes, values and perspectives. Further, as individual assumptions change, the agency itself is slowly transformed towards becoming more responsive and adaptive. Assumptions within agencies are central because they determine the patterns of how individuals and groups relate to each other, organise their work, set priorities and how to engage external constituencies (Drucker 1995). Second is the importance of addressing the cultural properties of agencies themselves. By changing the cultural properties of an agency, e.g. by encouraging an outward focus and leadership style, individuals also change and the prospects for adaptation and responsiveness become enhanced.

Revisiting the Trialogue Model

In the light of the foregoing discussion on organisational culture and its underlying basic assumptions as determinants of responsiveness and adaptation, the Trialogue Model is a very useful conceptual construct. It highlights the fact that the problems facing society are not amenable to the processes of any one of the clusters on its own. For example, science remains central to generating knowledge and must not relinquish its traditional role of seeking new knowledge. But science must be pursued differently in order to be relevant to current and future biophysical, moral and social challenges. Equally, government processes should not lose sight of their role in fostering accountability, transparency and equity. Legislative and policy responsibilities of government remain noble and significant, but perhaps they are not enough. Government must engage more strategically with both science and society to resolve problems. Society can no longer afford to take a passive role. Society must engage government and science proactively and strategically, helping to redefine their roles and shaping their goals. A major implication for public service agencies is the need to align their assumptions with those of society, government and science processes. Otherwise, public service agencies are likely to fail to adapt and respond appropriately, as their assumptions will be inconsistent with those in the society, government and science processes.

Conclusions

At its core, the Trialogue Model is about managing for change. This means responsiveness and adaptation, both of which are usually prevented by organisational culture. To this end, it is important for actors in public service agencies to understand the prevailing assumptions — as the core of organisational cultures – in their agencies. Adaptation and responsiveness, arguably the thresholds to change, are unlikely to occur in the absence of a specific review of prevailing assumptions. Therefore, those involved in leading resource management agencies would benefit from engaging in continual exploration of assumptions as the portal for the ideas, perceptions and beliefs that influence change. Making assumptions explicit, and aligning them with society, government and science processes will require the institutionalisation of an adaptive approach so as to re-examine assumptions continuously in a structured manner. Failing to do so predisposes public service agencies to experiencing change as a disruptive force, resulting from a failure to appreciate the assumptions at play in society, government and science processes.

References

Alexander JN (1999) Institutional constraints to protected area funding. Parks **9**

Argyris C (1993) Knowledge for action: a guide to overcoming barriers to organisation change. Jossey Bass, San Francisco

Argyris C, Schön DA (1978) Organisation learning: a theory in action perspective. Addison-Wesley, Reading

Bellah R, Madsen R, Sullivan WM, Swindler A, Tipton SM (1991) The good society. Alfred A Knopf, New York

Bellamy JA, McDonald GT, Syme GJ, Butterworth JE (1999) Evaluating integrated resource management. Society and Natural Resources **12**:337-353

Bensted-Smith R, Cobb S (1995) Reform of protected area institutions in East Africa. Parks **5**:3-19

Bolman LG, Deal TE (1991) Reframing organisations: artistry, choice and leadership. Jossey-Bass, San Francisco

Brinkerhorff DW (1999) The implementation of environment and natural resource policy in Africa: blending regulatory enforcement with community participation. International Journal of Organisational Theory and Behaviour (2)1-2:233-259

Burt RS, Gabbay SM, Holt G, Moran R (1994) Contingent organisation as a network theory: the culture performance contingency function. Acta Sociologica 37:346-370

Caldwell LK (1990) Between two worlds: science, the environmental movement, and policy choice. Cambridge University Press, New York

Conley A, Moote MA (2003) Evaluating collaborative natural resource management. Society and Natural Resources 16:371-386

Concise Oxford dictionary. (2004) 10th Edition. Oxford University Press, Oxford

Cortner HJ, Moote MA (1994) Trends and issues in land and water resources management: setting the agenda for change. Environmental Management **18**:167-173

Cortner HJ, Shannon MA, Wallace MG, Burke S, Moote MA (1996) Institutional barriers and incentives for ecosystem management: a problem analysis. General technical report PNW-GTR 354. Portland, OR, US Department of Agriculture, Forest Service, Pacific NorthWest Research Station

Drucker P (1995) Managing in a time of great change. Butterworth Heineman, Oxford

Etzioni A (1961) A comparative analysis of complex organisations. Free Press, Glencoe

Folke C, Carpenter S, Elmquist T, Gunderson L, Holling CS, Walker B, Bengtsson J, Berkes F, Colding J, Danell K, Falkenmark M, Gordon L, Kasperson R, Kautsky N, Kinzig A, Levin S, Mäler K-G, Moberg F, Ohlsson L, Olsson P, Ostrom E, Reid W, Rockström J, Savenije H, Svedin U (2002) Resilience and sustainable development: building adaptive capacity in a world of transformations. The Environmental Advisory Council of the Swedish Government. Available online at http://www.consecol.org/Journal [Accessed 13 April 2003]

Gordon G, DiTomaso N (1992) Predicting corporate performance from organisational culture. Journal of Management Studies **29**:783-798

Gunderson LH, Holling CS, Light SS (Eds) (1995) Barriers and bridges to the renewal of ecosystems and institutions. Columbia University Press, New York

Hampden-Turner CM (1994) Corporate culture: from vicious to virtuous circles. Piatkus, London

Hampden-Turner CM (1990) Charting the corporate mind: from dilemma to strategy. Blackwell, Oxford

Harberlain TA (1988) Improving interdisciplinary research: integrating the social and natural sciences. Society and Natural Resources 1:231-250

Hoekstra M (1995) AQUA – a framework integrated water policy analysis. Global Series Report 6. Ryksinstitrust voor Volkesgesonheid en Milieu, Bilthoven, Netherlands

Holling CS, Meffe GK (1996) Command and control and the pathology of natural resource management. Conservation Biology 10:328-337

Kotter JP (1996) Leading change. Harvard Business School, Boston, MA

Kotter J, Heskett J (1992) Corporate culture and performance. Free Press, New York

Lee KN (1999) Appraising adaptive management. Conservation Ecology (3)2:3. Available online at http://www.consecol.org/vol3/iss2/art3

Lubchenco J (1999) Entering the century of the environment: a new social contract for science. Science (279)23:491-497

McCool SF (2003) Integration of social science into protected area stewardship: challenges and opportunities. Paper presented at the George Wright Society Biennial Conference, San Diego, California, 14 April 2003

McLain RJ and Lee RG (1996) Adaptive management: promises and pitfalls. Environmental Management, vol 20, no 4, pp 437-448.

McNeely JA, Schutyser F (2003) Protected areas in 2023: scenarios for an uncertain future. IUCN, Gland

Nyambe N (2005) Organisational culture and its underlying basic assumptions as a determinant of response to change: a case study of the conservation sector in KwaZulu-Natal, South Africa. PhD Dissertation. CEAD, University of KwaZulu-Natal, Pietermaritzburg

Nyphus PJ (2002) A role for natural resource social science in biodiversity risk assessment. Society and Natural Resources 15:923-932

O'Reilly CA, Chatman JA (1996) Culture as social control: corporations, culture and commitment. In: Staw BM and Cummings LL (eds) Research in organisational behaviour. JAI Press, Greenwich

Ostrom E (1986) An agenda for the study of institutions. Public Choice 48:3-25

Parsons T (1960) Structure and process in modern societies. Free Press, Glencoe UK

Patterson M, Williams D (1998) Paradigms and problems: the practice of social science in natural resource management. Society and Natural Resources 11:279-295

Peters T, Waterman RH (1982) In search of excellence: lessons from America's best-run companies. Harper and Row, New York

Pettigrew AM (1985) The awakening giant: continuity and change at ICI. Blackwell. Oxford

Presber JS (1997) An institutional approach to protected area management performance. Conference readings. Political Economy Research Centre Forum, The politics and economics of protected area management, Emigrant Press, Montana

Rogers K, Roux D, Biggs H (2000) Challenges for catchment management agencies: lessons from bureaucracies, business and resource management. Water SA (26)4:505-511. Available online at http://www.wrc.org.za [Accessed 9 February 2003]

Salafsky N, Margoluis R, Redford K (2001) Adaptive management: a tool for conservation practitioners. Biodiversity Support Programme/WWF-US. Available online at http://www.BSPonline.org [Accessed 4 June 2002]

Schein EH (1985) Organisational culture and leadership. Jossey-Bass, San Francisco

Schein EH (1992) Organisational culture and leadership. 2nd edn. Jossey-Bass, San Francisco

Schlager E, Ostrom E (1992) Property-rights regimes and natural resources: a conceptual analysis. Land Economics 68:249-262

Scott WR (1987) Organisations: rational, natural and open systems. 2nd edn. Prentice Hall, Englewood Cliffs

Selznick P (1957) Leadership and administration. Harper and Row, New York

Senge P (1990) The fifth discipline: the art and practice of the learning organisation. Doubleday/Currency, New York

Senge P, Kleiner A, Roberts C, Ross R, Roth G, Bryan S (1999) The dance of change: the challenges of sustaining momentum in learning organisations. Nicholas Brealey, London

Stoll-Kleeman S (2001) Reconciling opposition to protected areas management in Europe: the Germany experience. Journal of Environmental Planning and Management 44:109-128

Turton AR, Hattingh J, Claassen M, Roux DR, Ashton PJ (2005) Towards a model for ecosystem governance: an integrated water resource management example. Paper presented at the International Symposium on ecosystem governance, 10-13 October 2005, Kwa Maritane Bush Lodge, Pilanesberg, South Africa

Walker B, Carpenter S, Marty Anderies J, Abel, NC, Janssen M, Lebel L, Norberg J, Peterson G, Pritchard R (2002) Resilience management in social-ecological systems: a working hypothesis for a participatory approach. Conservation Ecology 6:14

World Commission on Protected Areas (WCPA) (2003) A guide to securing protected areas in the face of global change: options and guidelines. (Draft) IUCN, Gland, Switzerland

Weber M (1893) Max Weber on capitalism, bureaucracy and religion: a selection of texts. Andreski S (ed) Allen and Unwin, Boston

Lessons from Changes in Governance of Fire Management: The Ukuvuka Operation Firestop Campaign

Sandra Fowkes

Abstract

Practical experience from the Ukuvuka Campaign, a short-term initiative in the governance of fire in the Cape Peninsula, Cape Town, South Africa offers insights that may have an application in enhancing ecosystem governance. Effective ecosystem governance requires many behaviour changes among citizens, both individual and institutional. Ukuvuka explored, facilitated and catalysed a number of these changes. The lessons learned are also used to look specifically at the interaction between government and society in ecosystem governance. Finally, the paper evaluates the model of ecosystem governance prepared as a basis for discussion at the international symposium on ecosystem governance held at KwaMaritane in October 2005.

Ukuvuka Operation Firestop Campaign was a four-year partnership set up to address the issue of fire in the Cape Peninsula. The initiative was triggered by a crisis caused by serious veld fires that burnt some 30% of Table Mountain National Park which Cape Town surrounds, and also damaged or destroyed some 60 homes and buildings. In addition, the campaign was challenged to focus some attention on the on-going problem of fires in informal settlements.

The governance lessons learned were that:

- Setting up demonstration projects in a "safe space" can facilitate the behaviour changes required for governance of ecosystems by testing a new behaviour in a low-risk environment;
- For institutional behaviour change, an internal champion is needed. However, the champion does not necessarily have to be very senior. Personal passion is potent;
- For the behaviour changes needed for ecosystem governance to be systemic and sustainable, initiatives using the model of interest-based, rights-based and power-based approaches look promising;

- Effective governance in an ecosystem needs diverse role players to work together. Drawing together such a wide range of participants, many of them often in conflict with one another, is challenging. Engaging reluctant participants may be more easily done in a short-term low-risk situation outside of the long-term institutions;
- Use of the special-purpose vehicle of a "non-owned" body can be a tool for government to undertake the research and development that can assist it to deal with the current challenge to society of rapidly evolving and intricate policy problems; and
- The experience gained from Ukuvuka indicated that science is a player with power unequal to that of government or society in the governance space. Science, or knowledge, may play a far more powerful and useful role in shaping the governance decision space by making itself available to all the role players.

Keywords: fire management, behaviour change; short-term initiative; low-risk safe-space for innovation; interest-, rights-, power-based interventions; invasive alien vegetation

Introduction

The models, tools and methodologies used in what has become known as the governance of ecosystems have undergone significant changes. These range from 1970s command-and-control regulation, through use of economic tools in the 1980s, to public participation and partnerships in the 1990s. It is recognised that in the 2000s "*society is faced with rapidly-evolving and intricate policy problems that demand complicated choices between possible solutions, often under conditions characterised by uncertainty. The successful implementation of sustainable ecosystem...policies is...dependent on...the organisation of efficient institutions, implementation structures and public support. It is therefore necessary to mobilise, integrate, and improve state-of-the-art models, tools and methodologies, activities and resources, to bridge the gap between 1) the scientific community; 2) decision makers; and 3) the public and build and develop competences*" (Gooch 2005). The extent of the challenge requires not only enhanced use of existing tools but also new models, tools and methodologies. The purpose of this chapter is to describe one such approach that emerged out of a response to a failure in ecosystem governance, and to share some of the relevant lessons learned from it. The failure was in governance of fire in the Cape Peninsula, of the City of Cape Town, South Africa in

January 2000. The new approach was a short-term partnership between government, business and civil society called the Ukuvuka Operation Firestop Campaign.

South Africa has excellent policies and laws relating to ecosystem governance – especially to water. However, the challenges lie in translating policy intent into implemented action on the ground. "A key reason for implementation failure seems to be the tendency to superimpose an adaptive approach on a non-adaptive (e.g. command-and-control or bureaucratic) decision-taking environment. Institutional capacity to adapt to and shape change is an important prerequisite for effecting adaptive management of ecosystems" (Roux et al. 2005).

There are significant behaviour changes required in both institutions and individuals in order to deliver effective ecosystem governance. For example, Gooch (2005) identifies that, "*Other contemporary complications within the field include the fact that serious flaws in the traditional technical approach to water management have been demonstrated to exist...Water managers now need more than just technical skills if they are to successfully address the issues before them.*" This comment applies to other sectors and certainly also to fire management.

There are many tools for policy implementation to give effect to ecosystem governance. Among them is the use of appropriate institutional forms. Mintzberg (1996) states that robust sustainable institutions are best achieved through the balances, and the implied checks, exercised by countervailing forces. He makes the point that, where these strong countervailing forces are absent, the societies are not sustainable. He cites the political breakdown in Eastern Europe as the result of very strong government without the countervailing checks and balances from other sectors of society. "*Indeed, there is a role in our society for different kinds of organisations and for the different contributions they make in such areas as research, education and health care.*" (Mintzberg 1996).

Mintzberg identifies four kinds of organisation – those owned publicly; privately; collectively and non-owned, and comments that "*unfortunately, we in the West have not come to terms with the full range of possibilities. Because capitalism has supposedly triumphed, the private sector has become good, the public sector bad, and the cooperatively owned and non-owned sectors irrelevant*" (Mintzberg 1996).

The initiative described in this chapter can be considered a non-owned institution, i.e. an organisation not-for-profit and controlled by "*self-selecting and often very diverse boards of directors*" (ibid). Examples of non-owned organisations are charities, volunteer and activist organisations and, in some countries, certain hospitals and universities. Well-known examples of non-owned institutions are the Red Cross and Greenpeace and, of particular relevance to the present topic, the World Commission on Dams (WCD).

One of the achievements of the WCD was "*to actively break through traditional boundaries of thinking – to look at familiar issues from a different perspective.*" The commissioners further commented that, "*the trust required to enable different sectors and players to work together must still be consolidated. ...We have conducted the first comprehensive and global review of the performance of dams and their contribution to development. We have done this through an inclusive process that has brought all significant players into the debate. And we believe we have shifted the centre of gravity in the dams debate to one focused on options assessment and participatory decision-making*" (World Commission on Dams 2000).

The WCD operated at a global scale and made a significant contribution to global ecosystem governance. Operating at a totally different scale, the local level, the non-owned institution, the Ukuvuka Operation Firestop Campaign, contributed to changing the behaviour of individuals and institutions involved in fire management and governance, and it offers lessons applicable to the wider field of ecosystem governance. It differed from other governance tools in having a short-term existence, some four years only, and it is argued that it was largely this feature that created an enabling environment for behaviour change in ecosystem governance.

The contribution made at such differing scales by short term non-owned institutions suggests strongly that those concerned with effective ecosystem governance should actively pursue the use of non-owned institutions and redeem this valuable tool from the fate of irrelevance described by Mintzberg.

Background

The winter of 1999 was particularly dry in the Mediterranean-climate area of the Western Cape, South Africa. A hot summer with searing winds followed and many areas of the province burnt fiercely in January 2000. From 16-21 January the Cape Peninsula region of the City of Cape Town was in flames. The fires were so much hotter and more extensive than before because of the considerable amount of extra fuel provided by the invasive alien vegetation that was flourishing in this area. The greatest impact was at the interfaces between the natural area and the urban areas. When the fires were eventually extinguished, 30% of the National Park had burnt and some 60 homes and buildings had been damaged or destroyed.

Those six days of continuous fire in a major South African city can be viewed as the outcome of a failure of governance, and particularly of ecosystem governance.

Graham et al. (2003) state that, "*We define governance as the interactions among structures, processes and traditions that determine how power and responsibilities are exercised, how decisions are taken and how citizens or others stakeholders have their say. Fundamentally, it is about power, relationships and accountability: who has influence, who decides, and how decision-makers are held accountable.*"

With hindsight, it was clear that the then prevailing structures, processes and traditions would exacerbate rather than mitigate such an extreme event as the January 2000 fires.

The Existing Governance Structures and Processes

In 2000 the City of Cape Town was divided into six Metropolitan Local Councils (MLCs), with a seventh overarching coordinating council. Three of these seven structures were involved in the area that burnt in the January 2000 fires. The two operational MLCs had policies and procedures that differed from one another. In addition, a new institution had entered the stage. In 1998 a National Park was created in order to conserve a key part of the Cape Floral Kingdom, the smallest of the world's six floral kingdoms. The South African National Parks (SANParks) organisation had taken over the running of a collection of fragmented areas. These were of such national and global conservation value that they merited a higher level of protection and coherent management by a single competent entity. Although the unique indigenous vegetation of the Cape Floral Kingdom is fire-adapted, one of the key threats it faces is fires that are too hot and too frequent. The invasive alien plants that have infested the area significantly, over many years, increase the amount of burnable fuel as well as the temperature at which fires burn.

Even though there was a Memorandum of Agreement between the MLCs, SANParks and various other bodies for dealing with emergencies such as fire, the structures in place were clearly not adequate for the enormity of the situation. More fundamentally, there was no policy that guided planners in incorporating proactive fire-prevention into the way that the urban-natural interface was planned and that guided decisions taken about land-use on the urban edges. In addition, the City Fire Services had a proud tradition of 150 years of fire-suppression which focused on the built-environment, but which did not cover the adjacent natural areas. The previous system of fire-breaks in the Peninsula was not well maintained. Finally, the nature of the terrain, i.e. a narrow, mountainous peninsula some 70 km long, meant that there was no robust and reliable emergency communication system in place.

Existing Governance Traditions

Responsibilities for minimising and managing the amount of fuel created by invasive aliens were not embraced by many landowners, either individuals or institutions. There was certainly no coordinating policy or guideline in the City at the time. Although there were laws on the statute book to control invasive alien vegetation, the mandated government department showed little will to use their legal instruments for prosecution. Such legal intervention was also politically unpopular, especially as much of the original introduction of invasive aliens was encouraged and funded by the previous government. It was also politically unpopular in the new post-1994 democratic dispensation to invest state resources on conservation issues when there were urgent social issues to be addressed. This changed when the innovative Working for Water programme forged linkages between ecological integrity and social justice by using removal of invasive alien plants as a means to pursue social goals such as employment creation, empowerment of marginalised people i.e. women, youth and the disabled, HIV/Aids education, early childhood learning as well as other societal transformation agendas.

South Africans have not had a tradition of litigating and environmental laws had been particularly weakly implemented. Ecological concerns had generally been of low status in the profit-dominated paradigm of business and also of low status in the socio-politically driven paradigm of the new democratic government. Finally, the concept of being responsible and accountable for imposing costs on others for acts of negligence has certainly not had a strong tradition in South Africa. With this hindsight overview of relevant aspects of the governance structures, processes and traditions, the crisis caused by the fires in January 2000 was not surprising.

Start of the Ukuvuka Campaign

It was against this background that the Ukuvuka Campaign was started. Within 10 days of the extinguishing of the fires a short-term (four-year) partnership between three spheres of government – national, provincial and local, and various role-players in society, primarily for-profit businesses and non-profit organisations, had been agreed in principle. The first draft of the business plan had been prepared. The partnership was called the Santam Cape Argus Ukuvuka Operation Firestop Campaign. The name of the campaign, an isiXhosa word meaning *"to wake up, or to rise up"*, reflected the need for action and change, and its desired outcome was that, by the end of the campaign, the conditions and behaviours that had led to these two very different kinds of fire would be so changed that fires of that level of severity would not happen again.

The role-players were to collaborate in a campaign to change the behaviour of citizens – both individual members of the public as well as institutions – with respect to the focal issue of the campaign, which was fire. The fires had occurred in both the natural areas and in the informal settlements of the Cape Peninsula. Informal settlements had expanded dramatically as a result of the flood of *"refugees"* from ecologically and economically unsustainable rural areas in the previous "homelands/Bantustans". The material used to build shacks is often highly flammable. The shacks are built in areas that are close to work opportunities and consequently have high densities of dwellings which also contribute to the high fire risk.

There were many lessons learned from Ukuvuka. This chapter selects those key lessons relevant to governance and tells some of the stories that taught the lessons. *"It is not surprising that attempts to pursue these synergies between biodiversity, ecosystem management, and human well-being are fraught with difficulty. At both the local scale and the global scale, these strategies are inherently multi-sectoral, multi-stakeholder, and multidisciplinary. And they are being pursued in a world with highly sectoral institutions, inequitable distribution of wealth, little experience with participatory processes, and little reward for multidisciplinary research. But these strategies do often succeed and the lessons from success and failure are guiding new initiatives today."* (Reid, in O'Riordan and Stoll-Kleeman 2002).

The Governance Lessons Learned

Setting up demonstration projects in a "safe space" can facilitate the behaviour changes required for governance of ecosystems by testing a new behaviour in a low-risk environment.

Crises shake up individuals, institutions and systems and create an opportunity to fast-track some of the changes that governance requires. The whole concept and setup of the Ukuvuka Campaign was a case in point of the speed with which the right blend of vision, political will and skilled communication can initiate change when there is a crisis. The commitment to support and fund Ukuvuka, plus the first draft of its business plan was underway within 10 days of the end of the fires. There is no doubt that people with the energy to exploit the opportunity provided by crisis can catalyse change. Harvard Business School Professor JP Kotter (1995) argues that the vital first step towards achieving successful organisational change is the creation of a sense of urgency and a need for change. Kotter cites examples of business leaders who have deliberately created major crises in their organisations primarily as a way of generating sufficient impetus to achieve major organisational change.

However, it was also the experience in Ukuvuka that change could also be facilitated by setting up a demonstration or pilot project. Such piloting needs a safe space that shields those who are risk-averse, especially civil servants, and encourages a willing suspension of disbelief for a negotiated period. Crucial to its success is a small, competent, well-resourced team of people working within a framework of good governance but outside the constraints and bureaucracy of long term institutions.

Case Study 1: Rapid Attack on Informal Settlement Fires

Rapid attack is the approach increasingly being used to control fires. The chances of suppressing a fire are greatly increased if the attack can be launched as soon as possible within the "*golden twenty minutes*" of its starting. Clearly, aerial delivery is an effective way to deliver that rapid attack. Within the Cape Peninsula, helicopters with quick-release buckets have been used to drop 3,000 litres of water on mountain fires to contain and cool them, so making it possible for ground forces to play their role in fire control.

One of the fire managers in the City of Cape Town wanted to use a similar approach to containing fires in informal settlements. Such an approach had, to his knowledge, not been used elsewhere. The idea of a helicopter "water-bombing" informal settlements was highly contentious. Many of his colleagues regarded the idea with considerable scepticism, giving a litany of reasons as to why it would not work. In that climate the likelihood of obtaining funding from the City, let alone permission to test out such an approach, was simply non-existent. Also, politically, it was a non-starter – the imagination could run rife with what the media might do with an initiative that dumped three tons of water on people who are already severely disadvantaged, vulnerable and in dire poverty. Nevertheless, he wanted to try it, given the lack of success of the alternatives and the depressingly common situation of hundreds of dwellings burning and thousands of people being sucked yet deeper into the vortex of poverty after each fire.

He saw in Ukuvuka an opportunity to test the approach, as Ukuvuka was tasked with changing the conditions that lead to fires. Ukuvuka had human and financial resources and a responsive enough governance structure to set up such a test relatively quickly. Perhaps most importantly, Ukuvuka was not risk-averse. Failure of an initiative would not threaten the partnership. Ukuvuka was intended, after all, to cease to exist at the end of its four-year term. Conducting the test through such an arrangement could insulate the City from the risk.

The first test took place in July 2003. Through its network of contacts, Ukuvuka arranged to construct a mock informal settlement of 16 shacks on some vacant state land. A helicopter company donated flying time and fuel for the test. The master-stroke in the design of the test was to enlist the active fire-fighters, especially the resident cynics and sceptics, to evaluate the test against criteria which they themselves had set up in advance.

The first test was a victim of its own success. After burning just a few of the mock shacks, the fire was put out very swiftly. The fire-retardant foam added to the water in the bucket was so effective that the fire service's trainer, who prided himself as being a champion "arsonist", was unable to re-start the blaze! The verdict of the evaluators was that the mock-up did not mimic reality sufficiently. Firstly, there were not the usual contents – mattresses, paraffin stoves, TV sets, clothing – that fuelled typical informal settlement blazes. And, secondly, in July the fierce winds typical of the summer fire season, were absent.

The second test, in November 2003, provided gale-force winds and enough fuel in the shack contents to cause the temperature to rise to nearly 1,000°C. Concern about dropping three tons of water on people had lead to a decision for the helicopter to use a smaller bucket with a capacity of only 600 litres. The test showed that this was not enough water to be effective, and that the larger-capacity buckets should be used. The reckoning was that anyone close to shacks burning at 1,000°C would be unlikely to be in a condition to care about being deluged with large amounts of water.

The evaluation showed the potential of the approach, but there was still internal resistance to use of helicopters and it took a further 18 months before helicopters were used to control a particularly vicious informal settlement fire. Sadly, the helicopters were only called in long after the 'golden 20 minutes' period, too late to make a real difference.

The person with the passion for testing this radically different approach to suppressing fires in informal settlement fires was appointed Chief Fire Officer for the City of Cape Town in July 2005. It will be interesting to see how a strongly-resisted approach, tested in the "safe space" provided by Ukuvuka, may be incorporated into standard operating practice in future.

Fig. 11.1. *Testing the efficacy of using a helicopter to control fire in a mock informal settlement*

For institutional behaviour change, an internal champion is needed. However, the champion does not necessarily have to be very senior. Personal passion is potent.

In the story of the rapid aerial attack on informal settlement fires, the proponent of the approach was a senior officer, but this is not a necessary condition. Change can be initiated by people of much lower status.

Case Study 2: The Fire and Life Safety Education Story

March 2001 saw two young fire officers coming to Ukuvuka with a slightly diffident request for some funding to test an idea. They had recently been exposed to the concept of proactively stopping fires from starting rather than only waiting to suppress them reactively. A course given by a visiting American fire officer had seeded the idea that they could use an entertaining, lively way to embed fire-prevention behaviours in the citizens of Cape Town, especially its youth. The R5 000 (about US$900) funding that they sourced from Ukuvuka, and the opportunity of "performing" at the annual Navy festival in April confirmed for them the potential of the approach.

From this small start the two young fire officers set their sights on a major initiative at the City of Cape Town's annual Youth Environmental School (YES) in the first week of June. A "retired" fire engine was converted to a vibrantly painted, music-filled focus for the "edutainment" sessions that were prepared to communicate messages like "stop, drop and roll", "crawl low under smoke and go, go, go." By the end of the YES week, some thousands of children had been exposed to the messages, delivered with high-energy fun around Noddy, the fire engine.

From its tentative beginnings in March 2001, the momentum of the initiative accelerated dramatically. The project not only motivated further funding from Ukuvuka but also worked with members of the Ukuvuka team who contributed their expertise in project design and management, communication, building media profile, and project evaluation. Within two years there were two converted fire engines, some 200 fire fighters trained in the foundation course of Fire and Life Safety Education, and tens of thousands of children exposed to the messages of the programme. Most impressively, all the outreach work was done on a voluntary basis by fire fighters in their off-duty time, as there were no formal posts for Fire and Life Safety in the Fire Service at the time. The ultimate compliment to the change that the passionate pair had started was that, after a presentation to the fire chiefs of the seven Metropolitan Local Councils, the City made provision for the creation of 23 posts for Fire and Life Safety work.

In September 2005 the Chief Fire Officer for the City of Cape Town commented that "we have seen a reduction in fires. I'm not sure if I can attribute the reduction to the Fire and Life Safety programme, but I would like to say that it has played some part...We have gone through a complete paradigm-shift, from reactive to proactive approaches." (P Smith pers. comm.)

It should be noted that, although the 23 posts exist on paper, the City has not yet formalised the step of providing a budget and appointing staff to the posts. There is seldom an end to a story of creating change. However, from a tentative, modest beginning, the passion of two young fire officers helped to start a major change in a 150 year-long tradition of a well-established institution.

In contrast, Ukuvuka also learned that a number of other worthy initiatives started by Ukuvuka team members did not take a life of their own and were not adopted by the long-term institutions. None of these initiatives had internal champions in the long-term organisations.

For the behaviour changes needed for ecosystem governance to be systemic and sustainable, initiatives using the model of interest-based, rights-based and power-based approaches look promising.

A model which was originally used as the basis for the design of dispute systems provides valuable insights that are relevant to ecosystem governance. As ecosystem governance requires so many role-players of such differing natures, there will inevitably be conflict and disputes. The goal of dispute systems design is to create a system that maximises benefits and minimises costs. Similarly, ecosystem governance aims to deal with the many trade-offs on the road to sustainability – however one may define sustainability.

An effective dispute-resolution system might be viewed as a pyramid: most disputes are resolved through reconciling interests – the base of the pyramid, some through determining who has rights, and fewest through determining who is more powerful – the apex of the pyramid. Ury et al. (1988) explain that working with "interests" typically costs less and yields better results than determining who is right, or has rights, or is more powerful.

Interests are those factors that shape people's behaviour or are the underlying reasons for what they say they want. Rights are generally created through a country's laws or through common law. Power, in Ury's model, can be political power, economic power, the power of the media to create awareness and shape opinion, or the power of protest, such as "lying in front of bulldozers."

In the Ukuvuka situation a somewhat different application of the dispute system design model was used. The initiative was designed to include all three elements: interest-, rights-, and power-based, but in appropriate proportions. In the case described below the use of all three elements, and the synergy between them, was essential for the positive outcome of increased compliance to remove invasive alien vegetation. The majority of resources were invested in the interest-based aspects. Far less was invested in rights-based aspects. The least amount of time, person-power and finances were invested in power-based aspects.

Fig. 11.2. *Enthusiastic fire officers use "edutainment" to impart Fire and Life Safety messages to children, and even to involve the Chief Fire Officer of the City of Cape Town*

Case Study 3: Using Interest-Based, Rights-Based and Power-Based Interventions to Promote Sustainable and Systemic Behaviour Change

To address the issue of damage and destruction caused by invasive alien plants around the urban edge, a significant behaviour change in citizens – both individual and institutional – was required. Ukuvuka wanted to change the behaviour of landowners so that they removed the invasive alien plants that led to fires being hotter, fiercer and more frequent. The core problem was the presence of too much fuel from too many invasive alien plants. The fuel load led to fires that impacted on both property and the unique biodiversity "capital" of the Peninsula.

The typical approach to promoting behaviour-change falls under the overarching heading of communication and includes awareness-raising, making information available, developing knowledge, influencing attitudes and building advocacy. Ukuvuka's communication efforts were directed at the general population of the campaign area, and particularly at those priority targets who owned or managed property around the urban edge.

Ukuvuka developed an innovative integrated communication campaign of:

- Ten different street-posters;

- Related radio plus newspaper and magazine print advertising – specifically designed to cross-promote the various media and so to leverage a small budget;
- Relevant editorial for print and radio as well as for the website;
- Detailed information delivered to each urban edge household in an engaging and novel way. An oversize matchbox, labeled "matches are not the only threat to our mountains" which contained:
  - One information booklet about the top 10 alien plants, and another on "*The Law and you*", which explained citizens' rights and responsibilities relative to invasive alien plants;
  - An invitation to public information sessions; and
- Public meetings.

Any healthy sceptic would correctly identify this worthy investment of time, effort and funds as likely to preach to the converted and to be sub-optimal. And so it would have been, had it not been part of the combined set of interventions informed by the insights of Ury's dispute system design model. In order to leverage the communication initiatives described above – collectively known as interest-based interventions – they needed to be undertaken in concert with rights-based initiatives that demonstrated the consequences of ignoring the legislation that set out the rights and responsibilities of citizens to control the spread of invasive alien plants.

Ukuvuka's rights-based initiatives took the form of doing all the preparatory legal work required to initiate prosecution in terms of the relevant law, the Conservation of Agricultural Resources Act (CARA). The department responsible for enacting and enforcing the law was the National Department of Agriculture and Land Affairs. All the communication efforts had provided the foundation on which the prosecution could proceed. No landowner could claim that they were unaware of the problems posed by invasive alien vegetation or that they were unaware of their rights and responsibilities relative to any alien vegetation invasions on their properties.

The first properties targeted for prosecution were situated in a catchment vulnerable to invasive alien plant infestation and in which Government had already invested significant resources to remove invasive alien plants. The selection of these properties was as a result of strategic analysis of the catchments of the campaign area.

Although the sceptic's cynicism may lessen, knowing that the communication efforts were in tandem with prosecution, it would again be appropriate to wonder if this would be a long-term solution to the issue. The final insight from the dispute systems design model is that judicious use of power-based interventions is an important contribution towards the changing of behaviour.

The Department of Agriculture did not show much enthusiasm for prosecuting in terms of its CARA legislation. However, Ukuvuka worked with the Deputy Minister of Justice and Constitutional Affairs, who was also an Ukuvuka Board member, to convene meetings with all the role-players needed to initiate prosecution. The demonstration of political will from that level was a crucial power-based contribution to breaking through the delays, stalling and reluctance to prosecute that had been the situation previously.

Ukuvuka was able to use another power-based intervention through its contacts with media to ensure that the developments in the prosecutions were well publicised. As soon as the news report of the first-ever prosecution for having invasive alien plants on properties in the Cape Peninsula hit the streets, the impact caused noticeable ripples.

Finally, Ukuvuka's ability to use the power-based approach of providing funding that deployed staffing and consultant skills was important in keeping the momentum of prosecution going.

Fig. 11.3. *A strategic package of inter-related communication initiatives contributed to raising awareness – to the extent of being the topic of a cartoon in a daily newspaper*

Measurement of the effects of the combined initiatives is not easy or direct. Some of the arguably most relevant indicators were indirect. For the first time, advertisements for property sales included wording such as "*Land recently cleared of alien vegetation*" "*Half a hectare to build on with spectacular 360°views R2,25 mill* (includes removal of alien vegetation)" (Ukuvuka Final Report 2004). If an initiative merits a cartoon in a daily newspaper, this suggests that that initiative has become significantly prominent in the awareness of the newspaper-reading population. Ukuvuka was the subject of such a cartoon (Cape Times cartoon by Tony Grogan Tuesday 30 March 2004).

By October 2005, out of the eight properties that were targeted for prosecution, the clearing of invasive alien plants has occurred on six. The environmental lawyer who acted as a consultant to Ukuvuka was of the opinion that the alien clearing was overwhelmingly motivated by the threat of prosecution (David Waddilove pers. comm.).

Effective governance in an ecosystem needs diverse role-players to work together. Drawing in such a wide range of participants, many of them often in conflict with one another, is challenging. Engaging reluctant participants may be more easily done in a short-term low-risk situation outside of the long term institutions.

Ukuvuka learnt the value of being a short-term body in encouraging "*a willing suspension of disbelief.*" Where participants would certainly not have been willing to work as partners in perpetuity, if their involvement was only for a short time, they were willing to try to collaborate. Ukuvuka also learned that the existence of an independent short-term body facilitating the collaboration of all the stakeholders in fire management helped to reduce real or perceived turf battles between the major organisations.

Use of the special purpose vehicle of a "non-owned" body can be a tool for government to undertake the research and development (R&D) that can assist it to deal with the current challenge to society of rapidly evolving and intricate policy problems.

The response of many businesses to the current rapidly evolving and changing context of society is to invest heavily in R&D. Monitoring and Evaluation expert Michael Quinn Patton shared the information that some 60% of global corporation 3M's after-tax profits are invested in R&D, and some 80% of their current product line did not exist as little as five years ago. (Patton pers. comm.) Government is in no way insulated from this rate of change experienced by business. However, its response is seldom to invest in R&D. Experience from Ukuvuka suggests that partnership in a "non-owned" body can be a useful way of testing innovative approaches and piloting ways of applying resources more effectively and efficiently, and injecting additional short-term human capital to build long term capacity within government.

Evaluation of the Trialogue model

In preparation for the international symposium on ecosystem governance, the organisers prepared a paper to use as a starting-point for discussions (see Chapter 1). The hypothesis presented suggested, in essence, that the success of ecosystem governance depends on a "Trialogue" – three key actor-cluster and their processes. In the diagram of the model, each of the three parties is represented by one angle of an equilateral triangle. The three players identified in the model are government, society and science.

Underlying the model there appears to be a rather more fundamental unstated assumption that the outcome of successful or good ecosystem governance is some greater likelihood of sustainability. In practical terms, the assumption seems to be that any robust conceptual framework that emerges from the symposium could give guidelines on ways to direct our limited ecosystem management resources more effectively and efficiently. The framework would also indicate the proactive steps that role-players, especially science, might take to influence and shape the governance space.

The part of the model with which this chapter deals is the interface between Government and Society. The lens of the Ukuvuka experience suggests that the model, and the way that it is diagrammatically represented, is not yet sufficiently accurate or textured to be useful in practical applications.

The Relative Power of the Players

In the model diagram, representing the three players as points of an equilateral triangle implies equivalence of influence or power. The experience from the Ukuvuka Campaign suggests that, in the governance of fire in the Cape Peninsula, the "science process" pole had significantly less influence than either the "government process" or the "society process" poles.

The "science process" had clearly established that fire was an ever-present reality in the indigenous vegetation of the region. The natural vegetation required appropriately-spaced fire events for its sustainable existence. This meant that a major metropolitan area with a population of some three million that surrounded some 24,000 ha of natural vegetation, would always bear some risk in the interface zone between the urban and natural areas. What "science" had also clearly established was that the dramatically increased severity of the fire events was due, in large part, to the dramatically increased amount of flammable fuel (Van Wilgen et al. 1992). This additional fuel load came from the proliferation of invading alien trees that so easily spread across the permeable urban edge and which overwhelmed the adjacent natural areas.

What the science process had not been successful in doing was to apply its influence to embed the insights of the huge societal costs of invasive alien plants into the minds, hearts and value systems of government and society, and hence into the enforcement of appropriate legislation. Thus the critical view, as seen through the lens of the Ukuvuka experience, suggests that the power of the three role-players in the "Trialogue" is not equal.

Indeed, the fact that crucial aspects of ecosystem governance such as systems thinking, and understanding of the goods and services provided by healthy ecosystems are not in the mainstream of political processes and societal values, is a further indicator that the science pole does not exercise the same influence as do government and society in Turton et al.'s ecosystem governance model. Myers (2002) challenges "Where are the leaders? Answer: they are the politicians, the policy experts, the economic planners, the business chiefs, the media, the churches, and what is known as civic society writ large. Few of them show signs of knowing how to supply the crucial lead. They are generally acquainted with the nature and size of the challenge, i.e. the threat of depletion to biodiversity, yet they offer scarcely a 'cheep'. In the recent national election campaigns in the United States and Britain no candidate gave the merest mention of anything remotely resembling an extinction spasm." So in spite of science creating awareness about threats posed to ecosystems by extinctions, science did not exert sufficient influence on political leaders to persuade them to catalyse any activism or behaviour change.

The single pole in the model described as "society process" is very richly textured and this is also highlighted in Myers' (2002) description of civic society. Within this texture are multiple resources to deploy to improve sustainability through our ecosystem governance. The model would be enhanced if it elaborated on this texture. Indeed, given the diversity of the role-players that seem to be grouped under the Society heading, there is an argument to be built, by those analysts looking at the exercise of power, to question if society should not be represented by more than one pole or angle. Society could be conceived of as including at least *two* groups, i.e. business and civil society. These two groups have significantly different value systems and means of exercising influence. It is arguable that, as science has so much less influence than government or society, it should not be considered as a separate role-player at all but, more appropriately, as a resource to be deployed by any or all of the players to enhance the quality of the decisions taken.

The practical challenge that is the consequence of these observations is to ask how the science process can strengthen its influence and so can better make its important contribution to governance (the 'push' aspect). In addition, to ask how the other players can better engage the science process (the 'pull' aspect) to shape their values and inform their decisions in the governance space.

With the diversity of role-players, and with their competing and conflicting interests, negotiation is an essential process at all the interfaces, and especially in the central governance space.

O'Riordan and Stoll-Kleeman (2002) comment that "*Political theorists like Gary Stoker regard governance as the emergence of new styles of governing, in which the boundaries between international responsibility, national responsibility and local action become blurred, where the 'market' is both privately and socially regulated into a varied mix of incentives and prescriptions, and where formal government no longer controls all the levers of power and authority. Thus the 'steering' of public affairs becomes more and more a matter of joint responsibility amongst a variety of actors.*" This is particularly relevant in the situation where external costs are imposed on one set of role-players through the negligence of others, e.g. as in the case of land infested with alien plants creating a fire hazard for neighbours.

Conclusion

Effective ecosystem governance requires new or enhanced models, tools and methodologies, as well as policies, laws and procedures. A crucial next step is to implement these changes. Such implementation frequently requires behaviour change, which can be triggered in a number of ways.

The Ukuvuka experience showed how crisis could be channelled to unlock political support and to make resources available, and ultimately lead to behaviour change. Further, Ukuvuka showed how Government, in partnership with society, i.e. both business and civil society, can catalyse internal change in long-term institutions.

In retrospect, it is possible to see how Ukuvuka undertook "research and development" by testing ways in which the governance of fire could be achieved differently. Generally, individuals and institutions do not change that easily, as there is an inexorable momentum to continue along their route of "business as usual". However, by demonstrating what change might look like from a low-risk safe space, it was easier to build support for incorporating new ways of approaching fire management in the long-term institutions.

In order to embed the changes in the long-term institutions effectively, Ukuvuka learned that it was necessary to have at least one champion within that institution. Ukuvuka had experience of constructive ideas developed from within the Ukuvuka team and demonstrated on the ground, which never took a life of their own in the long-term institutions, due to the lack of such a champion. This highlighted the importance of the existence of that internal champion.

Ukuvuka's experience suggests that there is merit in the use of the device of a **short-term** body. The short-term nature means that the body is less likely to be perceived as a threat to long-term institutions. It is less likely to generate competition for "turf". It is easier to create an enabling environment for innovation when disbelief is suspended. One factor that makes is easier to suspend disbelief is knowing that there is a clear end-point in sight. It is also easier for parties, who would not commit to working together in the long term, to collaborate in the short term.

The nature of the short-term body is that it should be small, adequately resourced with finances and skills, and should operate within governance structures composed of all the stakeholders. It should be able to respond flexibly and nimbly. Finally, it should be staffed with people with sound interpersonal skills and negotiation competence.

Ukuvuka, which existed only from 2000 to 2004, demonstrated some viable means for relatively swift behaviour change relating to the management of fire in the Cape Peninsula. There have been changes in the behaviour of both institutions and individuals that have been facilitated by the Ukuvuka campaign. The unanswered issue is how sustainable these changes will prove to be. This experience suggests that, in the challenge to implement effective ecosystem governance, the tool of using a short-term, issue-focused partnership between (at the least) government and business, could be of value.

Finally, in evaluating the Trialogue model, it is appropriate for the model to show a space in which the processes of governance take place. The space is appropriately determined by points representing actor-clusters. This chapter questions the way in which those actor-clusters are defined and also questions the appropriate number of actor-clusters. The governance space may well be shaped by three actor-clusters but, in certain circumstances, there may be more or less than three. The experience of practical applications strongly suggests that science does not hold equivalent power to government or society. Business, with its economic power, and civil society, through the power of religious institutions or trade unions, for example, exercise a far more direct influence as actor-clusters. The Ukuvuka experience also suggested that science is not an appropriate separate actor-cluster. There is no doubt that "science" has an important contribution to make to ecosystem governance. However, it is suggested that it would be more appropriate for that role to be in support of *all* the role-players involved in ecosystem governance.

References

Gooch G (2005) From dialogue to trialogue sustainable governance and civil society. Proceedings: International symposium on ecosystem governance. Kwa Maritane, South Africa, 10-13 October

Graham J, Amos B, Plumptre T (2003) Governance principles for protected areas in the 21st century. Prepared for Fifth World Parks Congress

Myers N (2002) Biodiversity and biodepletion: the need for a paradigm shift. In: O'Riordan T, Stoll-Kleeman S (eds) Biodiversity, sustainability and human communities: protecting beyond the protected. Cambridge University Press

O'Riordan T, Stoll-Kleeman S (2002) Deliberative democracy and participatory biodiversity. In: O'Riordan T, Stoll-Kleeman S (eds) Biodiversity, sustainability and human communities: protecting beyond the protected. Cambridge University Press

Kotter JP (1995) Why transformation efforts fail. Harvard Business Review 73:59-67

Mintzberg H (1996) Managing government governing management. Harvard Business Review, May-June:75-83

Roux D, Murray K, Van Wyk E (2005) Principles to create learning environments for good ecosystem governance. Proceedings: International symposium on ecosystem governance. Kwa Maritane, South Africa, 10-13 October

Turton AR, Hattingh J, Claassen M, Roux DR, Ashton PJ (2005) Towards a model for ecosystem governance: an integrated water resource management example. Paper presented at the International symposium on ecosystem governance, 10-13 October 2005, Kwa Maritane Bush Lodge, Pilanesberg, South Africa

Ukuvuka Campaign (2004) Final report of the Santam Cape Argus Ukuvuka Campaign, Cape Town: Ukavuka

Ury W, Brett J, Goldberg S (1988) Getting disputes resolved: designing systems to cut the costs of conflict. Jossey-Bass Publishers, San Francisco

Van Wilgen BW, Richardson DM, Kruger FJ, Van Hensbergen HJ (eds) (1992) Fire in South African mountain fynbos: ecosystem, community and species response at Swartboskloof. Ecological Studies 93. Springer-Verlag, Berlin-Heidelberg-New York, 325 pp

World Commission on Dams Report Overview (2000) Dams and development: a new framework for decision-making

Acknowledgement

The work for this chapter was supported by funding from the Water Research Commission as part of a wider research project entitled: "*Investigation of a potential institutional form to pilot innovations in water management by CMAs: Lessons from Ukuvuka an innovative partnership for natural resource management.*" The WRC's assistance is gratefully acknowledged.

Websites

www.ukuvuka.org.za
www.dwaf.gov.za/wfw/
www.dams.org/report/wcd_overview

Personal Communications

Mr Piet Smith, Chief Fire Officer, City of Cape Town. September 2005
Mr David Waddilove, Environmental Lawyer and Consultant to Ukuvuka. July 2005
Prof Michael Quinn Patton. May 2002

The Government-and-Society Challenge in a Fledgling Democracy – Ecosystem Governance in South Africa, with a Particular Focus on the Management of the Phongolo Floodplains and Reservoir

Barbara Schreiner

Abstract

South African political reform coincided with a global trend in the reform of water resource management. Consequently, South Africa is an excellent case study in the role of water in addressing historic socio-economic inequity while at the same time trying to legislate for sustainable development. Undoing the legacy of more than a century of institutionalised discrimination is a complex undertaking made even more problematic by the fact that water scarcity constraints limit the future South African economic potential. Getting it right is therefore a political necessity and also a necessary condition for the future wellbeing of a country that plays a vital role as the engine of growth on the African continent. This chapter charts the evolution of a fledgling democracy in which water resources management is seen as a vehicle for social and environmental justice as well as a tool for deepening the democratic experience. The case of the Phongolopoort Dam is used to illustrate these complexities.

Keywords: ecosystem governance, government-society, fledgling democracy, mature democracy, Integrated Water Resource Management (IWRM), participatory governance, Agenda 21

Introduction

Turton et al. (2005) have proposed a Trialogue Model through which to examine the processes at work in the governance of ecosystems. One side of the Trialogue is the government-society interface. This chapter examines that interface with particular reference to experiences derived from the management of the Phongolopoort Dam. This chapter examines some of the challenges of the government-society interface in a fledgling democracy, where power relations between and within stakeholder groups are extremely unequal. Challenges also exist in recognising and responding to the different needs of stakeholder groups while still protecting the aquatic environment. The chapter refers to the crucial role of government in ensuring that the voices of the disempowered are heard.

From Apartheid to Democracy

In 1994 South Africans of all races went to the polls together for the first time in the history of the country. Prior to this, black South Africans had been denied the vote, and many had been forced to become "citizens" of the so-called "independent homelands." Prior to 1994, a system of "apartheid" had been maintained by the white minority, under which, black South Africans were essentially second-class citizens: they were denied the vote, had restrictions placed on them in terms of mobility, job opportunities and education, and were systematically deprived of their access to natural resources, predominantly through forced resettlement and removal from the land. The 1913 Land Act is perhaps the most notorious of the several pieces of legislation that deprived black South Africans of land, and, by 1994, it resulted in 87% of the land being in the hands of the small white minority.

Although water legislation prior to 1994, i.e. the Water Act of 1956, was not in itself overtly racist, access to water was directly linked to access to land. Thus, through the promulgation of legislation that prevented blacks from owning land, black South Africans were simultaneously deprived of access to water for productive purposes. As a result, by 1994, access to water was firmly in the hands of the white minority, with white commercial agriculture the predominant water user. Similar disparities in economic status existed, with large numbers of black South Africans living in poverty in rural and peri-urban areas. In 1995, 61% of Africans were poor, while only 1% of whites were considered poor (May 2000).

The ANC government, voted into power in 1994, set out its intention to redress the racial inequities established under apartheid. This approach is enshrined in South Africa's Constitution, The Constitution of the Republic of South Africa, 1996 (Act 108 of 1996) which states in its Preamble that "*We, the people of South Africa, recognise the injustices of our past.*" The Constitution further, in the Bill of Rights, sets out the right of all in South Africa to equality, dignity, and life. It also gives all in South Africa the right to an environment that is "not harmful to their health or well-being" and the right to have the environment protected for the benefit of present and future generations".

While all South Africans have had the right to vote since 1994, democracy in South Africa is still in its early stages. It is a mere 11 years since the first democratic elections in South Africa. In the global context, mature democracies have been in place for many generations – for most South Africans of voting age, democracy has been in existence for less than half their lives. At the same time, while political democracy has been in place in South Africa for 11 years, economic democracy has still not been achieved. In 2004 29.8% of South African households lived below the poverty line (Hirsch 2005), unemployment in 2004 was 26%, using a definition that includes people who sought work in the last seven days, and 40% when including people who did not seek work in the last seven days (Hirsch 2005) and many South Africans, particularly in the rural areas, are still functionally illiterate. Access to the economy and to economic power is still largely in the hands of the white minority and a small black elite.

Thus, one of the challenges facing the Department of Water Affairs and Forestry (DWAF) is to implement participatory and consultative integrated water resources management in a context shaped by a young democracy and racially-biased access to economic power.

Post-1994 Institutional Context for Water Resource Management

Building on the mandate of the Constitution, the then Minister of Water Affairs and Forestry, Prof Kader Asmal, led a consultative process to develop the White Paper on a National Water Policy for South Africa (adopted by Cabinet in 1997), and the National Water Act, 1998 (Act 36 of 1998). The White Paper builds on the basis of the Constitution, stating that *"...it is, therefore, the duty of the government to make sure that...there is sufficient water to maintain the ecological integrity of our water resources and that...sustainable justifiable economic and social development are promoted."*

Both the policy and legislation are very clear on the need to redress the racial and gender imbalances in access to water that were developed under apartheid, as well as to ensure protection of aquatic ecosystems to ensure sustainable use of water resources. The purpose of the National Water Act includes ensuring:

- That the nation's water resources are protected, conserved, managed and controlled in ways which take into account, amongst other factors:

- Promoting equitable access to water;
- Redressing the results of past racial and gender discrimination; and
- Protecting aquatic and associated ecosystems and their biological diversity.

The Act makes provision for the declaration of a Reserve which must set aside, in any significant water resource, sufficient water of a suitable quality to provide for basic human needs and to maintain the functioning of aquatic ecosystems. The Reserve is given the status of a right, and this water cannot be used for other purposes. Only once the Reserve has been determined and protected may water be made available for economic and other purposes.

The Act also makes provision for a classification system that will enable the protection of water resources to different degrees, allowing for certain water resources to receive a very high degree of protection, while others may be more highly used.

The legislative context within which the Reserve may be determined or the Classification system implemented, is one of openness and transparency. Thus the Act requires, with regard to the development of the classification system and the determination of the Reserve, that the relevant information must be published in the Government Gazette, and that the Minister must "*consider what further steps, if any, are appropriate to bring the contents of the notice to the attention of interested persons, and take those steps which the Minister considers appropriate*."

The Act further sets out the institutional arrangements for management of water resources in South Africa, such as the establishment of Catchment Management Agencies (CMAs) that may manage water within particular water management areas. The governing boards of these Agencies are to include representatives of key stakeholder groupings within the water management area, including local government, provincial government, and representatives of water users such as agriculture, mining, power generation and communities. Nineteen CMAs are in the process of being established. The first, the Nkomati Catchment Management Agency has been formally established, and the Governing Board held its first meeting in September 2005. Four other CMAs are due for establishment in 2005/6. However, until such time as the CMAs exist, the Department of Water Affairs and Forestry must fulfill the role of the CMAs, including consultation with stakeholders on various water management issues.

Rationale for Protection of Aquatic Ecosystems and Participatory Governance Approaches

The rationale for the protection of aquatic ecosystems, as set out in the White Paper, is that water resources must be managed within their capacity to recover from human and other impacts, so that use can be maintained in the long term. The White Paper also refers to the importance of the "silent services" provided by water resources, i.e. the removal and purification of wastes, supply of food and plants, retention and storage of water, opportunities for tourism and recreation, and the conservation of biodiversity. The rationale is perhaps set out most succinctly in the National Water Resource Strategy (DWAF 2004b) which states that:

The quantity, quality and reliability of water required to maintain the ecological functions on which humans depend shall be reserved so that the human use of water does not individually or cumulatively compromise the long-term sustainability of aquatic and associated ecosystems.

The White Paper also sets out several reasons for the consultative approach that has been adopted by the Department. The first reason is captured in the White Paper which states that "*The challenge of water management in the 21st century...is to treat the development, use and protection of our water as a common endeavour in the interests of all, in the spirit of a new patriotism, rather than as a series of conflicts between different groups*" (DWAF 1997).

The second reason is that the White Paper grew out of an international discourse that included the UN Conference on the Human Environment (Stockholm 1972), the World Conference on Water and the Environment (Dublin 1992), the UNCED World Summit – Agenda 21 (Rio de Janiero 1992), the Global Water Partnership meeting (Stockholm 1996) and the First World Water Forum (Marrakesh 1997) all of which stressed participatory water management and devolution of functions to the lowest appropriate level. At the same time, the stress laid on consultation must be seen against a background of the pre-1994 South African regime in which people, particularly poor black people, were excluded from decisions that affected their lives. The approach to governance post-1994 was deliberately designed to establish participatory democracy in which people could participate in the decisions that affected their lives.

Thus, the White Paper directs that "*the process of balancing social and economic benefits as well as of determining environmental objectives should involve those affected, or their representatives, in weighing up the options on an informed basis*" (DWAF 1997).

This is in line with the approach to stakeholder involvement in natural resource management that was given prominence through Agenda 21, and which has increasingly been accepted as best governance practice in many parts of the world. The process of ensuring stakeholder involvement in water resources management decision-making processes includes the establishment of institutions, be they catchment management forums or Water User Associations. The intention is that, through these institutions, stakeholders will have a structured way of giving their input into decision-making. The Phongolo case study, below, will identify some of the benefits and challenges associated with this approach.

Public consultation and involvement does not, however, remove the responsibility of custodianship from the Department. The White Paper makes it clear that the public trust places the responsibility on government to ensure that environmental interests are taken care of and the resource protected effectively. This responsibility cannot be abrogated through consultation with the public.

Obviously, the capacity of the state will impact on how well policy is implemented. The implementation of the above policy is examined through the consideration later in this chapter of the management of the Phongolapoort dam and the interaction with stakeholders in connection with the management of this dam.

Kooiman and Warner (2000) suggest that a successful governance-mix brings together the governing capacity of the public sector, the private sector and civil society. However, the balance between parties, the various roles of the parties and the extent to which they should be consulted or be able to influence decisions must be considered seriously. As will be seen in the Phongolapoort dam example, civil society is not a homogenous group, and therefore one of the challenges facing government is how to balance the different interests and power-relations between and within stakeholder groups.

The Government-Society Interface with regard to Management of Flood Releases from the Phongolapoort Dam

An example of some of the challenges inherent in the government-society interface, as they pertain to ecological governance, can be seen in the processes around management of the Phongolapoort Dam and the implications thereof for the floodplain lying downstream of the dam.

The Phongolo floodplain stretches for about 150 km from the Lebombo Mountains in northern KwaZulu Natal, South Africa, to the Mozambican coast near Maputo. The floodplain is characterised by a number of pans that, prior to the building of the Phongolapoort Dam, relied on the frequent flooding of the Phongolo River to fill them. Since the building of the Dam in 1973 these pans have been filled by flood-releases from the dam. However, the timing of these flood releases is different and their intensity is, obviously, lower than the natural events would have been. Aligning the timing and intensity of flood releases to the requirements of people around and below the dam, and to the needs of the floodplain ecosystem, has been a major challenge for the Department of Water Affairs and Forestry (Mwaka et al. 2005).

Mwaka et al. (2005) state further that the floodplain has been settled for over three hundred years, with the communities living around the plain practicing subsistence agriculture that was highly dependent on flood irrigation, using the receding flood waters. This system allowed for a sustainable balance between human needs and the needs of the ecosystem. The communities are also dependent on the floodplain ecosystem for fresh water, fish, building materials and fuel wood. Thus the local communities traditionally had an integrated and sustainable relationship with the floodplain ecosystem.

Since the construction of the dam, however, the picture has changed. The construction of the dam, through limiting the flood waters on which the pans depend, had a negative impact on the livelihoods of those communities dependent on the flooding of the pans. On the other hand, the control of the floodwaters allowed other economic activities, such as the growing of cotton, to take place in areas that would previously have been inaccessible due to flooding. The growing of cotton has brought economic benefits to some sections of the community below the dam. These cotton farmers have requested that flood releases be scheduled in such as way as to minimise their impact on the growing season. Scheduling floods to meet the requirements of the cotton farmers has, however, raised concerns that the pans are not filling at the appropriate time to provide for the needs of either the ecosystem or the subsistence-agriculture communities.

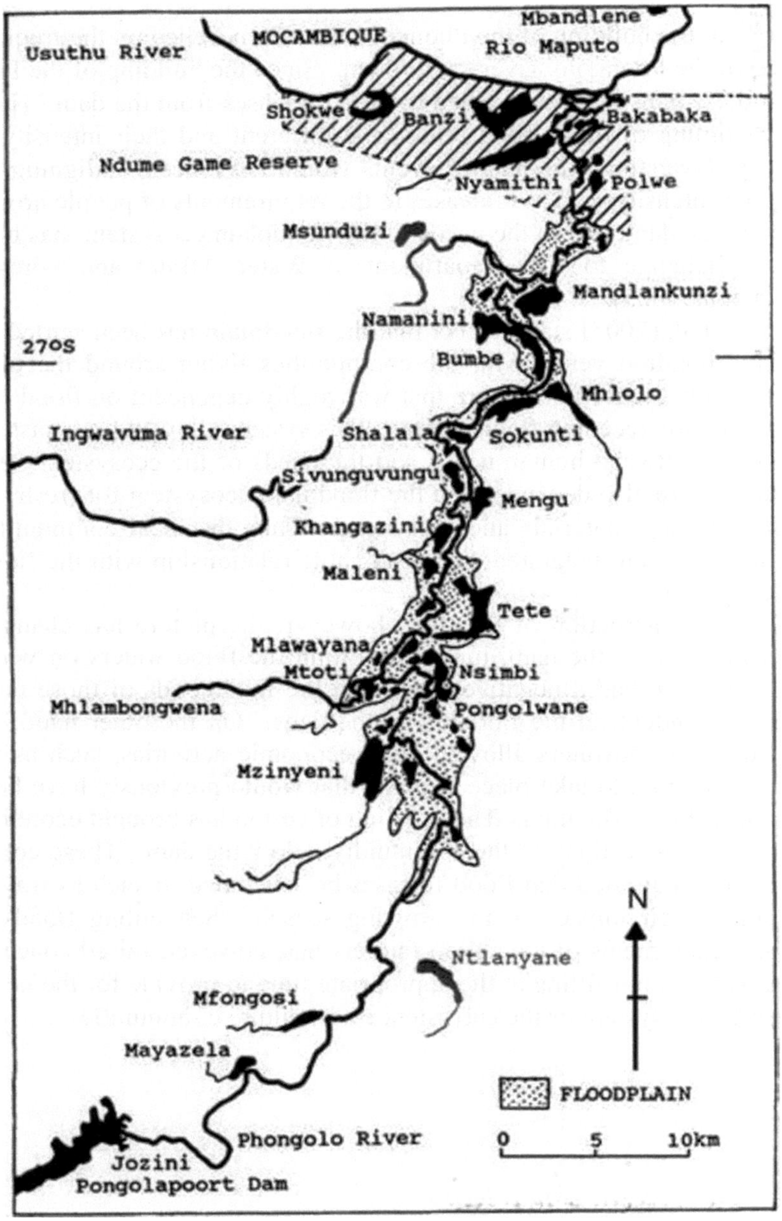

Fig. 12.1. Map of the Phongola floodplain in South Africa

Research by Heeg and Breen (1982) showed that the natural flood regime was characterised by several relatively short-duration flood peaks, not a constant flow or prolonged flood-flows. Flood intensity impacts on the level of inundation of the 90-odd pans in the floodplain. Fluctuating water levels result in certain areas being dry at some time, and inundated at others. The floodplain ecosystem is thus "a complex pattern of interacting flood-dependent components" (Heeg and Breen 1982).

Heeg and Breen also identified three major environmental impacts, deriving from the Phongolapoort Dam, that would be experienced by the people living on the floodplain, namely a reduction in water availability (the dam would capture all low flows), a deterioration in water quality, and a decrease in available land arising from a combination of population growth and the development of an irrigation scheme that would require the relocation of some of the subsistence farmers. They also recommended that the only reasonable development option that would ensure the continued existence of the floodplain as a viable ecosystem was a combination of agricultural development "in conjunction with a policy of floodplain preservation."

Consultation with Stakeholders

The history of stakeholder consultation in the Pongolapoort Dam area dates back to the mid-1980s. At that time, Clive Poultney and Zeph Nyathi were doing research into the relationship between livelihoods on the floodplain and the flood regime. After the construction of the dam in the early 1970s, the Department of Water Affairs and Forestry made decisions about the timing and intensity of flood releases *without* consulting with the floodplain communities. Flood releases were inappropriately-timed and often too long for the needs of the floodplain communities. Poultney and Nyathi, then involved in the Mboza Village Project in the area, worked with the community to establish a number of water committees on the floodplain. A range of categories of water-users were represented on the water committees, such as stock owners and traditional healers, and there was a specific category for women to be represented on the committees. Each water committee elected five members to sit on a central committee which was mandated to negotiate with the Department on the flood releases. These committees were active from 1986 to 1996 and were extremely well organised. However, the late 1990s saw the rise of an emerging group of cotton farmers on the floodplain, and of power struggles within the community and the water committees. At the same time, funding to NGOs dried up, as international donors redirected their funds to the new government. As a result, funds were not available to support the water committees and consultation over the large floodplain area became less effective (Poultney pers. comm. 2005).

The issue of the flood releases remained contested into the 2000s, and was compounded by the differing needs of various stakeholders around the impoundment itself. In order to deal with the needs of the latter groupings, in July 2001 the Department commissioned the compilation of a Sustainable Utilisation Plan (SUP) for Phongolapoort Dam, in terms of Section 113 of the National Water Act. The intention was to ensure sustainable, equitable and beneficial access to and use of the stored water.

The development of the SUP involved a wide stakeholder consultation process that included participation by local authorities, resource managers, and community facilitators, neighbouring landowners and politicians. The SUP covers issues such as management of the resource, the involvement of the private sector, community involvement and beneficiation, and monitoring and evaluation. While the SUP was not specifically aimed at managing flood releases from the dam, sustainable utilisation of the dam, as well as the regional ecosystem, obviously has implications for flood-release management.

During the process of consultation it was identified by stakeholders above the dam that a Water User Association (WUA) would be the most appropriate institution to implement the SUP plan. The WUA could be delegated the responsibility for managing access, utilisation and/or development of the resource, according to the SUP. This decision resulted in the establishment of the uPhongolo Dam Water User Association. The key categories of members of the WUA include upstream water-users, downstream water-users, women, disabled people, Mozambique and Swaziland, DWAF, local government, individual water-users, workers and trade unions, tourism associations and others.

The objectives of the uPhongolo Dam Water User Association are:

1. Conservation and protection of the water resource, as well as the state land surrounding the manmade lake;
2. Creation of opportunities for equitable access to the stored water;
3. Provision of socio-economic benefits to the region and its people;
4. Redress of past imbalances and inequalities;
5. Community involvement and accrual of benefits to the community; and
6. The monitoring and control of the use of the water surface and State Land surrounding the manmade lake.

During the process that led to the establishment of the uPhongolo Dam Water User Association, downstream users on the floodplain realised that their interests were different to those of the users above the dam and that they needed to form their own organisation, which they chose to call *Imfunda Yophongolo Dam WUA* (Phongola Dam Floodplain WUA). This Water User Association is still in the process of establishment.

Challenges of Involving All Stakeholders

The involvement of all of the stakeholders in the development of the SUP and the subsequent development of the two water user associations presented various challenges to the Department. The first challenge was that of identifying who the relevant stakeholders were.

The measures put in place to identify the stakeholders within the area of operation included:

- Notices in local and regional newspapers to advertise public meetings;
- Invitations to known organisations and individuals, with a request for them to refer the invitation to other possible stakeholders;
- Announcements on Maputaland Community Radio;
- Identification of further stakeholders at public meetings;
- Identification of stakeholders from surrounding communities by the Community Liaison Officer of Ezemvelo KZN Wildlife; and
- Using Local Government and Traditional Authorities to bring in other stakeholders.

Measures were put in place to bring in and involve marginalised groups. Upstream and downstream stakeholders, including small-scale farmers, traditional leaders, civil society representatives, farm workers and women were involved, so as to ensure appropriate liaison and communication in the absence of a "higher level" forum for integration of interests and addressing impacts.

The capacity of these stakeholders to participate in consultative processes varied considerably, and the interests of the stakeholders were very different. While some stakeholders had access to telephones, email, and private transport, other stakeholders were semi-literate and lived in communities without easy access to modern communications systems. Their access to transport was also limited by lack of private cars and lack of money to pay for taxis. Language was also a challenge, with English and isiZulu being the two dominant languages spoken in the area.

In order to deal with the issue of language, experienced facilitators were used and interpreters were present at all meetings. Where relevant, and where stakeholders requested it, meetings were held in isiZulu. Support was provided to ensure accessible venues and travel co-ordination, particularly via the Traditional Authorities. Venues were alternated to the north and south of the dam, to ensure accessibility for all stakeholders.

The Department made considerable attempts to ensure appropriate and timeous information was provided to stakeholders throughout the process, to ensure opportunity for involvement and to build capacity to participate. Practical process principles, including providing sufficient opportunity for involvement and comment, exchange of views, ongoing feedback, respect of cultural diversity and, importantly, special efforts for involving historically-disadvantaged communities were applied. Importantly, there was visible commitment from the Department of Water Affairs and Forestry throughout the process.

Despite the fact that the floodplain Water Committees were no longer active, it would appear that the experience of having been organised and having participated in structures focused on management of water had left its mark on these communities. Erstwhile members of these committees were particularly active in the processes around the establishment of the WUA and the decision-making process on flood releases (Mwaka pers. comm. 2005).

Ultimately, a plan was devised that covered issues pertaining to management of the resource, private sector involvement, community involvement and benefits and monitoring and evaluation. Page 8 of the SUP, under the Operational Guidelines for the Key Performance Area Natural Resource Management states that:

Water Releases: Input to the timeous releases of water will be provided by the Pongolapoort Dam management team through an open and consultative process with stakeholders around the Pongolapoort Dam. To enhance the natural resource management objectives of the dam it is essential that the management institution participate in negotiations regarding releases to ensure that the releases are environmentally sound and sustainable, based on ecological, social and economic considerations.

The Management Committee for the uPhongola Dam WUA was established only in August 2005, and thus has not yet impacted on the decision-making processes associated with releases. They will, however, participate in this process in future with a coordinated voice and formal mandate. The Imfunda Yophongolo Dam WUA, as mentioned, is still to be established. It remains to be seen what the impact of these two structures will be on the flood releases from the dam, and therefore on ecosystem governance in the floodplain. Currently, below the dam the demands of the cotton farmers have outweighed those of the subsistence farmers, who are more directly dependent on the functioning of the ecosystem. It is possible that the establishment of the Imfunda Yophongolo Dam WUA will see the capturing of the structure by the more vocal and organised cotton farmers, further marginalising the subsistence farmers. Two elements weigh against this.

The first is that the experience of the members of the previous water committees ensures that they are able to play an active and informed role in water-management decisions. The experience of having been involved in participatory structures previously is extremely valuable. In a fledgling democracy, where such experiences are limited, this is even more important.

The second is that the government has a particular role to play in terms of ensuring that the most marginalised elements of society are able to make their voices heard. Achieving this remains a challenge on the Phongolo floodplain.

In the case of the Phongolapoort Dam flood releases, it is particularly important that the voices of the most marginalised, the subsistence farmers, are heard, as they are the people most dependent on the healthy functioning of the ecosystem. Their voices are, therefore, in many ways a vicarious voice of the ecosystem. Nonetheless, the third point of the Trialogue, science – represented here by the Reserve determination for the floodplain – is important in order to ensure that the government-society interface is tempered by scientific evidence.

Conclusion

The policy and legislative context for the government-society interface in ecosystem governance is well developed in South Africa, from the Constitution through the National Water Act and the various strategies and guidelines developed to support implementation of the Act. The National Water Act, in particular, requires consultation with affected parties around water resources management issues. It also provides the legislative base for the establishment of institutions, such as Water User Associations, that enable users to come together in a formal structure for managing common water resources.

Implementation of the Act, however, brings with it significant challenges, particularly in the context of a society in which huge disparities in wealth, education and access to power still exist. The Phongolo case study shows how complex the process of interaction between government and civil society is, and how issues of access to power, gender differences, levels of poverty, impact on the ability of people and communities to interact with the process. In the case of a fledgling democracy, where significant inequities still exist in access to power and access to wealth it is necessary for government to be proactive in ensuring the full participation, in particular, of the most marginalised and vulnerable members of society. A number of elements must be deliberately and consciously put in place to oil the interface, particularly education and awareness around people's rights, water availability and allocation, protection of aquatic ecosystems, and the new water structures and institutions. An informed public is a powerful tool for ensuring that the needs of both the people and the environment are met. However, "society" is not a homogenous group and so a proactive stance is necessary from government to ensure that the voices of the most marginalised, including poor rural women, are heard and that the government-society interface does not become a channel for the powerful to capture access to the resource.

It is, however, also clear that a solid basis of scientific knowledge can contribute to achieving a rational and appropriate solution. The research that has been done on the Phongolo floodplain over the last fews decades has contributed towards the development of a body of knowledge on the functioning of the ecosystem that has supported the recent process.

The Phongolo experience is contributing to the development of more effective processes for strengthening the government-society interface in the interests of pro-poor integrated water resources management in South Africa.

References

Constitution of the uPhongolo Dam Water User Association, April 2005

Dept of Water Affairs and Forestry (DWAF) (1997) White Paper on a National Water Policy for South Africa. Government Printer, Pretoria

DWAF (1998) National Water (South Africa) Act. Department of Water Affairs and Forestry, Government Printer, Pretoria

DWAF (2004a) Sustainable utilisation plan for Pongolapoort dam. Department Water Affairs and Forestry, Unpublished draft

DWAF (2004b) National water resource strategy. Department Water Affairs and Forestry, Government Gazette, South Africa

Establishment proposal for the uPhongolo dam Water User Association. 20 October 2004. Letter from Establishment Committee Chairperson to the Regional Director: KwaZulu Natal

Heeg J, Breen CM (1982) Man and the Pongolo floodplain. Report of the Committee for Inland Water Ecosystems. National Programme for Environmental Sciences. South African National Scientific Programmes Report 56, June 1982. Council for Scientific and Industrial Research, Pretoria, South Africa

Hirsch A (2005) Season of hope: economic reform under Mandela and Mbeki. UKZN Press, South Africa

Kooiman J, Warner J (2000) Three is a crowd? reintroducing civil society in water governance. Abstract prepared for The Fourth International Research Symposium on public management, 10-11 April 2000. Erasmus University Rotterdam, The Netherlands. Available online at http://www.eur.nl/fsw/-english/news/irspm/kooiman.htm

May J (2000) (ed) Poverty and inequality in South Africa: meeting the challenge. David Philip, South Africa

Mwaka B, Mbuli C, Cai R, Zwane C (2005) Integrating Pongolapoort dam flood control operations with the livelihood of communities on the Pongola floodplain. Proceedings: 12th South African national hydrological symposium. SANCHIAS, Johannesburg, South Africa

Mwaka B (2005) Conversation about role and influence of members in the establishment and decision making of the Water User Association of the Phongola dam

Poultney C (2005) Conversation about research into the relationship between livelihoods on the Phongola flood plain and the flood regime

The Constitution of the Republic of South Africa, 1996 (Act 108 of 1996). Government Printer, Pretoria

Turton A, Hattingh H, Claassen M, Roux D (2005) Towards a model for ecosystem governance: an integrated water resources management example. Paper prepared for the International symposium on ecosystem governance, 10-13 October 2005, Kwa Maritane Bush Lodge, Pilanesberg, South Africa

Turton AR, Meissner R (Unpublished mimeograph) The hydro-social contract and its manifestation in society. African Water Issues Research Unit, Pretoria University, South Africa

Part 3

Cross-cutting Governance Requirements

Principles Enabling Learning Environments for Good Ecosystem Governance

Dirk J Roux, Kevin Murray and Ernita van Wyk

Abstract

Complexity and, by implication, change and uncertainty, are inherent features of ecosystems. In managing ecosystems, or linked social-ecological systems, decisions are often based on insufficient or uncertain data and information. Appropriate and sufficient knowledge, which essentially resides in people, is a critical factor for making informed decisions under such circumstances. Informed action is a function of what we know, and our knowledge is a product of what and how we have learned.

Because of the central importance of learning, this chapter proposes that the development of an appropriate learning capability should not be left to chance but should be the result of deliberate intervention to establish the conditions for an organisation to operate in a learning mode. Focusing on organisations or agencies with mandates for ecosystem governance, the chapter sets out to identify the principles that will enable the creation of such learning environments.

Firstly, the key concepts of knowledge, learning and ecosystem governance are defined. Secondly, the chapter identifies main issues of concern regarding (a) the type of knowledge that needs to be created for good ecosystem governance; (b) the desirable processes for learning or knowledge creation; and (c) the characteristics of good learners. Thirdly, these main issues (10 in all) form the basis for formulating nine principles intended to enable the setting up of appropriate learning environments for ecosystem governance.

The proposed principles are summarised as follows. Good ecosystem governance requires positively persistent and adaptive people with a culture of empathy for other knowledge systems and levels. Their knowledge must be trans-disciplinary, moulded by a common future focus, acquired by patiently engaging their prior knowledge and learning by doing, in an environment of social knowledge sharing.

It is concluded that good learning practice would promote the achievement of some of the principles underlying the practice of good ecosystem governance, notably effective stakeholder engagement, adaptability and transformability. The proposed learning principles could be used as a framework to assess the learning proficiency of ecosystem management agencies and to develop learning strategies for such agencies.

Keywords: learning environments, ecosystems governance, knowledge, social-ecological systems, complex adaptive systems

Introduction

A seminal paper by Lubchenco (1998) and related discussions (Kinzig et al. 2000; Holling 2001; Gunderson and Holling 2002) highlight the need for new types of relationship between society, science and government in order to achieve more resilient social-ecological systems. Turton et al. (2006) argue that the achievement of good ecosystem governance is a function of the Trialogue, or knowledge interfacing between society, science and government. Indeed, human dependencies on ecosystems have become so intense that people have to coordinate their decisions and actions to avoid destructive patterns of resource use (Maarleveld and Dangbégnon 1999). The way in which different domains of interest such as society, science and government generate and share knowledge to serve the requirements of ecosystem governance is of key importance. We simply cannot afford to create knowledge about, and policies for, these systems in an isolated or reactive manner.

By implication then, in order to collectively and strategically make sense of, and respond to, the feedbacks of our complex world, we need to create and foster large-scale learning systems with appropriate links between component learning systems and communities of practice (Wenger 2005; Snyder and Wenger 2004; Schein 2004). Within these systems the value and appropriateness of knowledge will be negotiated as part of a social process (Wenger 2005) and will be influenced by the dynamic nature of the value systems of participants. We believe that such learning systems will generate knowledge that is more appropriate to governing ecosystems in a world of change and intensified interdependency. However, the ways in which we proceed, and the structures and cultures we develop to support these ideas, will profoundly influence the type and style of governance that emerges.

The dilemma is that most organisations responsible for natural resource management are typically designed for effectively applying what they know rather than for learning. The issue is how we selectively un-learn or forget outdated habits, retain and nurture fitting habits, reflect on the most appropriate future direction and strategically acquire new knowledge – all at the same time. Only "learning organisations" can achieve this. These are organisations where people continually expand their capacity to create the results they truly desire, where new and expansive patterns of thinking are nurtured, where collective aspiration is set free, and where people are continually learning to learn together (Senge et al. 1994). Developing this kind of learning capability should not be left to chance. It should be the result of deliberate intervention to establish the conditions for an organisation to operate in a learning mode (Goh and Richards 1997). The assumption is that there is an organisational archetype that typifies an environment conducive to learning.

The aim of this chapter is to identify the principles that will enable the creation of such learning environments, in the context of managing complex social-ecological systems. The intent is to lay the groundwork for the development of a strategy for enabling effective learning, and not to develop the strategy as such. The primary focus of this chapter is on enabling learning in those agencies with the official mandate of governing ecosystems, which fall largely within the domain of government. However, part of learning enablement is to participate in relevant learning systems outside one's primary or internal learning system. In this sense, the identified principles provide a framework for achieving collaborative learning or knowledge interfacing (Roux et al. 2006b) between government and domains such as science and civil society.

Key Concepts

Knowledge

Knowledge is a mix of fluid experiences, values, contextual information and intuition that provides a framework for evaluating and incorporating new experiences and information (adapted from Davenport and Prusak 1997). *Knowledge* should not be confused with *information*, which is loosely defined as interpreted data or data endowed with some relevance and purpose (Drucker 2001). For the purpose of this chapter, we will consider that which is stored in books, libraries and on computer systems as information, while knowledge, as defined above, resides in people's heads.

Learning

Learning, for current purposes, is regarded as synonymous with "knowledge creation." It is defined as gaining knowledge, comprehension, or mastery of a specific field through experience and study. Miller and Morris (1999) suggest that knowledge comes about through the integration of (a) information derived from data; (b) theory that puts the information in the proper context; and (c) experience of how things work in the real world. This process of integration is also referred to as learning.

Ecosystem Governance

Ecosystems as Social-Ecological Systems

An ecosystem consists of plants, animals and micro-organisms that form biological communities that interact with each other and with the physical and chemical environment, with adjacent ecosystems and with the water cycle and the atmosphere (Odum 1989). Ecosystems play a fundamental role in supporting life on Earth at all hierarchical scales, e.g. ecosystems produce renewable resources and ecological services.

Although the above definition of an ecosystem is essentially ecological, most aspects of the structure, functioning and changes in Earth's ecosystems cannot be understood without accounting for the strong influence of humanity; at least in the present era. Social and ecological systems are co-evolving at both local and global levels – where human behaviour shapes nature and nature influences the development and behaviour of human society (Folke 2003). Our capacity to govern ecosystems effectively requires an understanding, not only of the ecological systems concerned, but also of the integrated social-ecological system (Folke and Hahn 2004). For this chapter we view humans as part of ecosystems, hence the use of the term social-ecological system.

Ecosystems as Complex Adaptive Systems

Linked social-ecological systems are complex systems, containing the inherent features of change and uncertainty (Gunderson and Holling 2002). The properties of these systems (e.g. vegetation cover and capital wealth) emerge from the interaction among the various components of the entire system and which themselves feed back to influence the subsequent development of those interactions (Levin 1998). Consequently, the behaviour of coupled social-ecological systems (a) is emergent rather than predetermined; (b) can rarely, if ever, be reversed to some exact prior state; and

(c) has a changing path that is often unpredictable (Stafford Smith and Reynolds 2002). As these systems change over time, any combination of external forces and intrinsic changes may result in slower or faster rates of change in each subsystem. When the rates of change in each subsystem become unsynchronised with each other, for example when ecological systems change faster than the ability of society to learn and respond, or vice versa, the overall social-ecological system becomes dysfunctional (Robbins et al. 2002).

Ecosystem Governance

Cooperative governance refers to a means for achieving direction, control, and coordination of individuals and organisations that have varying levels of autonomy, to advance the interests or objectives to which they jointly contribute. It also involves formal organisational structures, personal relationships, and judgement by those individuals working in the complex space of administering public programmes. It is inherently political and involves bargaining, negotiation, and compromise (Imperial 2004).

Good ecosystem governance can be defined as cooperative governance of functional social-ecological systems underpinned by the principles of (a) transparency (honesty and openness); (b) effective stakeholder engagement (including cooperation between all relevant government departments); (c) accountability of all role-players; (d) consistency (of application of procedures); and (e) adaptability and transformability. In essence, good governance is participative, adaptive and accountable management with integrity.

What Knowledge should be Created

The knowledge that needs to be created for good ecosystem governance has particular characteristics dictated by the complex nature of social-ecological systems. There must be depth and breadth of knowledge and, within this breadth, knowledge must be complementary.

Knowledge Breadth and Depth

The "absorptive capacity", or ability of an individual or social grouping to recognise the value of external information or knowledge, and assimilate and exploit it for benefit, is largely a function of the level of prior related knowledge (Cohen and Levinthal 1990; Zahra and George 2002). This prior knowledge may include basic skills, a shared language or knowledge of the most recent scientific or technical developments in a given field – each of which permit effective communication and transfer of knowledge across boundaries. Absorptive capacity determines the degree to which learning can take place based on exposure to external information or knowledge. It follows that diversity, or breadth of knowledge, increases the chance of external information relating to what is already known (Cohen and Levinthal 1990), and hence knowledge breadth enables an organisation to learn from, and respond to, a wider variety of situations effectively.

In the case of integrated water resources management, knowledge breadth should cover multiple levels. At the lowest level, the disciplines required include aquatic ecology, environmental chemistry, hydrology, geomorphology, sociology, economics, complexity science, etc. At higher levels the disciplines include politics, law, and even ethics. This multilevel view of interactions between hierarchical disciplines is proposed by Max-Neef (2005) and is termed "transdisciplinarity".

Knowledge breadth comprises not only substantive, technical knowledge; it also includes awareness of where useful complementary expertise resides, both inside and outside the organisation. Such meta-knowledge – knowledge of who knows what, who can help with what problem, or who can exploit new information – is essentially embedded in personal relationships and networks, both within and external to the organisation. Especially in times of rapid technological change, it may be wise to have a broad and even geographically dispersed knowledge base, as opposed to a more homogeneous and centralised knowledge base.

There may be limitations to the breadth within which any individual can still effectively create knowledge. Miller (1956) described a phenomenon in human mental processing where, given a list of numbers to remember, sounds (phonemes) to distinguish from one another, or a set of unrelated facts to recall, there is a critical change in performance at around seven items, and hence the magical number 7 ± 2. With up to this number, people can readily handle a variety of tasks. With more than seven, their processing ability is seriously hampered. One way of getting around this memory constraint is to cluster items; for example, clustering a group of 70 items into seven groups of 10 each. The seven groups can then be processed more easily.

The degree to which an organisation or institution is knowledgeable is not only a function of knowledge breadth, but also of the depth of knowledge in critical areas. Knowledge competency is based on understanding rather than memory, and understanding relates to a deeper process of connecting and organising knowledge around important concepts. Learning with understanding implies (a) a deep foundation of factual knowledge; (b) understanding of facts and ideas within a contextual framework; and (c) organisation of knowledge in ways that facilitate retrieval and application (NRC 2000).

Complementary Knowledge

The knowledge possessed by individuals and organisations involved in ecosystem governance is complementary only to the degree that their knowledge entities are different and, at the same time, related. In a governance system, the challenge is to balance knowledge diversity (to increase potential for acquiring new knowledge) with knowledge overlap (to enable effective communication and coordination). This can be achieved through knowledge interfacing (Roux et al. 2006b), which is an active process where diverse knowledge entities negotiate meaning and generate shared understanding.

Knowledge interfacing, or learning together, creates a healthy tension between knowledge diversity and common understanding (Wenger 2005). Knowledge interfacing also provides a link between an individual's knowledge and the capability of the larger governance system. This social dimension of learning is of particular relevance in a diverse institution, where it is important to align resources and actions behind a common high-level purpose.

A common obstacle to effective knowledge interfacing is the natural tendency for learning to take place within groups of knowledge homogeneity. Since existing knowledge influences both the ability to put new knowledge into memory, and to recall and use such knowledge, learning is most efficient when the object being studied relates to what is already known (Cohen and Levinthal 1990). Quite naturally, people adopt learning patterns that favour subject matter that relates to previously accumulated knowledge. The downside is that the more a person's knowledge is shaped by learning within a defined field, the harder it becomes to associate with knowledge that emerges from other fields. Miller and Morris (1999) refer to this tendency as "trained incapacity" – the conundrum in which the more we know about something the harder it is to learn to do it differently. If all the knowledge professionals in an organisation share the same specialised language, they will be effective in communicating with one another, but they may not be able to tap into diverse external knowledge sources.

How Knowledge should be Created

The processes of learning, or knowledge creation, are many and varied. However, again largely driven by the complexity inherent in ecosystem governance, some are more desirable than others in achieving a state in which the kind of knowledge discussed above can be achieved.

Learning as a Social Process

Although learning or cognitive development happens at the individual level, much of our learning takes place through social interaction. An individual experiencing an idea, behaviour or attitude in a social setting then internalises this as part of his or her frame of reference. This internalisation does not involve merely the transferring of reality between people. Rather, the learner actively processes the experience and then integrates this experience into his or her frame of reference – thereby changing or developing that frame of reference. The frame of reference is not merely absorbed or transferred verbatim between individuals, but actively constructed by the learner as the result of the social experience (Doolittle 1997).

In this section we highlight three dimensions of social learning, namely learning from more knowledgeable individuals, learning in communities of practice, and learning in teams.

Learning from More Knowledgeable Individuals

This form of social learning starts from childhood, where a child's interaction with more advanced peers or adults is key to their development of "higher mental functions such as language, logic, problem-solving skills, moral reasoning and memory schemas" (Doolittle 1997). Russian psychologist Lev Vygotsky (1896-1934) postulated that an individual's immediate potential for growth is bound by a lower and upper limit. The lower limit is that which he or she can accomplish independently. The upper limit is that which he or she can accomplish with the help of a more knowledgeable person such as a peer, mentor, tutor or teacher. Vygotsky referred to this region of immediate potential for learning, between the upper and lower limits, as the zone of proximal development. With practice, experience and increased knowledge, the learner's zone will move in the direction of the upper limit. Eventually the learner will be able to accomplish independently what he or she previously could only achieve with much assistance (Doolittle 1997). At this stage, a more skilled mentor or teacher is necessary to maintain the learner's immediate potential for learning.

Following from this, the potential for learning and knowledge development in a social system depends on:

- The quality of the social interaction between the less and the more knowledgeable individuals, and
- The upper limit of knowledgeable individuals available in a particular social system.

Learning in Teams

A team can be defined as "*a group of people with a high degree of interdependence, geared towards the achievement of a goal or the completion of a task*" (Parker 1994). Learning in teams comprises those activities through which team members acquire, share, and combine knowledge into a collective product through joint experiences (Argote 1999).

Doolittle (1997) presents five basic elements that are paramount to a team enjoying a cooperative learning experience, or working together in a social setting to solve problems:

- Positive interdependence: This element is achieved when each team member understands and values the need for group cooperation in the attainment of his or her personal goals, the other team members' goals and the goals of the entire team. Positive interdependence may take the form of goal interdependence, task interdependence, resource interdependence, role/function interdependence, or reward interdependence;
- Face-to-face interaction: This element involves individual group members encouraging and facilitating the efforts of other group members to complete tasks in order to achieve team goals. This interaction is characterised by the members 'providing each other with assistance, exchanging needed resources, and offering feedback';
- Individual accountability: This element is about holding each member accountable for mastering certain material. Since team members are likely to vary in their level of expertness, it is important that the team determine the level of mastery appropriate for each member; i.e. to maximise the learning potential of each member. Individual accountability prevents situations in which select group members do most of the work. Although several team members may be engaged in a collaborative effort on a particular task, the idea is that each member should grow and develop towards being able to do independently what he or she can only do in collaboration at present;

- Small group and inter-personal skills: This basic element is necessary to perform competently in association with others. The acquisition of social skills is essential for each member to mediate and navigate his or her interactions with others in the context of cooperative learning; and
- Group self-evaluation: The purpose of this element is to clarify and improve the productiveness of all group members as contributors to achieving the group's goals. Group self-evaluation relates to meta-cognition, a process whereby people evaluate their own levels of mastery and understanding. In a team context, each member's learning, or cognitive development, is not only the responsibility of that member but also of the other members of the team. Since the level of mastery and learning proficiency of each individual is dynamic, constant self-evaluation and monitoring is necessary for groups to continue to be successful, and for individuals to be constantly challenged according to their learning potential (within their zones of proximal development).

The composition of teams is important, although research has yielded contradictory findings. Williams and O'Reilly (1998) concluded that cultural heterogeneity in groups is more likely to have negative effects on group performance. Although diverse teams were more creative, they had more difficulty coordinating and implementing ideas than homogenous groups. Bland and Ruffin (1992) suggest that disciplinary diversity tends to promote productivity if balanced with common values and goals. Watson et al. (1993) found that, although the performance of culturally homogenous groups was initially superior to that of heterogeneous groups, the performance of the latter group improved at a faster rate than that of homogenous groups. At the conclusion of a study lasting several months, the heterogeneous groups were more effective at identifying problems and generating solutions. In general, productivity and cooperative learning proficiency increases with the age of a team (Bland and Ruffin 1992).

Learning in Communities of Practice

In complex and diverse knowledge domains – such as that of ecosystem governance – it is imperative for practitioners to learn how to participate in (i.e. not to take ownership of) broader learning systems such as disciplines, sectors, regions, alliances and consortia. Practitioners have to interact with other practitioners because the knowledge base in any given field is too complex for one individual to cover. The challenge is to learn how to be one part of a larger social process where the benefit is mutual. "*You know your part, but because your knowledge is engaged, in practice you know more than you know*" (Wenger 2005). This is where the notion of communities of practice plays a critical role.

Communities of practice are "*groups of people who share a passion for something that they know how to do, and who interact regularly in order to learn how to do it better*" (Wenger 2004). Examples are a peer group interested in the conservation of an endangered species, or a community of photographers. Since membership is based on participation rather than official status, these communities are not bound by organisational affiliations. They constitute webs of inclusive relationships in which people feel valued and inspired to share their knowledge.

Wenger (2004) lists three fundamental characteristics of communities of practice:

- Domain: A community of practice is about something. Its identity is defined not only by a task, as it would be for a team, but by an "area" of knowledge that needs to be explored and developed. The area of focus, or domain, provides a common purpose that gives the community its identity and defines the key issues that members need to address;
- Community: A community involves people who interact and who develop relationships that enable them to address problems and share knowledge. Community refers to the group of people for whom the domain is relevant, and the quality of the relationships among members. These relationships enable collective learning; and
- Practice: A community of practice is not merely a community of interest. It brings together practitioners who are involved in doing something. Over time, they accumulate practical knowledge in their domain. They also have a special connection with each other because they share actual experiences. They understand each other's stories and insights. This allows them to learn from each other and build on each other's expertise. Practice refers to the body of knowledge, methods, tools, stories, cases and documents which members share and develop together.

The self-organising nature of such communities is key to them realising their full potential. Since they are largely based on voluntary participation, "management" has a different set of rules to that of conventional organisational management. To be successful, communities must generate enough excitement, relevance and value to attract and engage members.

With knowledge as the currency of concern, a participant's status within a community of practice is determined by (a) the degree to which he/she is knowledgeable; and (b) a willingness to share. Wenger et al. (2002) identify three main levels of participation: A "core group" actively participates and largely determines the agenda and activities. An "active group" also participates actively, but with less intensity. A "peripheral group" (usually the majority) prefers to observe rather than contribute. There may also be an outside group of people who are not formally members, though they have a temporary interest. The core group (or community leaders), which in many instances comes down to an individual, will determine the degree of productivity and success.

Basing Learning on Prior Knowledge

People construct new knowledge and understanding based on what they already know and believe (NRC 2000). This assumption is well illustrated by the children's book *Fish is Fish* (Lionni 1970), which describes a fish keen to learn about the world outside its pool. It befriends a tadpole who grows into a frog, eventually leaving the water to explore terrestrial life. The frog returns a few days later and tells the fish about the birds, cows and people that he has seen. In the mind of the fish, all these creatures are fish-like: birds are fish with wings; cows are fish with udders and horns; and people walk upright on their tailfins. This tale demonstrates both the creative opportunity and dangers inherent in the fact that people construct new knowledge based on their current knowledge and worldview.

The worldview, or frame of reference of an individual, is moulded by a combination of factors such as: relatively generic experiences attributable to developmental stages through which learners may have passed; the learning of individuals based on their personal and idiosyncratic experiences; and knowledge acquired through fulfilling specific social roles, which may be specific to a race, class, gender, culture or ethnic affiliation; the era in which an individual grew up (age-related), environmental or geographic influences; or a particular disciplinary/theoretical background. The frame of reference of an individual provides the context within which information is processed, whereby two individuals with divergent worldviews may draw radically different conclusions from the same information. This context-specific nature of knowledge often constrains an individual or organisation's ability to understand, replicate or exploit external knowledge (Zahra and George 2002).

Newstrom (1983), cited by Becker (2005), refers to the lack of recognising previous learning as the "clean slate fallacy", whereby it is assumed that learners represent a clean slate or empty vessel waiting to acquire new knowledge without the interference of previous knowledge. In fact, if the initial understanding of individuals is not engaged, they may fail to grasp new concepts and information that are presented; or they may memorise these but not fully understand them. Since knowledge is dynamic, the changing conceptions of learners should be monitored as learning proceeds. As an example, when teaching children about the fact that planet earth is round, one can start with their conceptualisation, which may relate more to a pancake than a sphere – i.e. it is round but at the same time flat so that people can walk on it (NRC 2000).

Due to the cumulative nature of learning, learning efficiency is greatest when the object of learning is related to what is already known. The implication for organisations is that, by having already acquired some knowledge in a specific area, an organisation may more readily accumulate what additional knowledge becomes available and hence more fully exploit a particular wave of technological development. It also follows that it will be difficult for a group to pursue a completely new direction without at least some prior knowledge of the new field (e.g. in times of radical technological change).

Time Demands of Learning and Reflection

Modern society tends to value action and not reflection, to the extent that we are often too busy to learn. Learning is often not seen as a valid activity at work and, in most organisations, staff admit that they learn in their "free" time. As a result, group learning in the workplace, which builds collective understanding, is neglected.

In an era of information overload and a multitude of responsibilities for every individual, it is difficult for learners to organise knowledge meaningfully. Often there is only superficial coverage of facts before moving on to the next topic; there is insufficient time to develop important, organising ideas. The cognitive activity of information integration requires time and the time required for learning complex subject matter can be substantial. For example, world-class chess masters require between 50,000 and 100,000 hours of practice to reach that level of mastery. They rely on a knowledge base containing some 50,000 familiar patterns to guide their selection of moves (Chase and Simon 1973; Simon and Chase 1973, cited in NRC 2000).

Expertness comes with time. People who have developed expertise in particular areas are, by definition, able to think effectively about problems in those areas. It is not merely general abilities, such as memory or intelligence, nor the use of general strategies that differentiate experts from novices. Instead, "*experts have acquired extensive knowledge that affects what they notice and how they organise, represent, and interpret information in their environment. This, in turn, affects their abilities to remember, reason and solve problems*" (NRC 2000). They don't simply attempt to do the same things more efficiently; they attempt to do things better.

Contextualised and Abstract Knowledge Creation

All learning integrates thinking and doing (Senge et al. 2005). Practical experience plays a critical role in learning proficiency and in creating knowledge competence. Experience relates to "doing". For example, "doing science" involves activities such as testing hypotheses through experimentation, observation and engaging a peer-review process. One can only become an experienced scientist if you engage these activities over a long period.

However, knowledge that is over-contextualised can inhibit knowledge transfer. For example, some Brazilian street children could perform mathematical calculations when making sales in the street, but were unable to solve similar problems presented in a school context (Garraher 1986, cited in NRC 2000).

On the other hand, abstract representations can facilitate transfer. Exposure to multiple contexts allows learners to identify core concepts and represent them abstractly (transcending the specificity of particular contexts and examples). The higher the levels of abstraction, the more flexible knowledge transfer can be.

The above acknowledges the need for theory and that this should be clearly related to practical experiences. This duality can be achieved through action research, where research is achieved through intervening in what is being researched. In fact, unless we intervene, we will not learn what some of the essential dynamics of the system really are. Through action research the development of a theoretical discourse enables new ways of conceptualising, understanding and doing, while reflective practice becomes a source of theorising (Roux et al. 2006a).

A Culture for Knowledge Creation

Knowledge creation has as much to do with the nature of the people as with the nature of the knowledge itself. Culture is seen as the shared or commonly held beliefs, assumptions, values and norms, and behaviours that govern organisations (Schein 2004).

Adaptability and Unlearning

When it comes to balancing robustness and flexibility in the management of linked social-ecological systems, "adaptive management" is considered to be the most appropriate management approach; for example, it will outperform optimisation approaches that seek stable targets (Holling 2001). Whilst there is general consensus that adaptive management can facilitate functional feedback loops between social and ecological systems (e.g. Olsson et al. 2004), and thereby enhance the resilience of these linked systems, the institution of such adaptive processes has proven problematic (Walters 1997). A key reason for implementation failure seems to be the tendency to superimpose an adaptive approach on a non-adaptive (e.g. command-and-control or bureaucratic) decision-making environment (Rogers et al. 2000). Institutional capacity to adapt to and shape change is an important prerequisite for effecting adaptive management of ecosystems (Berkes et al. 2003).

Adaptability is the capacity of actors in a system to influence the resilience of that system, e.g. whether they can intentionally avoid crossing into an undesirable system regime (Walker et al. 2004). Giampietro (2004) provides a complementary definition of adaptability, namely the ability to adjust our own identity to retain fitness in the face of changing goals and changing constraints. Fitness means the ability to maintain congruence among (a) a set of goals; (b) the set of processes required to achieve them; and (c) constraints imposed by boundary conditions.

In order to be adaptable, organisations need employees with ability to adapt and handle change. This translates to an ability to change our frames of reference, and relates to the emerging concept of "unlearning". Becker (2005) quotes Delahaye (2000) as follows: "*it is interesting to reflect that the concept of unlearning only recently has become a phenomenon worthy of consideration in adult and organisational learning. Centuries ago, an individual's knowledge would last a lifetime, indeed knowledge would be passed down generations and still be highly useful. This has changed during this century until, as we pass into the new millennium, knowledge becomes rapidly obsolete – hence the need to consider the unlearning process.*"

Becker (2005) defines unlearning as the process by which individuals and organisations acknowledge and release prior learning (including assumptions and frames of reference) in order to accommodate new information and behaviours. The process of unlearning should not be seen as the total "removal" of past learning, but rather the relinquishing of past learning as a result of new learning that reveals new ways of choosing a response to a particular situation. As shown in other parts of this chapter, an individual's prior knowledge is valuable to any learning process. Prior knowledge (which determines absorptive capacity) increases the ability of an individual to acquire related knowledge. However, where the new information or knowledge is unrelated to prior knowledge, or creates dissonance, prior knowledge may also inhibit unlearning – a phenomenon known as proactive inhibition. Proactive inhibition protects knowledge already acquired by an individual by disregarding conflicting information (Lyndon 1989).

It is generally assumed that the more expert people are in a particular field, the more difficult it becomes to relinquish that experience and knowledge, which was acquired and reinforced over a long period of time and which may have become deeply-entrenched beliefs. More-recently-acquired knowledge may be easier to unlearn, because the individual may have less of an emotional attachment to this knowledge (Becker 2005). From the above it seems that, apart from the possible cognitive hurdles to unlearning, there is also an attitudinal dimension, namely whether an individual is willing to unlearn.

A further constraint to adaptive learning is being in a state of fear and anxiety. Under such circumstances, individuals and organisations are likely to revert to "habitual ways of thinking." In a state of fear our dominant learning mode is reactive, through which we mostly reinforce pre-established knowledge and frames of reference. We need to perceive our working environments as safe to "see" alternative futures and to learn along new and dynamic trajectories towards such futures (Senge et al. 2005).

Organisational Memory

The collective ability of individuals to learn and unlearn eventually translates into an organisational ability. A related concept is that of organisational memory, defined as "*the means to retain and transmit information from past to future members*" (Stein 1995), or "how organisations encode, store, and retrieve the lessons of history despite the turnover of personnel and the passage of time" (Levitt and March 1988). These definitions highlight the need for some degree of institutional stability (maintaining memory through people staying in the system long enough to develop a sense of shared responsibility and ownership of outcomes) as a balance for adaptability (relinquishing outdated knowledge and adjusting to more appropriate responses).

Secondly, the above definitions recognise that organisational memory is not only based on explicit information, but also the tacit knowledge that resides in people. The dynamic interplay between tacit knowledge and explicit information is key to the processes of learning and knowledge sharing. This can be achieved by moving from a view of knowledge as a "thing" that can be transferred, to viewing knowledge as a "process of relating" that involves negotiation of meaning among partners. This latter view requires establishment of a trusting relationship between all parties involved to enable them to share and compare the various interpretations of their messages when sending and receiving.

The social process of sharing and passing knowledge on is of critical importance to ensure that individual knowledge does not remain with the individual, but becomes available to all concerned. We identify three complementary forms of knowledge sharing, each contributing to organisational memory in a different way:

- Explicit sharing: Knowledge, which is tacitly possessed, is codified or made explicit in the form of numbers, words and equations. In this explicit form, "knowledge" can relatively easily become part of organisational artefacts (tools, documents, procedures, etc.), and can be transferred and "owned" by organisations. Such explicit knowledge is stripped from its human context and strictly speaking does not represent true knowledge, but rather information. Once codified, control is lost over the subsequent use of that information. Identical information will always provoke different meanings for different individuals. Secondly, the loss of content, and particularly context, means that such explicit forms are only ever a partial representation of what we know (Snowden 2002). Furthermore, the more intricate the knowledge, the less effectively it can be codified (James and Minnis 2004).
- Tacit sharing: Tacit knowledge can often be transmitted more thoroughly through narratives. This allows the exchange of context associated with the content of interest that may be difficult, or even impossible, to write down. This context may be conveyed in subtle forms such as emotion or body language. The construction of stories, anecdotes or metaphor is a particularly useful technique for storing and sharing knowledge. The creation of space for strategic conversations and storytelling is increasingly recognised as good knowledge sharing practice.
- Subtle sharing: Knowledge that cannot be written down or verbalised exists only within and between the minds of people. This deeply tacit knowledge firstly accounts for that which we don't know that we know – until we need to know it. To the observer, resultant actions can appear to be based on "gut feel" or intuition. In the collective, the value of this knowledge manifests, for example, in the efficiency with which established teams, such as sport teams, operate. There is an unwritten and unspoken code between members that allows such teams to achieve results that are not possible for beginner teams. This does not come easily, and is often the result of a long process based on a combination of peer- and self-selection, getting to know each other intimately and developing trust.

Motivation for Learning

Motivation determines the amount of time that people are willing to devote to learning. Challenges, however, must be at the proper level of difficulty to be, and to remain, motivating. Tasks that are too easy become boring, and tasks that are too difficult frustrate. The tendency of learners to persist in the face of difficulty is affected by whether they are motivated for performance or for learning. People motivated for learning like new challenges; those that are performance motivated are more concerned about their status and making mistakes.

Meta-cognition refers to people's ability to monitor their current levels of mastery and understanding. In essence, they take control of their learning when they are able to recognise when they understand and when they need more information. A meta-cognitive approach to learning includes a focus on sense making, self-assessment and reflection on what worked and what needs improving. These practices have been shown to increase the degree to which learners can apply their learning to new settings and events (NRC 2000).

Appreciative Inquiry

Inquiry within a group context will form one of the major processes within the broader learning process. The way in which we inquire into diverse perspectives is a cultural phenomenon and will strongly influence the 'flavour' of the learning process, as well as determine what knowledge is generated.

Appreciative inquiry (Cooperrider et al. 2003, Senge et al. 1999) involves a cooperative and evolutionary search for the best in people, their organisations and the world around them. The approach strives to enable creativity, knowledge and spirit in the community of learning and focus on a common future vision. This vision should intimately determine what knowledge is created.

Senge et al. (1999) note that appreciative inquiry also requires bringing empathy into day-to-day practice. Empathy means developing an understanding of another so intimate that the feelings, thoughts and motives of a person are readily comprehended. To be empathetic means to "try on" different perspectives and assumptions, temporarily suspending your own in the process, so that you can inquire into the reasons why people hold them.

Approaches such as appreciative inquiry (Cooperrider et al. 2003) and systems-based and consensus-seeking inquiry (Churchman 1971) have been promoted as a means for making multiple perspectives and truths explicit. Making multiple perspectives explicit will be an important process if science, society and government are to co-evolve and reconcile these diverse perspectives in order to produce appropriate knowledge to serve a common purpose.

This philosophy challenges the notion of 'expert'. Although experts and novices differ in the nature and depth of their knowledge competence (NRC 2000), an appreciative approach promotes humility and respect in the inquiry process, regardless of level or domain of knowledge. Rather, multiple perspectives are respected and sourced with appreciation and humility to produce a robust and dynamic joint perspective over time.

Guiding Principles

The complexity and uncertainties typically associated with ecosystem governance place high demands on ecosystem managers. In order to provide basic guidance on what learning is needed, how learning can be achieved most effectively, and who are the most appropriate learners, a set of principles has been distilled from the above discourse. The principles define the Dos and often, by implication, the Don'ts, of what is required. A principle is regarded here as providing guidance on what should be strived for, typically acknowledging underlying assumptions.

The principles to be strived for in any endeavour define, in effect, the "rules of the game." They differ from the strategy, which is the "game plan" itself. There can be a number of strategies, e.g. suited to different organisations, that apply the principles and which may all achieve the desired end result.

Common Future Focus

Assumptions: (a) Organisational learning not relating to, or overlapping with, a common purpose can be fragmented or non-complementary, inefficient and potentially ineffective; and (b) Individuals, and different organisations, typically have fundamentally different knowledge (or reference) systems that intimately determine their interpretation of new information and of the best way forward.

Principle: "Common future focus" strives to ensure that all stakeholders have agreed to a well-defined vision of the future, and that this actively determines what is learned.

Strategies: Rely heavily on catchment-visioning processes and active engagement with all stakeholders. We use this to define a desired future state of the resource and to focus all stakeholders on striving for it.

Prior Knowledge Engagement

Assumptions: The capacity for knowledge acquisition ("absorptive capacity", or learning efficiency) increases in proportion to the level of accumulated prior knowledge. This existing knowledge is intimately related to an individual learner's knowledge (or reference) system (e.g. age, culture, language, formal education, etc.).

Principle: "Prior knowledge engagement" strives to ensure that knowledge creation acknowledges, monitors, is tailored to, interacts with, adapts to, and builds on what learners already know.

Strategies: Encourage open discussion on individual perceptions of the vision of the future and how to achieve it. Ensure individuals have, or are given, an adequate basis on which to base further learning. Tailor, as far as possible, a variety of knowledge creation methods to individual and organisational preferences.

Transdisciplinarity

Assumptions: (a) The integrated nature of decision-making associated with ecosystem governance (across a hierarchy of technical, ecological, political, economic, social and institutional disciplines) requires a significant breadth of knowledge at all levels; and (b) The complex, and often technical, nature of decision-making associated with ecosystem management sometimes requires a significant depth of knowledge.

Principle: "Transdisciplinarity" strives to ensure that the knowledge that is created (in individuals and in organisations) is appropriate and adequate at each level in a hierarchy of disciplines (e.g. from technical through political to ethical) and, where necessary, adequately detailed (i.e. based on deep understanding).

Strategies: Identify potential champions and develop them into "Jacks of all trades and masters of some", but first basing this on a mastery of one discipline. Ensure the necessary detailed technical and scientific expertise to support decision-making exists within the organisation or can be effectively outsourced. Encourage learning outside "comfort zones" while ensuring that appropriate psychological support is provided to those who may feel threatened (e.g. by using role models).

Learning by Doing

Assumptions: (a) Effective learning involves practical experience. Learners must “get their feet wet and hands dirty” to fully “know”; and (b) Reflective practice and the development of theory can be mutually reinforcing.

Principle: “Learning by doing” strives to ensure that knowledge is also created through hands-on practical experience.

Strategy: Extensive fieldwork should be mandatory.

Social Knowledge Sharing

Assumptions: (a) Learning is a social process (i.e. it is rooted in social interactions and heavily influenced by individual and organisational knowledge systems); (b) Knowledge can be difficult to extract – it often exists only in people's heads; (c) Knowledge sharing can occur both explicitly and implicitly (subconsciously); (d) Mutual understanding across different knowledge systems arises out of debate and negotiation; and (e) Potential for cognitive development or learning is supported through access to and quality interaction with more knowledgeable individuals.

Principle: “Social knowledge sharing” strives to facilitate freely interactive sharing, inquiry, debate and negotiation of new information between learners and those with the knowledge that should be shared.

Strategies: Create many opportunities for working with experts, particularly under field conditions. Encourage secondments. Encourage discussion.

Patience

Assumption: (a) Ecosystems and their governance are complex. The time required to become sufficiently knowledgeable to be able to recognise emerging patterns and to solve complex problems can be substantial; and (b) Transdisciplinary learning, ensuring its effectiveness, gaining practical experience and allowing time for reflection, are all time consuming.

Principle: “Patience” strives to ensure that adequate time is allowed for absorbing appropriate knowledge and that, during the learning process, the expectations of all concerned are realistic.

Strategies: Ensure learning takes place over an adequate time period that encourages the creation of sound knowledge while simultaneously ensuring that expectations of practitioners who are learning are aligned with their current capabilities.

Positive Persistence

Assumption: The demands of good ecosystem governance require a positive attitude towards problem solving and decision-making.

Principle: "Positive persistence" strives to ensure that learners have determined yet positive and enthusiastic attitudes to acquiring new knowledge.

Strategy: Ensure that people with this character trait are recruited into critical ecosystem governance positions.

Empathy

Assumptions: Knowledge is widespread and diverse. Showing respect for the knowledge of others promotes respect in return and hence more effective engagement and learning.

Principle: "Empathy" strives to stimulate co-creation of new knowledge by nurturing a culture in learners to interact and share with other knowledge systems (cultural, political, scientific, etc.) and knowledge levels (novice, expert) with understanding and an ethic of mutual respect for knowledge (in all its forms), wisdom, culture, language, abilities, concerns and inputs of all stakeholders.

Strategies: Expose learners directly to fundamentally different knowledge systems. Ensure they understand that there can, and should, be vastly different points of view. Apply the widely-used slogan "connecting through culture."

Adaptability

Assumptions: Ecosystems present "moving targets", to which management approaches must be able to adapt. This is partly due to their complexity.

Principle: "Adaptability" strives to ensure that learning processes adapt to the learners and that the learners themselves learn to adapt to and handle change.

Strategies: Adapt the learning process to the reference systems of individual learners. Encourage "unlearning" of inappropriate knowledge. Provide a safe supportive environment for both learning and unlearning.

Synthesis

The proposed principles can be summarised as follows:

Good ecosystem governance requires **positively persistent** and **adaptive** people with a culture of **empathy** for other knowledge systems and levels. Their knowledge must be **transdisciplinary**, moulded by a **common future focus**, acquired by **patiently engaging their prior knowledge** and **learning by doing** in an environment of **social knowledge sharing**.

These proposed principles are intended not only to facilitate a more appropriate learning process for good ecosystem governance, but also to promote the achievement of some of the proposed principles underlying the practice of good ecosystem governance, such as effective stakeholder engagement and adaptability and transformability. These interrelationships are indicated in Figure 13.1.

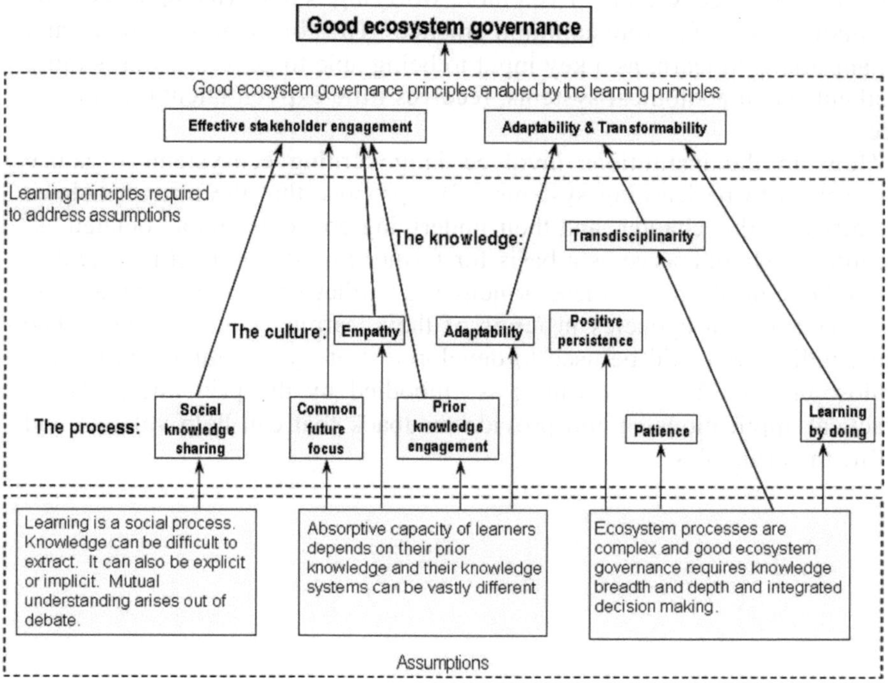

Fig. 13.1. *Illustration of the relationship between some assumptions, the proposed learning principles and the good ecosystem governance principles enabled by them*

Conclusion

The dynamic interplay between societal behaviour and ecosystem responses results in unforeseen interdependencies and ever-emerging patterns. Consequently, continuous adaptation of our management practices is required to ensure sustainable management. Yet, our capacity to do better is completely dependent on our capacity to learn, and our present understanding of issues, and ability to respond to such issues are functions of our past learning. Furthermore, bridging organisational, disciplinary, cultural and functional boundaries will be central to promoting learning within the social system of the water institution.

Breen et al. (2003) identified the promotion of individual and organisational learning in a cooperative context for river management as a strategic implementation priority. However, there seems to be a general lack of empirical studies in the area of learning as related to the management or governance of ecosystems. Even more worrying is that what appears to be common sense is far from common practice. The ability of individuals and organisations to learn, as a key input to being able to govern towards more resilient social-ecological systems, receives little explicit attention in practice.

How do the institutions involved in governing ecosystems organise themselves to be learning systems? We propose that the nine principles identified in this chapter, and their underlying concepts and associated assumptions, should serve as a basis for further investigation. In particular, it would be useful to use these principles as indicators for auditing ecosystem management agencies in terms of their learning environments. The same indicators could be used to develop a strategy for implementing the philosophy of effective learning as embodied by the principles. These practical applications should provide feedback that can help validate and refine the principles.

References

Argote L (1999) Organizational learning: creating, retaining and transferring knowledge. Kluwer Academic Publishers, Boston

Becker K (2005) Individual and organisational unlearning: directions for future research. International Journal of Organisational Behaviour 9:659-670

Berkes F, Colding J, Folke C (eds) (2003) Navigating social-ecological systems: building resilience for complexity and change. Cambridge University Press, Cambridge, UK

Bland CJ, Ruffin MT (1992) Characteristics of a productive research environment: literature review. Academic Medicine 67:385-397

Breen CN, Cox D, Dickens C, Mander M, Roux DJ, Turton A, Van Wyk E (2003) Strategic review of river research. WRC Report 1198/1/03. Water Research Commission, Pretoria

Churchman CW (1971) The design of inquiring systems: basic concepts of systems and organisation. Basic Books, Inc., New York

Cohen WM, Levinthal DA (1990) Absorptive capacity: a new perspective on learning and innovation. Administrative Science Quarterly 35:128-152

Cooperrider DL, Whitney D, Stavros JM (2003) Appreciative inquiry handbook. Lakeshore Publishers, Bedford Heights

Davenport TH, Prusak L (1997) Working knowledge: how organizations manage what they know. Harvard Business School Press, Boston

Doolittle PE (1997) Vygotsky's zone of proximal development as a theoretical foundation for cooperative learning. Journal on Excellence in College Training 8(1):83-103

Drucker P (2001) Management challenges for the 21st century. Harper Business Press, New York

Folke C (2003) Freshwater for resilience: a shift in thinking. Philosophical transactions of the Royal Society of London 358:2027-2036

Folke C, Hahn T (2004) Social-ecological transformation for ecosystem management: the development of adaptive co-management of a wetland landscape in southern Sweden. Ecology and Society 9:2. Available online: at http://www.ecologyandsociety.org/vol9/iss4/art2

Giampietro M (2004) Multi-scale integrated analysis of groecosystems. CRC Press, London

Goh S, Richards G (1997) Benchmarking the learning capability of organizations. European Management Journal 15:575-583

Gunderson LH, Holling CS (eds) (2002) Panarchy: understanding transformations in human and natural systems. Island Press, Washington, DC

Holling CS (2001) Understanding the complexity of economic, ecological and social systems. Ecosystems 4:390-405

Imperial MT (2004) Collaboration and performance management in network settings: lessons from three watershed governance efforts. IBM Center for the Business of Government, Washington, DC

James CH, Minnis WC (2004) Organizational storytelling: it makes sense. Business Horizons 47:23-32

Kinzig AP, Carpenter S, Dove M, Heal G, Levin S, Lubchenco J, Schneider SH, Starrett D (2000) Nature and society: an imperative for integrated environmental research. Report from a workshop held June 2000, Arizona. Available online at http://lsweb.la.asu.edu/akinzig/report.htm

Levin SA (1998) Ecosystems and the biosphere as complex adaptive systems. Ecosystems 1:431-436

Levitt B, March JG (1988) Organizational learning. Annual Review of Sociology 14:319-340

Lionni L (1970) Fish is fish. Scholastic Press, New York

Lubchenco J (1998) Entering the century of the environment: a new social contract for science. Science 279:491-497

Lyndon H (1989) I did it my way! an introduction to "old way/new way" methodology. Australasian Journal of Special Education 13:32-37

Maarleveld M, Dangbégnon C (1999) Managing natural resources: a social learning perspective. Agriculture and Human Values 16:267-280

Max-Neef MA (2005) Foundations of transdisciplinarity. Ecological Economics 53:5-16

Miller GA (1956) The magical number seven, plus or minus two: some limits on our capacity for processing information. The Psychological Review 63:81-97

Miller WL, Morris L (1999) 4th Generation R&D – managing knowledge, technology, and innovation. John Wiley & Sons, Inc., New York

NRC (2000) How people learn: brain, mind, experience, and school. National Academy Press, Washington, DC

Odum EP (1989) Ecology and our endangered life-support systems. Sinauer Associates, Sunderland, Massachussets

Olsson P, Folke C, Hahn T (2004) Social-ecological transformation for ecosystem management: the development of adaptive co-management of a wetland landscape in southern Sweden. Ecology and Society 9:2. Available online at http://www.ecologyandsociety.org/vol9/iss4/art2

Parker G (1994) Cross-functional teams: working with allies, enemies, and other strangers. Jossey-Bass, San Francisco

Robbins PF, Abel N, Jiang H, Mortimore M, Mulligan M, Okin GS, Stafford-Smith DM, Turner, II BL (2002) Desertification at the community scale: sustaining dynamic human-environment systems. In: Reynolds JF, Stafford-Smith DM (eds) Global desertification: do humans cause deserts? Dahlem Workshop Report 88, pp 326-355. Dahlem University Press, Berlin

Rogers K, Roux D, Biggs H (2000) Challenges for catchment management agencies: lessons from bureaucracies, business and resource management. Water SA 26:505-511

Roux DJ, MacKay HM, Hill L (2006a) Consolidation and transfer of knowledge and experience gained in the development and implementation of water and related policy in South Africa. WRC Report 1295/1/06. Water Research Commission, Pretoria

Roux DJ, Rogers KH, Biggs HC, Ashton PJ, Sergeant A (2006b) Bridging the science-management divide: moving from unidirectional knowledge transfer to knowledge interfacing and sharing. Ecology and Society 11:4. Available online at http://www.ecologyandsociety.org/vol11/iss1/art4/

Senge PM, Kleiner A, Roberts C, Ross RB, Smith BJ (1994) The fifth discipline fieldbook: strategies and tools for building a learning organization. Doubleday, New York

Senge P, Kleiner A, Roberts C, Ross R, Smith B (1999) A fifth disciple resource: the dance of change. Nicolas Brealey Publishing, London

Senge P, Scharmer CO, Jaworski J, Flowers BS (2005) Presence: exploring profound change in people, organizations and society. Nicholas Brealey Publishing, London

Schein EH (2004) Innovative cultures and adaptive organisations. Chapter 7 in: Creating a learning culture: strategy, technology and practice. Conner ML, Clawson JG (eds) Cambridge University Press, Cambridge

Snowden D (2002) Complex acts of knowing: paradox and descriptive self-awareness. Journal of Knowledge Management 6:100-111

Snyder WM, Wenger E (2004) Our world as a learning system: a communities of practice approach. Chapter 2 in: Creating a learning culture: strategy, technology and practice. Conner ML, Clawson JG (eds) Cambridge University Press, Cambridge

Stafford Smith DM, Reynolds JF (2002) Desertification: a new paradigm for an old problem. In: Reynolds JF, Stafford Smith DM (eds) Global desertification: do humans cause deserts? Dahlem Workshop Report 88, pp 403-424. Dahlem University Press, Berlin

Stein E (1995) Organizational memory: review of concepts and recommendations for management. International Journal of Information Management 15:17-32

Turton AR, Hattingh J, Claassen M, Roux DJ, Ashton PJ (2005) Towards a model for ecosystem governance: an integrated water resource management example. Opening chapter of this book

Walker B, Holling CS, Carpenter SR, Kinzig A (2004) Resilience, adaptability and transformability in social-ecological systems. Ecology and Society 9:5. Available online at http://www.ecologyandsociety.org/vol9/iss2/art5

Walters CJ (1997) Challenges in adaptive management of riparian and coastal ecosystems. Conservation Ecology 1(2):1. Available online at http://-www.consecol.org/vol1/iss2/art1

Watson WE, Kumar K, Michaelson LK (1993) Cultural diversity's impact on interaction processes and performance: comparing homogenous and diverse task groups. Academy of Management Journal 36:590-602

Wenger E (2004) Knowledge management as a doughnut: shaping your knowledge strategy through communities of practice. Ivey Business Journal January/February 2005

Wenger E (2005) Learning for a small planet: a research agenda. Available online at http://www.ewenger.com/research/index.htm

Wenger E, McDermott R, Snyder W (2002) Cultivating communities of practice: a guide to managing knowledge. Boston Business School Press, Boston

Williams KY, O'Reilly CA (1998) Demography and diversity in organizations: a review of 40 years of research. Research in Organizational Behaviour 20:77-140

Zahra S, George G (2002) Absorptive capacity: a review, reconceptualisation, and extension. Academy of Management Review 27:185-203

The Role of Communication in Governance: The River Health Programme as a Case Study

Wilma F Strydom, Liesl Hill and Estie Eloff

Abstract

Within South Africa, active public participation in decision-making processes and policy development is a requirement for governance to be successful. Similarly, ecosystem governance requires active public participation in environmental issues. For the public to become progressively more involved in decision-making, an increase in public understanding of, active involvement in and engagement with science, is needed. This emphasises the importance of sound science communication strategies that will lead to informed responses to environmental issues.

This Chapter describes the role of communication between the components of the ecosystem governance Trialogue Model, namely science, society and government. The focus is directed particularly towards the role of communication in the society-science interface, highlighting the lessons learnt and communication challenges arising from the South African River Health Programme's State-of-Rivers reporting component.

Keywords: science communication, river health, society-science, information distribution

Introduction

The 1992 Rio Declaration on Environment and Development imposes on all human beings the obligation to involve themselves in environmental issues and to participate in decisions relating to the environment (UNCED 1992). In addition, the Global Water Partnership acknowledges the importance of governance in solving global water-related issues (GWP 2000). The need for broader stakeholder participation and more inclusive decision-making is increasingly recognised and encouraged (Santi and Grenna 2003) in many democratic societies, including South Africa. This in turn requires accurate and frequent communication that can ensure that the needed information is provided to stakeholders, to enable them to make informed decisions.

Concerns about the relationships between government and science, and between science and society, were raised in the early and mid-1900s in South Africa and the United States of America, respectively. Even though the establishment of a formal scientific institution in South Africa was recommended in 1921, the CSIR was established only in 1945 (Kingwill 1990). During the early 1950s the American Association for the Advancement of Science (AAAS) called for a broader public understanding of science (Gregory and Miller 1998). It was recognised that, as science support was moving more and more towards national resources, the engagement of the public became essential to science, which has to meet society's needs. Despite this commitment to improve the ability of the public to understand science, appropriate communication remained somewhat of a foreign concept to many scientists.

Efforts to improve the management of the natural environment and to stop environmental degradation have received increasing attention and more and more people have become aware of environmental issues and seek to influence the future through education, individual actions, legislation and other initiatives in order to put pressure on those with influence and power (Trudgill 1990). Trudgill also acknowledged the importance of having the ability to communicate appropriate knowledge so as to improve environmental practices. However, it was only fairly recently that the broad body of scientists has accepted the importance of communicating scientific information to non-scientists effectively (Gregory and Miller 1998).

Environmental decision-making, including water resource management, faces enormous challenges in South Africa due to increasing pressure on natural resources. Adding to these challenges is the diversity of stakeholders that have to be informed and that take part in decision-making processes. Stakeholders range from well-educated individuals, who are fluent in several languages and who have access to various channels of communication, to those who have not had formal schooling and who cannot read.

This chapter examines the relationship between communication and governance. The focus is specifically on the importance of communication between scientists and society, and uses the South African River Health Programme (RHP), with emphasis on State-of-Rivers reporting, as a case study to illustrate the importance and challenges facing this bidirectional communication between scientists and society.

What is Communication?

Communication is not easy to define. Clevenger (1991) rightly notes that communication is "*one of the most overworked terms in the English language*". More than 120 definitions of communication exist. Littlejohn and Foss (2005) give a summary of various renowned authors' definitions of communication. According to Hoben (1954), communication is "*the verbal interchange of a thought or idea*". Saywell and Cotton (1999) define communication as the "*transmission of data, information or knowledge between two or more points*", which focuses on communication between provider(s) and recipient(s).

Littlejohn and Foss (2005) also noted that the value of understanding is increased when value statements are added to the definition. Littlejohn et al (2005) added to the communication definition: communication is a "*level of observation*" that links discontinuous parts of the world, but a communication should also have intent – a conscious intent to affect behaviour – and a judgment of whether the information was received, understood and acted upon. Sometimes, communication can also be viewed as a process of negotiation, where negotiation is a two-way process of exchange (Gregory and Miller 1998). For example, to ensure that a communication addresses public needs, the public must first express clearly what those needs are. Communication is clearly an interactive process, which requires a two-way exchange of information, knowledge and experience.

To summarise: communication must link people, groups or parts of the world together. Communication should also have an intention or purpose, for example to change behaviour. A communication would not have any value if it was not received, understood and acted upon by the target audience.

Criteria for Good Communication

Various authors have made explicit the criteria that are required for good communication. The views of Lewenstein (1992), and Saywell and Cotton (1999) are listed below:

- Identify and understand the audience;
- Address user needs and requirements;
- Ensure that objective information is available and readily accessible;
- Ensure that users of the information are able to comprehend/understand the information provided;
- Adapt findings to the local context – use intermediaries as interpreters of research results where needed;

- Maximise exposure through output in various media and through several communication channels; and
- Create an environment for interactivity and feedback between the sender(s) and the receiver(s) of the information.

The number of stakeholders that have to be engaged in communication processes and also be involved in the decision-making process is wider now than ever before. In addition, these stakeholders have different levels of understanding, perceptions, interests and attitudes towards environmental issues, which mean that they need to be approached in different ways (Santi and Grenna 2003).

What is Governance?

In a democratic society, governance encompasses the processes and interactions between government and the public, as well as the products or conditions arising from that interaction. A country's government, as its elected leadership, is held accountable for aggregating the public's interests and of finding ways to satisfy those needs.

Good Governance

Based on the above description, good governance can be described as a transparent process between government and civil society, where the interests of society are addressed and implemented by government, using the available government tools (UNDP 1997; Van Kersbergen and Van Waarden 2001; Santi and Grenna 2003; Hattingh et al. 2004a, 2004b). Hattingh et al. (2004a) identified good governance, together with government instruments, such as policies, strategies, legislation, as well as adequately-resourced implementing institutions and mechanisms of implementation, as important ingredients for successful democratic system functioning. Good governance should also be effective, participative, dynamic, inclusive, responsive, integrative, accountable, and based on trust.

For the purposes of this chapter, and based on the Trialogue Model given in the introductory chapter, governance can be described as the relationships between government, society and science. The bidirectional relationships, or interfaces, between government and society, government and science and between science and society, as well as the relationships between these interfaces determine governance. Good governance is the degree to which these relationships and interfaces are functional and effective (Turton et al. 2005).

For the Society-Science relationship or interface to be dynamic and able to meet its expectations, the pillars of environmental governance as a whole, namely government, society and science, have to be equally supportive of each other. The relationship between science and society, and specifically the role of communication – the main focus of this chapter – is one aspect of ecosystem governance that has become more prominent over the last few years.

The Link between Communication and Good Governance

To be effective, government and the processes and products of governance must be closely interlinked. In addition, effective communication between these processes and products provides a vital link that helps to build good and effective governance. Rowe and Fudge (2003) argue that accountability of the governance process is largely facilitated through mechanisms of adequate communication. This has been echoed by Hattingh et al. (2004) who identified a lack of access to appropriate information and inadequate communication as obstacles to good ecosystem governance. Together, these emphasise the importance of communicating information to society, and ultimately the responsibility of scientists to communicate science in a manner that would improve society's understanding of science and that would lead to informed responses to environmental issues.

Whatever way one looks at it, it is clear that environmental governance poses a significant communication challenge.

Interfaces between the Science, Government and Society Processes in Ecosystem Governance: The Role of Communication

Recognising the importance of the groups that form the "cornerstones" of the governance process, namely the science, government and society components, this section concentrates on the interfaces between these groups and the role that communication plays in these interactions.

The Government-Society Interface

In South Africa, the national government is the custodian of the nation's water resources. As public trustee, national government must ensure that the development, apportionment, management and use of those resources are carried out prudently and wisely (DWAF 1997). The Constitution of South Africa (RSA 1996) requires consultation on all matters that can impact on people's lives. The public (society) therefore has the right to participate in, and influence, decision-making that may affect them in any manner. In order for the public to participate in a meaningful way and to have equal opportunity to contribute in any consultative or participatory process, sufficient accessible information is required.

Government, therefore, has the responsibility of creating awareness and communicating accurate and relevant information to decision-makers in order to inform them and to add value to their decision-making processes. In addition to government's responsibility to create awareness amongst society, it is their responsibility to provide support to marginal groups, where necessary (MacKay 2003). Government therefore not only has a role to play in creating a public understanding of environmental issues, but has to encourage society to appreciate the natural environment and to take responsibility for their actions. Similarly, government needs to provide the public with information that could improve their wellbeing and livelihoods.

The Government-Science Interface

In order for government to achieve its aims and objectives and live up to its responsibilities with regard to societal issues, sustained investment in research and continuous communication between the science and government sectors is crucial. The South African government recognises the importance of science and therefore invests in a number of national research facilities. Many development interventions would have been much less successful, or would have failed, if research had not been done and the resulting technologies implemented.

The River Health Programme, as an example, was initiated by the Department of Water Affairs and Forestry (DWAF), which, together with the Department of Environmental Affairs and Tourism (DEAT) and the Water Research Commission (WRC), are the national custodians of the Programme. The Programme has a strong scientific foundation, which is strengthened by continuous improvement and refinement of the various programme components. These costs are shared and co-funded by the three partner organisations. The CSIR was also involved from the start and plays an important role as science partner. Ongoing research strengthens the scientific rigour of the programme.

Research results, however, need to be communicated effectively to both government and society. Conditions or "enabling environments" (Meyer 2002) for communication need to be created within the relevant departments and organisations, to ensure that decision-makers can successfully implement the research findings such as, for example, the proposed management recommendations that should lead to improved water resource management.

The Society-Science Interface

The importance of making science accessible to the general public was stressed by Warren Weaver, a board member of the American Association of the Advancement of Science, during the 1950s when he stated: "*it is absolutely essential that science...be better understood by government officials, business-men and, indeed, by all people*" (Gregory and Miller 1998). Apart from achieving an improved public understanding of science, the science community is also responsible for accounting for environmental and social acceptability of science outcomes, and for enhancing the quality of life of the people (Trudgill 1990).

Communication within the science community is well-developed. Research projects are often concluded by the writing of articles for journal publication. This is the most common and preferred method of communicating to colleagues (Metcalfe 2002; Winter 2004). Although there may be strong incentives for this, the role of a journal in providing broad-based information transfer is limited.

Although there is now a stronger move towards communicating scientific information to non-scientists, gaps still exist and communication within the society-science interface is in many instances still less than sufficient. It requires a conscious effort from scientists to improve their communication and distribution strategies to society and government.

Case Study: Communicating through State-of-Rivers Reporting

Background

The River Health Programme (RHP) is a national monitoring programme that measures and reports on the ecological state of rivers in South Africa. The Programme was initiated in 1994 by DWAF and came at a time when the Department's focus with regard to the management of water resources shifted towards a more integrated ecosystems approach. The National Water Act (Act No 36 of 1998) also stipulates that DWAF, as the national custodian of South Africa's water resources, is responsible for the monitoring and assessment of these water resources (RSA 1998).

The Programme adopted a model of national development and coordination with provincial or local implementation (Roux 2004). The Department of Water Affairs and Forestry, the Department of Environmental Affairs and Tourism and the Water Research Commission are the Programme's national partners. Members of the provincial implementation teams are typically drawn from provincial government departments, conservation authorities, universities, metropolitan councils, and water supply utilities. Unique partnerships have developed, which have resulted in the sharing of resources and technical expertise across various organisations to monitor, report and communicate information on the state of rivers in South Africa.

Broadening Understanding

The ultimate purpose of the RHP is for scientists to communicate the ecological condition or health of river systems, and the consequences of human activities on those systems, to water resource managers and planners to ensure that river systems are effectively managed. The Programme further aims to inform and educate the people of South Africa regarding the health of their rivers. One of the key challenges of the Programme is to communicate technical information in an effective and creative manner to a wide audience, ranging from scientists and water resource managers to politicians and the general public (Roux 1997). The River Health Programme's communication strategy (DWAF 2001) identified several communication channels. For the purpose of this chapter, we will focus on State-of-Rivers reporting, which addresses two of these channels, namely printed media and electronic media.

River health information is communicated with the ultimate purpose of changing the behaviour and attitudes of people, who utilise and benefit from our precious water resources, and of ensuring that these resources are developed and used in a sustainable manner. Changed behaviours would typically relate to the degree to which resource managers incorporate river health information in their decision-making processes and implement it in their strategies. Similarly, a positive change in civil society's perception and appreciation of rivers would testify to effective communication (Roux 2004).

Not all people, however, understand, appreciate and accept that aquatic ecosystems are protected to ensure a sustained provision of certain goods and services on which they rely for their livelihoods and well-being. This issue is discussed in more detail in the section on "*Lessons learnt and communication challenges within the River Health Programme context.*"

In order to communicate river health information that is intended to make a difference to people's values and belief systems, or to how resources are managed, three key aspects need to be carefully considered:

- The target audience, including the engagement with the audience during the communication process;
- Information packaging; and
- Information distribution.

Communication plays a central role within each of these three aspects, and is equally important in the different components of the governance Trialogue Model, as well as at the interfaces between these components.

The Audience

The concept of broader stakeholder participation, a more inclusive decision-making process, and subsequently the effective communication of science to a broad audience, have become increasingly recognised, both nationally and internationally (Santi and Grenna 2003). The number of stakeholders to be reached and to be involved in environmental decision-making processes, has become wider than ever. Apart from the responsibility of communicating river health information to a broad stakeholder audience, the River Health Programme also aims to create a general awareness regarding environmental issues that would, in turn, enable stakeholders to participate in decision-making processes that don't necessarily relate to the Programme. Since environmental governance relies on accurate, consistent and continuous communication, the onus on scientists and government in particular to communicate the right information to the right people at the right time, is highly relevant.

Different stakeholders have different levels of perception, knowledge, interests and attitudes towards environmental issues and need to be approached in different ways (Saywell and Cotton 1999; Santi and Grenna 2003). Not only is it necessary to have a clear understanding of stakeholders' information requirements, but it is also necessary to understand the kinds of decisions that have to be made, as well as the day-to-day responsibilities of those stakeholders.

Experience has shown that engaging with water resource managers at the start of a river survey is invaluable to the scientists and the RHP reporting team. Not only does this create the opportunity for the managers to share their knowledge of the catchment area, but scientists start to develop a better understanding and appreciation of management issues. Informal conversations next to a river and the opportunity for the managers to share their knowledge during active participation in technical workshops, add tremendous value to data interpretation. Apart from developing more of an understanding and appreciation of each other's issues, the interaction between managers and scientists also increases awareness of, and respect for, each other's views and beliefs, which in turn lead to improved teamwork.

Information Packaging

When communicating scientific information to stakeholders, great care must be taken in deciding what information to share and how best to achieve a clear understanding of it in the target audience. Designing an effective reporting format to communicate river health information was seen as a key priority of the RHP (Strydom 2003). State-of-Rivers reporting, now seen as the Programme's flagship product, was developed with two key considerations in mind:

- To produce a report in a non-technical and uncomplicated style and format that would be appealing to its audience – comprising mainly water resource managers, politicians and the general public; and
- To ensure that the information that is provided remains objective, credible and conveys a clear message.

State-of-Rivers reports are written in simple, accessible language, in full-colour brochure-style reports of about 40 pages that contain photographs, graphics and diagrams. Other products that also convey the river health message are: posters translated into several of South Africa's indigenous languages; non-verbal fun posters; and activity books. Since these products are mostly aimed at environmental education and creating public awareness, they are becoming more and more in demand and the numbers of these outputs are subsequently on the increase. These products are typically displayed at various water-related events such as Water Week activities, conferences and symposia and general awareness days.

Information Distribution

An information product can only have a desired impact if it reaches the audience that it is intended for. The importance of effective distribution strategies, to ensure that knowledge and information do not stay where they are generated (Saywell and Cotton 1999), but reach the intended target audience, cannot be over-emphasised. It is therefore necessary to know what information to distribute, to whom, and how. Roux (2004), however, points out that it may be ideal, for example, to hand-deliver State-of-Rivers reports to key recipients, but this is by no means a guarantee that they will read the reports and internalise their information, let alone initiate a required management intervention.

It is therefore necessary for scientists to investigate and make use of a variety of media for disseminating science information. The choice of media will largely be determined by the target audience itself, as well as by the purpose and nature of the message to be communicated. This is clearly accomplished in the River Health Programme communication efforts. Apart from being hand-delivered, reports and other information products are posted to the respective target audiences, sent electronically or downloaded from the River Health Programme website (RHP 2005). Provincial State-of-Rivers reports are usually launched during high-profile events, which are attended by provincial political leaders and, in some instances, by the Ministers or deputy Ministers of DWAF and the Department of Environmental Affairs and Tourism. The popular media, such as local newspapers, radio interviews as well as television, are another route to distribute river health information, but these have not yet been fully utilised.

An aspect that is often neglected is the fact that the dissemination of information, or information transfer, has to be an interactive process, which requires a two-way exchange of information (Denisov and Christoffersen 2001), creating a stimulus to the process of mutual learning, rather than just the linear transfer of knowledge from scientists to society (Saywell and Cotton 1999). Attempts therefore have to be made to collect user-feedback on a regular basis (Denisov and Christoffersen 2001), to find out from the different target audiences whether the information provided suited their particular needs. Within the River Health Programme an effort is made continuously to improve on the relevance of information provided in State-of-Rivers reports and related products, and to address the evolving information needs of report recipients. The approach followed is to send out questionnaires, or to do telephonic or personal interviews that prompt readers of the report to provide feedback on aspects such as the style of the report, as well as content and value of the information provided (Strydom 2003). Satisfaction reviews on two of the initial reports have indicated an increase of 27% in the proportion of readers that read more than 60% of the content. This improvement could largely be ascribed to adjustments related to content, presentation format and style (Strydom et al. 2002).

Lessons Learnt and Communication Challenges within the River Health Programme

As with environmental governance, the communication challenges of the River Health Programme are complex and diverse and several valuable lessons have been learnt. Some of these are shared below. Although this section focuses on the River Health Programme's State-of-Rivers experience, the learning from other science communication programmes, drawn from the literature, is also mentioned. These are used to emphasise other communication challenges that, although not encountered by the River Health Programme, are relevant to the discussion and could point out possible pitfalls.

Understanding User Needs and Requirements

To be truly useful, environmental information must improve understanding and knowledge and also contribute to effective decision-making (Saywell and Cotton 1999; Denisov and Christoffersen 2001). For information to add any value to decision-making processes, scientists must have a clear understanding of how the information they produce will be used, or how they want it to be used, and by whom. It is, however, extremely difficult to ensure that the information will actually meet these needs.

Over and above the challenge to understand what it is that users want, society finds itself living in an era of information overload, where people are overwhelmed by information. It is impossible to read everything and so the user has to prioritise by making decisions on the perceived value of information. This was a reality which was soon realised in the River Health Programme. State-of-Rivers reports compete with many other information products for reader's attention. Although they are professionally designed, these high-quality full-colour reports, containing user-friendly layouts and catchy graphics, are still not read by many of the intended target audience. Just over 85% of the managers that responded to a questionnaire noted that they have read more than 60% of the second State-of-Rivers report, the Letaba and Luvuvhu River Systems (Strydom et al. 2002). This was a big improvement from the first report, the Crocodile, Sabie-Sand and Olifants River Systems report, where more than 60% of the report was read by 50% of the managers (Strydom et al. 2002). Through improved stakeholder involvement processes during the assessment and reporting of river health, the RHP has increased stakeholder buy-in.

The value of involving stakeholders throughout a project, from start to finish, cannot be emphasised enough. In the River Health Programme context, this addresses the society-science as well as the government-science interfaces and is contributing more and more to managers' and scientists' understanding of each other's frustrations and challenges on the ground. Not only does this enhance working relationships, but scientists start to develop a better understanding of what information needs to cater for, and how often information should be provided. Managers, on the other hand, are realising the value of environmental information to support their decision-making processes – to such an extent that biomonitoring is being included more and more as part of water-use licensing conditions.

Financial Challenges

A critical aspect to think through carefully when planning a project is the communication and distribution of outputs to stakeholders. When communicating to a broad stakeholder-base, it is particularly important to carefully consider how the information is packaged (different target audiences require different approaches) and how the stakeholders will access and use the information (which varies among stakeholders). Both these aspects result in additional workloads and have financial implications. The science communication component, therefore, has to be carefully planned for at the start of a project.

Generally, the communication component of a project can use up anything between two and 10% of the total project budget. This usually depends on the size of the project and the stakeholders to be reached. The more information products there are, the higher the communication cost. Also, the cost of packaging technical information into an easier and more understandable format for non-technical stakeholders should not be underestimated.

River Health Programme posters, activity books and brochures are currently being produced in English and translated into one or two other languages that are indigenous to a particular region. Distributing this information to an even wider public in South Africa would require a multitude of translations, plus additional printing, considering the fact that we have 11 official languages. Although the fun posters with a river health message are aimed at people and children who cannot read, one aspect that has not yet been investigated is the communication of river health information to visually-impaired people. All of this adds to the overall project costs.

In regions or provinces where partnerships between participating organisations are well-established, funding is either sourced from sponsors or costs are shared among partner organisations. In other instances, provinces look to DWAF head office for funding. Because government departments work on budget-cycles that require allocations to be made to specifically-approved projects in advance, it is either very difficult, or impossible to obtain funding from the Department if the activity in question is not specifically planned for within an existing budget cycle. Either that, or a laborious time-consuming process has to be followed to request funding from the Department. It is therefore imperative that the distribution of information is properly planned and budgeted for as an integral part of a project, and not dealt with in an ad hoc manner.

Communication, Awareness Creation and Attitude Change

Even though information gained through research may be accurate and relevant, the knowledge gained from it may be lost if these research findings are not effectively communicated or disseminated. Ideally, creating awareness amongst decision-makers should expand their environmental knowledge-base, and thereby ensure improved decisions based on best judgment and new information (Denisov and Christoffersen 2001). Decision-makers should understand the trade-offs between, for example, conservation and development as well of the long-term consequences for the provision of goods and services.

Awareness creation is the means of conveying messages to one or more audiences with the explicit purpose of establishing a change in the knowledge, attitudes and ultimately the behaviour or practices of those audiences. According to Denisov and Christoffersen (2001), it is difficult to understand what happens when people become aware of an issue, since there is no explicit link between awareness and actions taken. This presents an important challenge to effective communication.

One aim of State-of-Rivers reporting is to raise awareness amongst the users of river health information regarding the state of South Africa's river systems, what their environmental problems are and the consequences thereof on these ecosystems. If such awareness is created, for example amongst water resource managers, it should in theory support informed environmental decision-making.

Five years have elapsed since the first State-of-Rivers report, that on the Crocodile, Sabie-Sand and Olifants rivers, was produced. To date, 10 reports have been produced for selected river systems across the country. It has been a steep learning-curve for the reporting teams, changing and adapting the reporting format, the type of information and how it is presented. Despite this, and acknowledging that it is very difficult to measure, there is no hard evidence that State-of-Rivers reporting has or hasn't had the desired impact on the different target audiences. It is our impression, though, that river health information is now slowly starting to form part of water resource managers' decision processes.

Realising that communication guarantees neither a change in attitudes nor a change in behaviour; it is appreciated that State-of-Rivers reporting, in order to facilitate change, should be communicated in a more holistic way. Whereas in the past these reports focused mainly on the consequences of human behaviour for the environment, it is increasingly realized that environmental issues cannot be resolved without putting environmental information in a larger context, which includes economic and social information. Added to this, are the benefits of healthy ecosystems to society. It is also realised that mandates have to be addressed and roles and responsibilities of the various stakeholders of the Programme made explicit, when management actions are proposed.

The extent to which water resource managers incorporate river health information in their decision-making processes and implement it in their strategies and policies will provide an indication of the degree to which behaviours have changed.

Shared Understandings and Perceptions

Society does not necessarily share the concept, or appreciate and accept that aquatic ecosystems are protected so as to ensure a sustained provision of certain goods and services upon which people rely for their livelihoods and well-being. One would reason that this might be due to the fact that people are uninformed and do not realise the value of healthy river ecosystems. It can therefore be assumed that, if people are informed about the consequences of certain impacts on rivers and other aquatic ecosystems, they will adopt an environmentally-friendly behaviour pattern. However, as mentioned above, communication of or knowledge about aquatic ecosystems does not necessarily guarantee attitude- or behaviour-changes (Denisov and Christoffersen 2001), let alone people's perceptions about the value of aquatic ecosystems.

Poor communication, on the other hand, can create negative perceptions which can result in misunderstandings and mistrust amongst scientists, decision-makers and the public.

In the River Health Programme experience, examples of these types of negative perceptions, which can be ascribed to insufficient communication include:

- Many **water resource managers** believe that aquatic ecosystem protection is only about ensuring the survival of 'bugs' and fish. This can be due to their uncertainty as to how to incorporate aquatic ecosystem information into their decision-making and management frameworks;
- **Scientists** want to protect aquatic resources at all costs, irrespective of social and economic needs, and want to be prescriptive to managers regarding the levels at which particular ecosystems should be protected. This is largely due to a lack of trust, and because managers are perceived to be biased towards water resource development as opposed to ecosystem protection;
- **Politicians** allow developments that address short-term needs to take place without considering their impact on aquatic ecosystems, or the long-term consequences that this may have on the provision of ecosystem goods and services; and
- Some members of the **public** still view rivers purely as channels carrying water for their use and pleasure, and do not regard rivers as living ecosystems.

Improving communication between scientists and the different target groups will help refine our understanding of which factors have the greatest impact on people's perceptions and why. Communication efforts can subsequently focus on addressing these issues and, in doing so, can lead to an improved understanding of, and appreciation for, aquatic ecosystem protection.

Mental Models

Environmental knowledge is rarely the sole factor causing people to take decisions about environmental issues. Other factors, such as economic conditions, social issues, tradition, culture and academic training all act together in a complex way, either to weaken or strengthen, environmental information (Denisov and Christoffersen 2001). This in turn determines mental models, or a person's worldview, which Dawson (2000) defined as "the internal representations that individual cognitive systems create to make sense of the external environment." Our worldview, therefore, forms the basis of our behavioural responses and regulates the way we see and experience things around us, and the way we subsequently act.

Both water resource managers and decision-makers prefer to operate within their "comfort zones" – doing things the way they are used to. Similarly, scientists and society have their own "comfort zones" and mental models that they protect. Changes in thinking do not come easily. Sustained effort is needed to shift mental models.

This does not imply that people with different mental models cannot communicate. Scientists must just be aware that, when communicating environmental information to stakeholders, a deliberate effort is required to ensure that the message is conveyed accurately, since people's interpretations of a piece of information may differ, and their resulting responses may be entirely the opposite to what was anticipated.

Science Credibility

Scientific processes frequently deal with abstract concepts and systems approaches to complex ecosystem interactions and processes. Answers to research questions are often indefinite. Scientists are regularly required to provide expert opinions on certain matters; this may be because of the uncertainties that exist around data, or the lack of adequate data. Resource managers, on the other hand, need firm sets of facts and figures to base their decisions on, since it is difficult to interpret and accommodate uncertainty in decision-making frameworks. If scientists do not communicate their research findings realistically, their audience will question the credibility of scientists and science alike. This makes it even more difficult for scientists to gain legitimacy for their ideas and information.

The case study of sheep farming in the English Lake District, after the Chernobyl incident in early May 1986 (Wynne, in Lewenstein 1992), should be a warning that the credibility of science should never be compromised and that the process of monitoring and reporting on findings should be transparent at all times.

Six days after Russia's Chernobyl disaster, in early May 1986, the main cloud of radioactive contamination passed westward over the United Kingdom. Heavy thunderstorms disturbed this cloud and rained radioactive particles across the country. Radioactivity readings showed that the Cumbrian area in northern England was the worst affected and a ban was subsequently placed on the movement and slaughter of sheep in the area. Thereafter, experts wrongly and prematurely predicted that the threat would only last a few weeks and that the sheep farming community should not be concerned since the ban would have been lifted by the time that that season's lambs were ready for the market. However, communications from the government were delayed by several weeks, resulting in the farmers suffering financial losses. Compensation payments were inadequate and were also terminated without proper communication. No account was taken of the fact that it would take several weeks of manual labour to collect sheep from the steep hills. Eventually, some five years later, the same experts still declined to make any predictions of how long the residual radioactivity would last. In addition, scientific data derived from rainfall radioactivity measurements indicated that other areas in the United Kingdom should have been equally affected. Since this was not the case, the Sellafield nuclear fuels reprocessing complex, which is situated on the Cumbrian coast, was implicated to be contributing to radioactivity levels in the area. This nuclear plant was previously under suspicion for radioactivity fallout in the area, but allegations were denied by scientists and government.

This well-documented case study explains how audiences received and comprehended certain scientific information. It is a typical example where the perceived credibility of scientists and the government influenced the effectiveness of scientific communication to the farmers. From this example, it is clear that the main factors that have the potential of affecting science credibility include (Wynne, in Lewenstein 1992):

- The abstract nature of scientific research: the public can be left with the impression that scientists are hiding important information;
- Contradictory information: different sources may have opposing viewpoints, which causes public confusion and leads to a total mistrust of scientists;
- Insensitivity to local conditions and conventions: this is caused by a lack of understanding of local culture and knowledge, and the way things are done; and
- Perceived scientific arrogance and lack of reality: the assumption that lay people cannot handle uncertainty and risk, which may result in the scientists withholding selected information from the people.

Public Trust and the Role the Media Play

The print media play an important role in the society-science interface. The challenge posed by having to express opinions and judgments about scientific issues becomes complicated by the need also to report on the uncertainties. Priest (2004) indicated that the public prefers to receive well-balanced science information in order for them to draw their own conclusions and to make decisions. In contrast, people also seem to appreciate sensationalistic accounts (Priest 2004), because they lack an understanding of the issues that would allow them to discriminate between 'fact' and 'fiction'. Contrary to the end result of the one-sided information that the public often receive through the media, balanced communication will encourage broad public discussion (Priest 2004). Within the biotechnology controversy context, public opinions and reactions are attributable to trust rather than to knowledge (Priest 2001, 2003). Priest (2004) also wrote: "...individuals make decisions on the basis of the extent to which they trust those espousing different points of view...Opposition to biotechnology can be understood as a crisis of trust – trust in science, trust in industry, trust in regulation, trust in the credibility of critical voices, even trust in media to bring out the full story."

The involvement of the press, e.g. newspapers, radio and television, in communicating river health information has not yet been fully explored. This may be attributed to a widely-shared concern in the scientific community that the popular media seldom communicate the intended (i.e. correct) message to the public. The difficulty of *controlling* the message eventually conveyed or portrayed by the media leaves most scientists feeling uneasy, and as a result, most scientists do not trust the media.

As an example: a documentary programme (50/50 2005), aimed at communicating the health of our rivers – showing both the good and the bad – was filmed recently. However, when broadcast, the intended message was distorted and sensationalised, portraying the unhealthy rivers out of context and ignoring the role that the Department plays in rectifying the situation and blaming them for not doing their job. Another department had a similar experience, in which their original message was also distorted and, when they insisted that it should be rectified, they had to pay for "advertisement" space in the newspaper.

Social and Political Barriers

Trudgill (1990) points out that economic, social and political factors, rather than environmental considerations, are the determining factors when environment-related decisions are made. In the River Health Programme experience thus far, we are not convinced otherwise, in light of the following:

- South Africa's National Water Policy (DWAF 1997) and law (RSA 1998) provide a clear mandate for the protection and conservation of the country's water resources, and the River Health Programme supports this mandate through the use of scientifically-sound and cost-effective tools for monitoring and assessing river health. This is, however, by no means a guarantee that the scientific information that the

RHP generates, and which it communicates to society and water resource managers, will be taken into account in relevant decision-making processes;

- It is a reality that the sustainable management of water resources requires trade-offs to be made between social, economic and environmental imperatives in order to find an appropriate balance. It may, therefore, at times be necessary to trade-off protection of the ecosystem against short-term imperatives for economic or social development and, in other instances, it may be necessary not to allow any development at all along certain stretches of a particular river (MacKay 2003). There is, however, the notion that political pressure is all too often excised to allow developments offering short-term economic returns to take place, in spite of their obvious long-term consequences for the environment; and
- Although the Trialogue Model suggests that all the three actors, i.e. science, society and government, are equally important, practice suggests otherwise. It is clear that decision-making currently favours the social considerations and, linked to them, the economic considerations.

Conclusion

This chapter focused on the importance of communication within the society-science interface of the governance Trialogue Model, and used the State-of-Rivers component of the South African River Health Programme as an example to demonstrate and discuss the role of communication between scientists and society, and the implications of inadequate communication. This example highlights both the lessons learnt, as well as the communication challenges that scientists face.

The role of science communication is gaining prominence, not only in South Africa, but also world wide. This relates directly to broader stakeholder participation, and more-inclusive decision-making relative to the environment. The importance of communicating the right information to the right people at the right time, and also in the right way, can not be over-emphasised. The communication challenge within environmental governance – particularly at the society-science interface is, however, complex and diverse and requires a thorough understanding and appreciation of the issues involved for it to be able to contribute to informed decision-making. In short, science communication must not be underestimated.

The three key aspects that have to be carefully considered in the communication process, in order to make a difference to people's values and belief systems, or how resources are being managed, are:

- The target audience, including the engagement with the audience during the communication process;
- Information packaging; and
- The distribution of information.

Although many stakeholders consider State-of-Rivers reporting to be a good example of how to communicate scientific river health information successfully, within each of the above points lie several communication barriers and challenges that need to be addressed. These include:

- 'Enabling' environments, and financial assistance for improving science communication;
- Further exploration of the bidirectional communication process – focusing on user needs and communicating the benefit to society of ecologically-sustained rivers;
- The need to expand holistic thinking about water resource use and the communication thereof;
- Continuous assessment of the impact of State-of-Rivers reporting on water resource management and on society in general. Are there any changed attitudes and behaviours, and do they lead to the desired actions; and
- Continuous examining of current science-communication practices and the improvement thereof to achieve the desired impact.

References

50/50 (Television programme) (2005) Rivers to die for – dying rivers. SABC 2 Programme, Sunday 22 May 2005. Transcript available online at http://-www.5050.co.za/inserts.asp?ID=6362 [Accessed on 15 March 2006]

Clevenger T Jr (1991) Can one not communicate? a conflict of models. Communication Studies **42**:351

Denisov N, Christoffersen L (2001) Impact of environmental information on decision-making processes and the environment. UNEP/GRID-Arendal Occasional Paper 01. Arendal, Norway

Dawson R (2000) Developing knowledge-based client relationships – the future of professional services. Butterworth Heinemann, Boston, pp 275

DWAF (1997) White paper on a national water policy for South Africa. Department of Water Affairs and Forestry, Pretoria

DWAF (2001) River Health Programme communication strategy. Draft internal report. Department of Water Affairs and Forestry, Pretoria

Gregory J, Miller S (1998) Science in public, communication, culture, and credibility. Basic Books, pp 294

GWP (2000) Towards water security: a framework for action. Global Water Partnership, The Hague

Hattingh J, Maree G, Van Wyk E, Turton AR (2004a) Environmental governance and equity in a democratic South Africa. Proceedings of the AWRA/IWLRI International Conference on water law governance, 29 August-2 September 2004. Dundee, Scotland

Hattingh J, Maree G, Claassen M, Oelofse S, Turton AR, Leaner J, Strydom W, Van Wyk E (2004b) Understanding the relationship between government and governance for effective implementation of government tools. CSIR report ENV-P-Conf 2004 016, CSIR, Pretoria

Kingwill DG (1990) The CSIR – the first forty years. Scientia Printers, CSIR, pp 352

Lewenstein BV (ed) (1992) When science meets the public. Proceedings of a workshop organised by the American Association for the Advancement of Science Committee on Public Understanding of Science and Technology. 17 February 1991, Washington, DC, pp 164

Littlejohn SW, Foss KA (eds) (2005) Theories of human communication. 8th Edition. Thomson Wadsworth

MacKay H (2003) Water policies and practices. In: Reed D, De Wit M (eds) (2003) Towards a just South Africa – the political economy of natural resource wealth. WWF, Macroeconomics Program Office

Metcalfe J (2002) Workbook on communication management and media skills for scientists & engineers. Foundation for Education, Science and Technology, Econnect, Australia, pp 1-30

Meyer H (2002) Communicating science to rural communities requires boundary crossing. Proceedings of the 7th International Conference on Public Communication of Science and Technology, Cape Town, 5-7 December 2002. Available online at http://www.saasta.ac.za/scicom/pcst7/meyer.pdf [Accessed May 2006]

Priest S (2001) Misplaced faith: communication variables as predictors of encouragement for biotechnology development. Science Communication **23**(2):97-110

Priest S, Bonfadelli H, Rusanen M (2003) The "trust gap" hypothesis: predicting support for biotechnology across national cultures as a function of trust in actors. Risk Analysis **23**(4):751-766

Priest S (2004) Biotechnology, media and public opinion across national boundaries. Ecquid Novi **25**(1):80-93

Republic of South Africa (RSA) (1996) The constitution of the Republic of South Africa (Act No 108 of 1996). Government of the Republic of South Africa, Pretoria

Republic of South Africa (RSA) (1998) National water act, 1998 (Act No 36 of 1998). Government of the Republic of South Africa, Pretoria

River Health Programme (RHP) (2005) The river health programme. Available online at http://www.csir.co.za/rhp/ [Accessed September 2005]

Roux DJ (1997) Overview of the design process and guidelines for implementation. River Health Programme Report Series 6. Available online at http://www.csir.co.za/rhp/reports/reportseries6.html [Accessed September 2005

Roux DJ (2004) From monitoring design to operational program: facilitating the transition under resource-limited conditions. In: Wiersma GB (ed) Environmental monitoring. CRC Press, Boca Raton, pp 631-648

Rowe J, Fudge C (2003) Linking national sustainable development strategy and local implementation: a case study in Sweden. Local Environment **8**(2):125-140

Santi E, Grenna L (2003) Environmental communications assessment – a framework of analysis for the environmental governance. Paper presented at the 7th Biennial Conference on Communication and the Environment (COCE 2003). Available online at http://siteresources.worldbank.org/EXTDEVCOMMENG/Resources/cocefinaldraft.pdf [Accessed September 2005]

Saywell DL, Cotton AP (1999) Spreading the word. practical guidelines for research dissemination strategies. interim findings. Water, Engineering and Development Centre. Loughborough University, United Kingdom

Strydom WF (2003) Compiling state-of-rivers reports and posters: a manual. River Health Programme Report Series 17. Available online at http://www.csir.co.za/rhp/reports/reportseries17.html [Accessed May 2006]

Strydom WF, Van Wyk E, Maree G, Maluleke TP (2002) The evolution of state-of-rivers reporting in South Africa. Proceedings of the 7th International Conference on Public Communication of Science and Technology, Cape Town, 5-7 December 2002. Available online at http://www.saasta.ac.za/scicom/-pcst7/strydom.pdf [Accessed May 2006]

Trudgill ST (1990) Barriers to a better environment. what stops us solving environmental problems? Belhaven Press, London, pp 151

Turton AR, Hattingh J, Claassen M, Roux DR, Ashton PJ (2005) Towards a model for ecosystem governance: an integrated water resource management example. Paper presented at the International Symposium on ecosystem governance, 10-13 October 2005, Kwa Maritane Bush Lodge, Pilanesberg, South Africa

UNCED (1992) Agenda 21 earth summit agreement. Report of United Nations Conference on Environment and Development: Rio declaration on environment and development, UN GAOR 47th Session. Annexure 1, Agenda Item 21, art 10. Available online at http://www.uneco.org/ECA/-Documents.html [Accessed June 2005]

UNDP (1997) United Nations Development Plan. Defining core characteristics of good governance. New York, NY: United Nations Development Programme, Management Development and Governance Division

Van Kersbergen K, Van Waarden F (2001) Shifts in governance: problems of legitimacy and accountability. The Hague, MAGW (Social Science Research Council), for: NWO (Netherlands Organisation for Scientific Research) Strategic Plan, 2002-2005, pp 77

Winter E (2004) Public communication of science and technology. German and European perspectives. Science Communication **25**(3):288-293

An Institutional Perspective on Governance – The Evolution of Integrated River Management in Victoria, Australia

Jane M Doolan

Abstract

Integrated river management in the state of Victoria, Australia has evolved considerably over the last fifteen years on a range of fronts, including the policy framework, level of community involvement, the technical base and the institutional arrangements. The Victorian River Health Programme is now regarded as the most successful river management and restoration programme in Australia. This chapter describes that programme, looks at how it has evolved over the last fifteen years in particular, and analyses some of the key factors influencing this evolution. It then examines the Trialogue Model for ecosystem governance in the light of the Victorian experience, concluding that, whilst the Trialogue hypothesis is correct at a very high level, it needs to build in a third dimension – that of *time*. This recognises that action at any one point in time can only occur to the extent supported by the three components taken together, and that evolution will occur only incrementally, building on past achievements and knowledge.

Keywords: ecosystem governance, institutional arrangements, river management, Australia, community involvement

Introduction

Victoria's rivers and streams are showing significant signs of degradation. Only 22% of Victoria's major rivers and tributaries are in good or excellent environmental condition, and one-third are in poor or very poor environmental condition. In addition, two thirds of the State's wetlands are in poor or very poor condition and nearly half of all its estuaries have been damaged. Many of these may still be on a downward trajectory. The seriousness of this situation, and the likely long-term impacts of this on regional economies and the wellbeing of communities, has been recognised by the Victorian Government who have set a target:

*To achieve significant improvements in the health of Victoria's rivers, floodplains and estuaries by 2010 to ensure that they are capable of delivering a wide range of services to the community (*Government of Victoria 2004*).*

The Victorian River Health Programme is the means for achieving this ambitious target.

The Victorian River Health Programme

The Victorian River Health Programme (VRHP) is a major government programme that is aimed at protecting healthy rivers and enhancing those in poor condition. The programme is an integrated one in two significant ways. Firstly, it aims to tackle the key causes of poor river health through integrated management – by collectively dealing with the problems of changed flow regimes, declining water quality and degraded riverine and floodplain habitats. Secondly, it is integrated in an institutional sense – integrating statewide policy and investment, regional planning and service delivery programmes, resource condition monitoring and reporting, and research and capability building.

The programme is built on a foundation of community involvement, recognising that it is only with long-term support, commitment and a willingness to undergo change from both the community and Government that any improvement in Victoria's rivers can be made. In line with this, the programme is based on the following philosophical principles that recognise this need for community support and involvement:

- That a healthy economy and society are dependent on a healthy environment.
- That, in the protection and restoration of rivers, trade-offs will need to be made between human use and the environmental condition of rivers and streams.
- That these trade-offs need to be made in open and transparent decision-making processes involving regional communities and stakeholder groups.
- That these decision-making processes are best undertaken at the regional level where they will have to be implemented (in line with the principle of subsidiarity), but within strong statewide policy frameworks.
- That these decision-making processes are based on the best available science within an adaptive management framework, improving in re-

sponse to knowledge gained through monitoring and research (DNRE 2002, Government of Victoria 2004).

The VRHP is effectively a programme where the Government aims to work in partnership with the community to achieve its goal and vision for river health, based on a continually-improving scientific and technical information base.

From an institutional perspective the VRHP has four key components, which work together to achieve the above aim:

- **Statewide policy and investment frameworks,** which embed the concept of working in partnerships with the community within an integrated catchment context. These statewide policy frameworks are outlined for the most part in two key documents – the *Victorian River Health Strategy* (DNRE 2002) and *Our Water Our Future* (Government of Victoria 2004). These frameworks:
  - Outline the statewide vision and targets for integrated river protection and restoration,
  - Detail policy for regional planning, priority and target setting in integrated river management,
  - Describe policy for specific management actions/functions which impact on river health (such as water allocation and the establishment of Environmental Water Reserves, management of water quality, riparian land, river channels and floodplains), and
  - Provide direction for the allocation of funds for regional integrated river management work programmes.
- **Regional institutional arrangements,** which are community-based and which deliver:
  - Integrated regional planning and priority setting for integrated river protection and restoration,
  - Integrated river protection and restoration work programmes, and
  - Achievement of regional river protection and restoration targets.
- **Resource condition monitoring, assessment and reporting.**
- **Innovation and continuous improvement programme**, with investment in knowledge generation and sharing, tools to assist in integration and decision-making and capability building in each of the above components.

The programme also has a funding base provided by the Victorian and Australian Governments and additional funds provided through an environmental component of water pricing. These components work together to form an adaptive, statewide programme.

Institutional Arrangements Supporting Integrated River Management

Whilst there are many agencies, groups and landholders in a catchment whose activities impact on river health, the key groups responsible for delivering the VRHP are:

- At the statewide level, the Victorian Department of Sustainability and Environment (DSE) whose role is to:
  - Set statewide policy and strategic directions for river restoration and for catchment and environmental protection,
  - Establish legislative frameworks and effective catchment/regional institutional arrangements,
  - Invest to achieve State and regional priorities,
  - Provide relevant advice, research and monitoring, planning, communication and some referral and enforcement functions, and
  - Participate in intergovernment processes and national approaches.
- At the regional level, Catchment Management Authorities (CMAs) – community-based statutory authorities – whose role is to:
  - Undertake strategic planning and community-engagement for land and water throughout their catchments, and
  - Act as a caretaker of river health with service-delivery responsibilities for:-
    - Integrated regional river planning,
    - Waterway management including riparian and instream habitat restoration, erosion control, management of sand and sediment, control of exotic species,
    - Operational management of the Environmental Water Reserve and provide river health input to water allocation decisions,
    - Coordination of water quality management,
    - Floodplain management,
    - Coordination of regional drainage,
    - Licensing works,
    - Community engagement, and
    - Monitoring and reporting of river health.

The full set of roles and responsibilities for the management of river health is outlined in DNRE (2002), whilst the full suite of functions of CMAs are detailed in Government of Victoria (2004), which also outlines the slightly different arrangements in place in metropolitan Melbourne.

In establishing CMAs across the State, two of the main aims of Government were to create:

- Community-based organisations that would ensure a high level of community input and engagement in the development and implementation of catchment management and river health programmes; and
- Organisations whose role was to provide leadership and effectively 'care' about rivers in their region. The CMAs are intended to fill a major institutional gap that commonly occurs in the management of rivers around the world; that is, whilst many groups are responsible for activities that impact on rivers, no one group has had the overall responsibility for the resultant environmental condition of the rivers. The CMAs have been given responsibility for a range of functions that impact directly on rivers (as listed above) and are expected to integrate the management of these functions to provide the highest level of river health outcomes possible for the funds invested. Most recently – in 2005 – the function of operational management of the Environmental Water Reserve was added to facilitate a truly integrated river management and restoration programme.

These two key groups, the DSE and the CMAs, develop partnerships with other key agencies and groups who have a role in river health, such as the Environment Protection Authority, rural and urban water authorities, local government, industry and landholders. There is also a strong relationship between the two groups, with the people working in each of these areas having a shared understanding of the programme, their role within the bigger picture of the programme, and the roles and requirements of the other players. This relationship is facilitated through one standing forum – the Waterway Managers Forum – with representation from the state policy group and regional service deliverers, and which discusses all matters of common policy, legislation and operational issues.

Linkages with Science

These two key groups also have close relationships with scientists working in the area of river health and enter into a range of research partnerships that support the overall programme. Most of the policies and practices have been developed with the input of formal scientific reference groups, representing a range of disciplines. The relationships between scientists and managers from both groups have been greatly facilitated by the creation of Cooperative Research Centres (CRCs), which are partnerships co-funded by the Australian Government between management agencies and research institutions. The VRHP has strong linkages with two river-related CRCs.

The VRHP is considered successful because it has:

- A strong, clear policy framework;
- Regional service deliverers with strong links to their communities;
- Funding;
- An evolving scientific/technical base; and
- Real government commitment to improving river condition.

The Path of Evolution

Twenty five years ago the picture was very different. In the period from 1950 through to 1980 there was no statewide policy or clear direction on river management. Regionally, there were 34 very small River Improvement and Drainage Trusts, which generally managed short reaches of rivers with the input of riparian landholders, on extremely limited budgets (Standing Consultative Committee of River Improvement 1983). The focus of these Trusts was extremely limited, concentrating on localised erosion control to protect public and private assets, flood protection and the drainage of arable land (DWR 1989). This reflected a very limited understanding of the importance of rivers to the community, which saw them often as a threat to be managed rather than a community asset or a simple water source. Work was undertaken by these trusts with little understanding of the river processes involved and the interdependencies between land management in the catchment and river condition, and often caused further environmental degradation. Water allocation processes did not recognise the environmental water requirements of rivers and generally did not provide water for the environment unless there were riparian rights involved. River scientists were not in the picture at all, working within their own disciplines, oblivious to the issues and challenges facing managers.

The picture changed somewhat in the 1980s with a review of river management across the State (Public Bodies Review Committee 1983, Standing Consultative Committee of River Improvement 1983). This set a general direction of catchment-based management arrangements. In response to this, between 1983 and 1996, a small statewide policy group was established within Government which facilitated the evolution of River Improvement Trusts into whole-catchment Waterway Management Authorities (WMAs). These WMAs had river catchments as their geographic areas of responsibility, rather than merely river reaches. They were community-based but, rather than just involving riparian landholders, they were encouraged to involve as many people as possible within the catchment and provide a leadership role on river issues. They were able to impose fees, under a tariff, across the catchment to support their activities. These authorities still had a significant focus on erosion control but had the capacity to treat the problem at the cause, wherever that occurred within the catchment. They also started to broaden their interest into other aspects of river condition, most notably into riparian condition and management of water quality issues. By 1996, WMAs covered approximately 40% of the State, leaving 60% without any river management service delivery at all, although at that stage proposals had been put forward by community steering groups for new authorities which would have increased the State coverage to 70%.

During this period some interaction between scientists and river managers started to occur, with the encouragement of the statewide policy group. A number of WMAs undertook geomorphological studies of the rivers in their catchments and investigations of some aspects of river ecology. River scientists started to operate in multidisciplinary teams, with particular interactions between river ecologists and water chemists and between river engineers and geomorphologists. Water allocation decision-making processes, whilst not involving the WMAs, did start to take into account environmental issues, although mostly as a secondary consideration. However, even this was limited by the very basic nature of the science of assessing environmental flow requirements. The results produced by these initial studies did not engender great confidence.

A further significant step in the evolution of integrated river management occurred in 1996 with a statewide review of catchment management arrangements. As a result of this review, the CMAs were established across the State in 1997, combining a strategic planning and coordination role for land and water management and the service delivery role of integrated river management as described above. In the period since 1997 the statewide policy group developed the *Victorian River Health Strategy*, the statewide policy framework for integrated river management (DNRE 2002), which has consolidated the role of CMAs as the 'caretaker of river health.' This role was further developed with the additional function of operational management of the Environmental Water Reserve in 2004/05, conferred in a review of water resource management (Government of Victoria 2004).

During the period 1997 to the present day, water allocation decision-making has considered the needs of the environment more seriously, with CMAs sitting around the table as key stakeholders. In 2004 this culminated in the establishment of the Environmental Water Reserve to provide a legal share of water for the environment.

In addition, the number interactions between scientists and managers escalated, with a number of scientists in multidisciplinary centres working with both regional managers and State policy officers on a range of management issues and challenges.

Factors Influencing Evolution

A range of factors have significantly influenced the evolution of river management in Victoria over the last 25 years to the integrated VRHP that we see today. These include:

- The development of **community ownership and commitment** towards rivers which has created the political impetus for action. This is the result of –
  - Improved understanding of the full range of services that rivers provide, that is, their importance for environment, recreation, tourism and social wellbeing, as well as for water provision,
  - Improved understanding of the interdependencies between the environmental condition of a river and the capacity of a river to provide the full range of these services,
  - Studies that show the economic value provided to the regional communities of healthy rivers, for instance the economic income generated by recreational fishing, estimated to be AUS$400M in Victoria, and
  - Personal observations of environmental degradation.
- The development of a **better technical understanding of river condition and river processes**. Whilst there have been many areas of advancement in these knowledge areas, those that have most influenced the evolution of integrated river management have included –
  - The development and application of integrated river assessment tools, such as the benchmarking of the environmental condition of rivers using the Index of Stream Condition (ISC) (DNRE 2000), which assessed 950 river reaches in 1999 representing 18,000 km of major rivers and tributaries. The ISC is an integrated measure of river condition aggregating 18 individual river-related variables into five sub-indices related to hydrology, water quality, aquatic biota, riparian health and the physical form of the river. The ISC was extremely important in an educative sense, in that it did not just focus on one variable related to a particular academic discipline, but described the major key factors contributing to river health that could be intuitively understood by community members. The statewide benchmarking

provided the first statewide picture of river condition, describing for the first time the extent and magnitude of the problem,

- The development of catchment modelling and decision support tools. These tools allowed the integration of best available science into a form that was readily useable by river managers, and enabled the development of 'what if' scenarios and options to improve regional river planning exercises and communicate with their communities,
- The development of river asset/value databases (DNRE 2002), which take data from the statewide benchmarking of river condition and other sources and describes, in a consistent, statewide fashion, the environmental, social and economic assets/values associated with each of the major river reaches across the State. This changed the focus of regional planning and provided a true vehicle for integrated management. These databases allow an assessment of the overall community value of the reach, the identification of priority reaches for protection and restoration, analyses of all the issues threatening these assets and the development of integrated works programmes to address them, and
- The development of methodologies for assessing the environmental flow requirements of rivers, which provided the information to input into water allocation decisions, showing the environmental risks and consequences associated with various water extraction scenarios. This has provided the technical basis to fully integrate river management with water allocation decisions and ultimately to enable the establishment of Environmental Water Reserves for all rivers across Victoria.

• The development of **improved regional institutional structures**. As mentioned above, these structures have evolved from the scattered, narrowly-focused, self-interested River Improvement Trusts in the period 1950s-1970s, to whole- catchment Waterway Management Authorities in the period 1980s to 1996 covering 40% of the State, to the statewide coverage by CMAs as caretakers of river health that we see today. The development has occurred particularly in the breadth of river-related issues that are the responsibility of these authorities and in the status and capacity that they now have. This growth was clearly incremental with each institutional stage building on the previous one. The growth was enabled by –
  - The increasing awareness, in both the community and the government, that with each institutional interaction there were still relevant and related issues that had not been covered which were likely to impact on river health if not addressed,

  - The increasing awareness of both the community and the government that all these river-related issues were most effectively dealt with in an integrated way, and
  - The authorities starting to develop a track record in river restoration and catchment management that engendered the development of general, though not total, confidence by community and government in them and in their capacity to take on new functions.
- The development of **statewide integrated policy frameworks** for river restoration and protection, which outlined clearly the Government's direction, management approaches, targets and specific policies. These became the focus for additional government funding and for communicating Government's commitment to river health to the community. They were also incorporated into performance agreements with the Australian Government.

How does the Evolution of the VRHP fit with the Trialogue Hypothesis?

The Trialogue hypothesis states that the degree to which governance is successful depends on six essential elements – the science process, the Government process, the society process and the interfaces between these three processes. At a broad level, this is undoubtedly true with respect to the VRHP. Each of these elements can be seen in its path of evolution. However, acknowledging this does not help a great deal. To gain some real insight, it is important to understand how each of these interact with each other, as well as together, particularly over time.

As mentioned above, the evolution of the VRHP has occurred incrementally, with development occurring strongly in each of the following key areas:

- Community understanding and ownership (the society processes);
- Technical underpinning (science processes);
- Regional institutional structures (government and society processes and their interface); and
- State-wide policy frameworks (government processes).

However, at any point in the path of evolution the actual institutional sophistication and construction of the VRHP reflected not just where each of the key elements (i.e. government, society and science) were in their development, but how well that development had been communicated and believed by the other key groups, prompting them to act or not.

For example, the decision by Government post-1983 to move to whole-catchment WMAs was taken after it had been established by scientists that river condition *was* degrading, and that the former narrow focus on bed- and bank-erosion and flood-protection was not helping and, in some cases, was accelerating the degradation. The need to take a whole-catchment approach to the management of rivers was supported by the scientific understanding of the time. This conclusion resonated with those members of the community who cared about rivers. It supported their own observations, and thus Government could begin to act, and consequently moved to establish WMAs. However, it should be noted that these did not spring up overnight. They could only be established where there was a high level of community support. In those areas where this was not present, further studies and discussions were undertaken with the community.

Similarly, the decision by Government to create the Environmental Water Reserve, and to have it managed as part of integrated river management, reflected the scientific understanding that flow was one of the key drivers of river ecosystems and that you couldn't successfully manage river health whilst flow remained outside the management loop. This was a more difficult proposition to implement, in that the numbers of community groups affected by the proposition grew to include water-users. They had to be convinced of the merit and need for the proposition before it could be introduced.

These decisions by Government, implemented with community acceptance, then threw up further technical challenges. In many ways, science provided the basic evidence for change, but actually did not provide anything particularly useful to help manage rivers by means of the new models. For example, the decision to manage within an integrated catchment management (ICM) context threw up the need for integrated modelling tools. These were not available before, but became an immediate challenge and need when managers were trying to operate within an ICM framework. Similarly, the decision to create and manage the Environmental Water Reserve has generated the need for integrated river operations tools, and these have now been accepted as a high priority for the next 'wave' of river research.

The experience with the evolution of the VRHP has shown that action could be instigated by any of the three groups, but could only be implemented once the other two groups had been convinced of the need for it. Moreover, the extent of action or programme development actually taken was limited to that supported by the information available and believed by the community.

Evolution only occurred at the pace at which Government could act on the basis of credible technical information and with the support of the community. There were no quantum leaps into the unknown. This is not surprising, as any action will have economic consequences on some if not all sectors of the community – either in increased taxes or charges or personal losses. Moreover, observations elsewhere, in regions where Government policy has outstripped community acceptance or where the community have tacitly accepted a Government direction without fully understanding the consequences for them, suggest that, in democracies at least, these policies will not be implemented but will either remain as rhetoric or be rejected by succeeding governments.

In summary, whilst the Trialogue hypothesis is correct at a very high level, it needs to build in a third dimension – that of time. This recognises that action at any one point in time can only occur to the extent supported by the three components taken together, and that any evolution will occur only incrementally, building on past achievements and knowledge.

Future Challenges

Whilst, in our view, the VRHP is now a successful programme, there are a number of key challenges that will have to be met in order to ensure both its continuation and any future progression. These have implications for all three components of governance – government, science and society. They include:

- Demonstrating and communicating river restoration successes to the community in the short term to maintain commitment over the long-term;
- Retaining community interest in urban areas who mostly pay for river restoration when action is mostly in rural regions;
- Increasing community's understanding, so that they can tackle some very difficult trade-off decisions, particularly in the area of water allocation, water recovery to enhance environmental flows;
- Retaining Government interest in the long-term against the plethora of competing demands;
- Understanding fully the implications of new risks, such as climate change or new exotic species; and
- Being able to describe the end-point of river restoration, i.e. ecologically healthy rivers, particularly in systems that are fundamentally changed by human activities, such as downstream of large dams.

It is only with the continued commitment and involvement of all three groups that we will achieve our targets and our vision for rivers in Victoria.

References

Department of Natural Resources and Environment-Victoria (DNRE) (2000) Victorian rivers: an environmental report card. Department of Natural Resources and Environment, Victoria

Department of Natural Resources and Environment-Victoria (DNRE) (2002) Victorian river health strategy. Department of Natural Resources and Environment Victoria, Australia

Division of Water Resources (DWR) (1989) Water Victoria: an environmental handbook. Department of Water Resources Victoria, Australia

Government of Victoria (2004) Our water our future: securing our water future together. White Paper. Department of Sustainability and Environment Victoria, Australia

Public Bodies Review Committee (1983) Future structures for water management. Report to Parliament. Government Printer, Melbourne

Standing Consultative Committee of River Improvement (1983) The state of the rivers. Report. Department of Conservation, Forests and Lands and State Rivers and Water Supply Commission, Victoria, Australia

Conclusion

Ecosystem Governance and the Trialogue Debate: An Overview of the Trialogue Relationship and the Engagement along Interfaces

Linda Godfrey

Abstract

Sustainable development is recognised as being core to the concept of good ecosystem governance, as is the amity of the relationship between government and society in the co-management of the environment. This relationship between government and society is a complex one, influenced by certain factors, including political and socio-economic systems, societal culture and science and technology. Science is believed to play a fundamental role in (i) understanding the relationship between government and society; (ii) capacitating society to enable them to engage effectively with government; and (iii) supporting government in the development of scientifically-sound policies and programmes, which aim to find a balance between development and ecosystem protection.

The Trialogue that develops between government, society and science and the engagement of the three partners along the Trialogue interfaces, is a dynamic and complex interaction, influenced by the political system of a country, the maturity and age of its democracy, the culture of the government departments and the conditions of society. Three models which look at the strength and rate of engagement along the Trialogue interfaces in an undemocratic society, a fledgling democracy and a maturing democracy are briefly discussed.

Keywords: Trialogue, ecosystem governance, stakeholder engagement, fledgling democracy, mature democracy

Ecosystem Governance

What is ecosystem governance? Governance, and in particular ecosystem governance as it relates to integrated water resource management, has been the topic of much debate in the literature, particularly as it relates to a clear definition (Rogers and Hall 2003; Imperial 2005). Ecosystem governance is seen by the author as the interaction between government and society, which enables socio-economic development and thereby poverty alleviation while ensuring ecosystem protection, i.e. the co-management of the environment by government and society to ensure sustainable resource utilisation for both current and future generations (Figure 16.1).

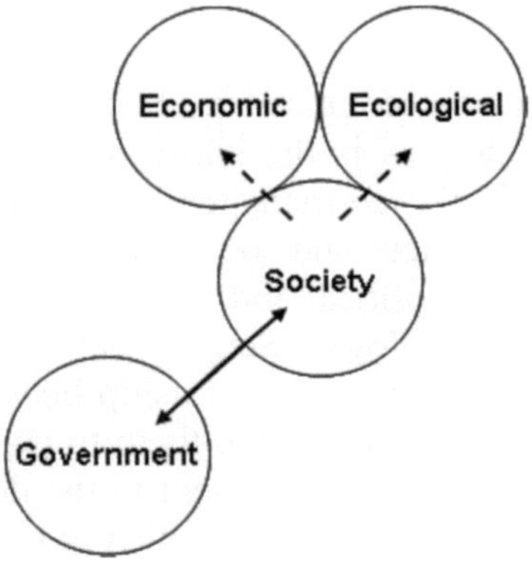

Fig. 16.1. *Governance seen as the interaction between government and society and the environment within which society functions*

Inherent in this definition is the need to identify, establish and implement "*mechanisms, processes and institutions through which citizens and groups articulate their interests, exercise their legal rights, meet their obligations and mediate their differences*" (UNDP 2001 in: Rogers and Hall 2003:7).

This need to co-manage is based on government's belief that it can no longer solve societal problems acting alone, particularly socio-environmental ones, and the private sector's belief that it cannot address the problems of the poor and the environment on its own (Imperial 2003; 2004; 2005, Rogers and Hall 2003).

The joint role of government and society in ensuring sustainable resource utilisation is central to the principles of the Rio Declaration on Environment and Development (UNCED 1992), in particular Principles 10, 20, 21, 22 and 27, which identify the roles of citizens, women, the youth and indigenous people in ensuring sustainable development. "*One of the fundamental prerequisites for the achievement of sustainable development is broad public participation in decision-making*" (UNCED 1992:219). This need for the co-management of a country's resources in ensuring sustainable resource utilisation is based on the ever increasing complexity of issues facing societies (Holmberg and Karlsson 1992 in: Robèrt et al. 1997), which require firstly a greater level of awareness and understanding by society and, secondly, a greater involvement by society in the decision-making process in order to find solutions that are acceptable to all parties (Gooch 2005a, Imperial 2004). For many developing countries, this approach to co-operative governance means a fundamental shift in the way in which a country manages its resources, from one of government governing (i.e. centralised command and control) to one of partnering with society, who have a vested interest in the sustainable management of the country's resources. This represents a change in environmental decision-making from one of "decide-announce-defend" to a "more inclusive and deliberative approach 'meet-understand-manage'" (Parr et al. 2003:5).

The sustainable utilisation of South Africa's water resources is fundamental to the National Water Act (1998: Chapter 1) which recognises "*the need to protect water resources, and the need to promote social and economic development through the use of water.*" As a developing country, a large percentage of South Africa's population rely on aquatic resources and their associated goods and services for their basic survival. The alleviation of poverty and hunger in a country such as South Africa involves ensuring continued access to such primary resources while simultaneously promoting economic development and the access by society to secondary goods and services. According to the South African national Department of Water Affairs and Forestry; "*Sustainable utilisation requires achievement of a balance between an acceptable level of long-term protection of water resources and water users, and society's present requirements for economic growth and development*" (DWAF 1999:5).

This inter-relationship between social, economic and ecological systems is a sound one, fundamental to the concept of sustainable development (UNCED 1992, Mebratu 1998), as is the notion of an inter-relationship between these three systems, the 'environment' in its broadest sense, and the government or political system of a country (Ashton 2005 (Figure 16.2a), Domoto 2001, Gooch 2005a (Figure 16.2b)).

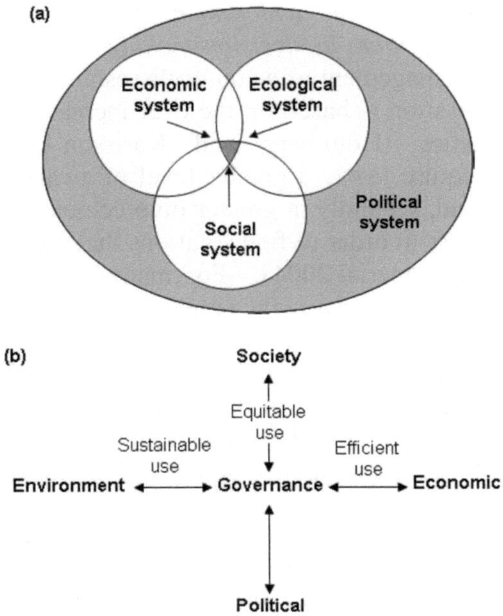

Fig. 16.2.(a)(b) *Conceptual diagram showing the four interacting systems – **(a)** (From Ashton and Chonguica 2003), **(b)** (From Gooch 2005a)*

From this interpretation of ecosystem governance, it is believed that governance can be interpreted as the relationship between government and its environment, defined in its broadest sense as comprising three systems, namely: social, economic and ecological (Mebratu 1998) (Figure 16.3).

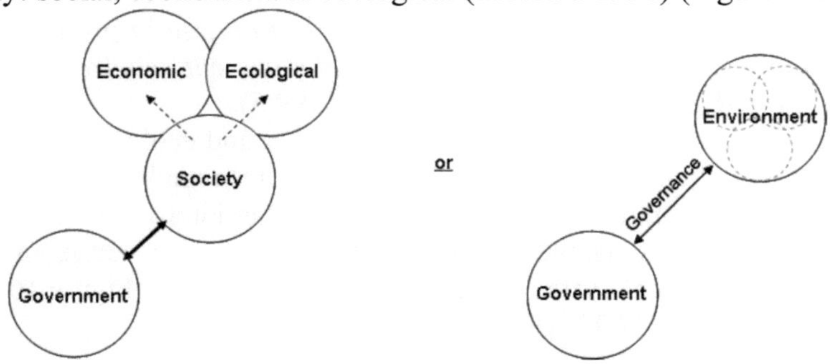

Fig. 16.3. *Governance and the government-society relationship*

This 'dialogue', defined as "*an exchange of views in the hope of ultimately reaching agreement*" (Chambers 1983:343) between government and society concerning their environment (Figure 16.3), is fundamental to the principle of 'good governance', which requires enabling conditions such as inclusiveness, accountability, participation, transparency, predictability and responsiveness (Rogers and Hall 2003).

The Governance Trialogue Model

The concept of a governance 'Trialogue' between government-society-science was put forward at the United Nations Conference on Environment and Development (UNCED) and summarised in Agenda 21 (UNCED 1992:240-242). Particular emphasis was placed on the need to strengthen the role of the scientific and technological community in "*decision-making processes concerning environment and development*" (UNCED 1992:240). The objectives to "*extend and open up the decision-making process and broaden the range of developmental and environmental issues where cooperation at all levels between the scientific and technological community and decision makers can take place*" and "*improve the exchange of knowledge and concerns between the scientific and technological community and the general public in order to enable policies and programmes to be better formulated, understood and supported*" were set by Agenda 21 (UNCED 1992:241).

This Trialogue discussion between government-society-science was again raised by Hattingh et al. 2005 and Turton et al. 2005 (Figure 16.4) and formed the basis for further discussion at a recent International Symposium on Ecosystems Governance held in South Africa in October 2005. At this meeting particular attention was paid to how the Trialogue relates to governance in fledgling versus mature democracies.

Based on the preceding discussion, the presence of government and society in the proposed Trialogue Model is not debated, other than to raise the point that society is part of a larger system which includes economic and ecological systems. In terms of this proposed government-society-science Trialogue model, the nature of the 'third corner' (currently occupied by science) would appear to be dependent upon the intended purpose of the Trialogue.

This '*third corner*' of the Trialogue is believed to provide a key viewpoint to understanding the underlying '*drivers*' of governance and, in particular, good governance. There would appear to be a number of important drivers to facilitating, understanding and supporting the relationship between government and society (Ashton 2005, Imperial 2004).

Such drivers include, amongst others:

- Administrative, political and legal systems;

- Finance and business;
- Societal culture; and
- Science and technology.

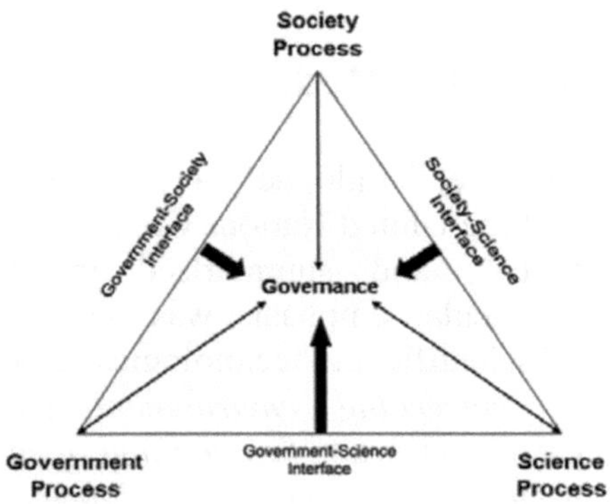

Fig. 16.4. *Proposed Trialogue Model (Hattingh et al. 2005)*

Each of these aspects has been addressed elsewhere in the literature and, with the exception of 'science', will not be explored in more detail here, other than to say that each one plays a critical role in understanding and fostering the relationship between government and society.

The focus of the following section is to explore the role that science plays in the governance relationship. Science is seen here in its broadest sense, to include:

- Formal science, housed by universities, research organisations, consultants and government, with specific sub-disciplines supporting the various components of the governance relationship, including:
  - Social,
  - Political science,
  - Environmental science,
  - Resource economics, and
  - The inter-relationship between these specialist domains.
- Informal, indigenous or citizen knowledge, typically housed by society.

The Role of Science in Governance

Science and technology were identified in Agenda 21 (UNCED 1992) as one of four means of implementing the programme areas aimed at achieving a sustainable level of protection of the quality and supply of freshwater resources. The way in which science was seen to achieve this was through research aimed at improving the understanding of the hydrological cycle and its inter-relationship with society, in particular through the development of information and expert systems, hydrological models, development of new and alternative low-cost technologies, development of endogenous capacity, monitoring and assessment of aquatic systems and ensuring integrated, long-term water resource planning (UNCED 1992:168-172).

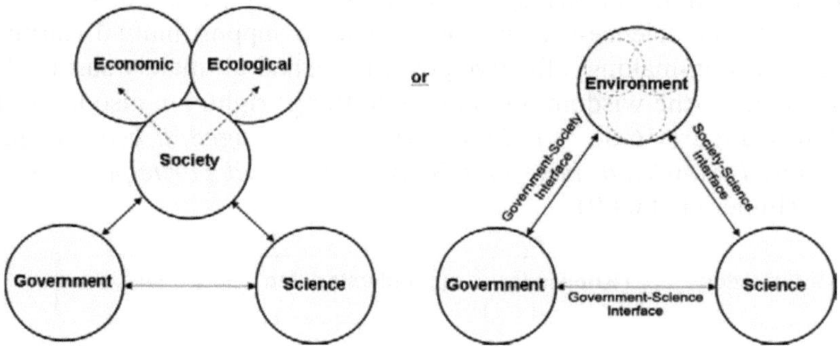

Fig. 16.5. *Role of science in the government-society relationship*

It is the author's opinion that the principle role of science, in the relationship between government and society (Figure 16.5), is to:

- Provide the scientific and technical understanding of this inter-relationship and its dynamics;
- Through appropriate means of communication, capacitate and support society so as to enable them to engage meaningfully with government; and
- Through a leadership or guiding role, use this understanding or knowledge to support government in balancing development and ecosystem protection through the development of sound policy and programmes, and technologies.

These roles are based on the stated need for governance systems, policies and decision-making processes to be '*informed by science*' (Magadlela 2005), '*based on science*' (Imperial 2004:1) and to be '*scientifically robust*' (UNCED 1992:258). It is believed that "*Government can typically only act when it is supported by credible technical information and/or upon support by the community* (society)" (Doolan 2005a). However, according to Imperial (2003:3), "*science is typically most influential in identifying problems, framing options, and evaluating progress but it rarely tells decision makers what to do.*"

Science is therefore believed to provide a sound and reliable knowledge base from which government and society can draw to support informed decision-making with respect to ecosystem management. The challenge therefore is to get government and society to see the value in this knowledge base, and then to engage with science along the Trialogue interfaces.

The Trialogue is therefore seen as a mechanism for facilitating the sharing of information, knowledge and wisdom (Imperial 2004, Parr et al. 2003) between science-society-government to support mutual learning, informed decision-making, effective governance and ultimately sustainable development, i.e. the wisdom not only to do things right, but also to do the right things (Figure 16.6). "*As information is exchanged, it becomes part of the shared knowledge base that is 'owned' by all participants in the process*" (Imperial 2004:9).

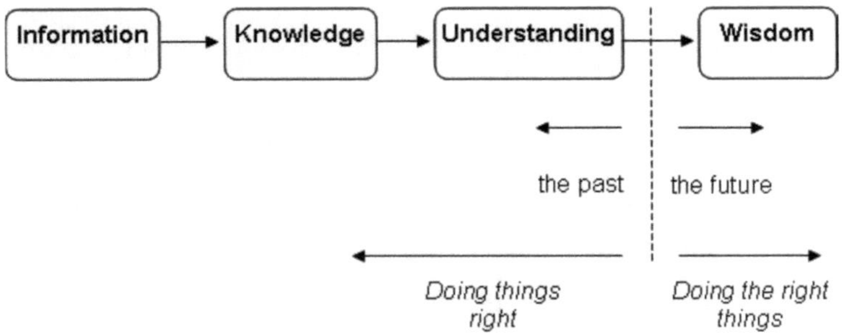

Fig. 16.6. *The path to wisdom (adapted from Miller and Morris 1999:87)*

For this Trialogue model to work, the principles of good governance require decision makers to create an environment conducive to independent research (UNCED 1992:240), buy-in by government and society to new approaches and technologies provided by science (Doolan 2005a), government and science to listen and respond to societal demands (Gooch 2005b:14), science to respond to government-society interventions (Doolan 2005a), and science to recognise the need to engage proactively with government and society. A trust relationship between the three partners, built on a common vision (Robèrt et al. 1997) is seen as vital to the Trialogue success.

The building of this common vision requires an investment from science in sharing existing knowledge. This should however not be done at the expense of the "*extension of that knowledge*" (Robèrt et al. 1997:80). The

United Nations identified the need continuously to "*enhance scientific understanding, improve long-term scientific assessments, strengthen scientific capabilities in all countries and ensure that the sciences are responsive to emerging needs*" (UNCED 1992:257).

The Trialogue Interfaces

The need for science to engage with government and society to ensure good governance and to support the principles of sustainable development are recognised. However, the strength of this engagement between the three, and the rate at which this engagement happens, needs to be looked at more closely to understand the implications of the interaction for good ecosystem governance.

The strength and activity of the interface between the three Trialogue partners appears to be dependent upon a number of factors, including:

- The political system of the country (i.e. democracy);
- The maturity and age of the democracy;
- The culture of the specific government departments engaged in the process; and
- The socio-economic conditions of the society.

Each of these four points is touched on in the following sections, relating specifically to the strength of engagement along the interfaces and the rate of engagement between players.

Strength of Engagement

Three possible models reflecting the strength of engagement along the Trialogue interfaces are presented in the following section, based on the political system and the relative maturity of democracy. These models are based on the author's personal experience working with government departments in a young democratic society, South Africa. It must be stressed that the inter-relationship between the three partners is neither as simple nor as clear-cut as is presented here. There are many nuances that influence the Trialogue interfaces which, in order to understand governance truly, need to be examined in much greater detail than can is possible here. It is also recognised that every country has its own unique set of physical, political, socio-economic, institutional and environmental characteristics and issues which influence the models presented here.

In an undemocratic society (Model 1) (Figure 16.7) the interface between government and society and between government and science is typically weak, with governance processes usually restricted to government.

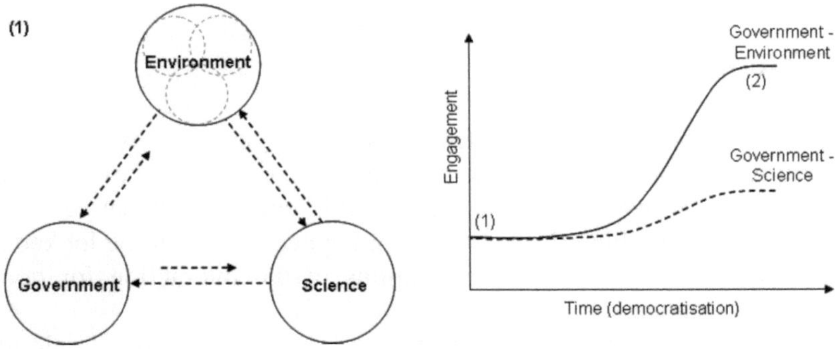

Fig. 16.7. *Model 1 – inter-relationship in an undemocratic society (weak-weak-strong)*

Here a strong society tries to engage with government, but without equal and opposite engagement, and typically with little success. This often leads to disputes and conflict between society and government. The same applies to science when trying to engage with government. In such a situation, science may take on an activist role, often using society to place pressure on government in order to be heard. This can result in science losing its independent, unbiased position, which may later reflect poorly on science when government does begin to engage with society in a more democratic environment. When science takes on the role of activist, society may feel used while government may feel threatened. Scientists and researchers who have played an active role in campaigning against government may typically break away from the formal science institutions to support non-governmental organisations (NGOs) as so-called "watchdogs" of government.

As the political situation of the country changes to one of a young democracy, as experienced in South Africa over the past 10 years, the strength of engagement along the Trialogue interfaces takes on a new dynamic, i.e. a shift from a purely government focus (1) to one of wider governance (2).

In the second model (Figure 16.8), that of a young democracy, there is typically a strong interaction between government and society, in ensuring that policy, legislation, strategy and programmes are developed and implemented in an open and transparent manner, i.e. less of a centralised command and control approach by government.

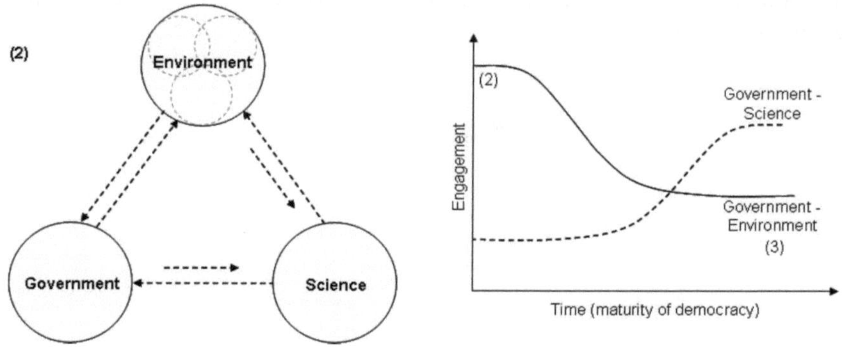

Fig. 16.8. *Model 2 – inter-relationship in a young democracy (strong-weak-weak)*

This is typically done to redress past discriminations and ensure that the voices of all people (society) are heard when outlining the requirements and principles for the future governance of the country. This is often done without the inclusion of, or consultation with, the scientific community.

There are several possible reasons for this lack of inclusion of science in governance in this model; these include:

- The value that science can bring to the discussion is not fully understood by government or society;
- Science typically does not communicate with government in a manner that is easily understood, particularly for government officials with a purely political rather than a technological or scientific background;
- In a non-democratic society, such as that which prevailed in South Africa pre-1994, the scientific community was seen to have abetted the apartheid regime (Model 1) and could therefore not be fully trusted in the new democracy; and
- The issues facing society are a lot more complex and the environment in which society lives a lot more uncertain. A decline in the level of confidence with which science can predict changes in and to the environment, and the resultant risks of such changes, has resulted in society being sceptical of the role of science (Gooch 2005b:2).

According to Domoto (2001:10) "*a damaging communication gap continues to lie between those who understand environmental problems (scientists) and those who have the political wherewithal to do something about them (policy makers).*"

In the third model (Figure 16.9), engagement between government and society in a maturing democracy is seen to diminish; however this is accompanied by a strengthening relationship between society-science and government-science.

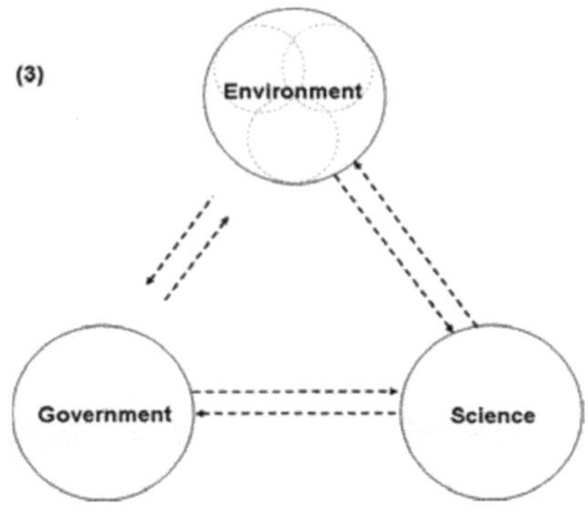

Fig. 16.9. *Model 3 – inter-relationship in a maturing democracy (weak-strong-strong)*

Government now see themselves as being in a position of greater understanding and control, as it relates to the needs of society. Government may also have been 'burnt' by previous stakeholder engagement processes and instead now seeks to avoid interaction with society, so as not to compromise or delay policy development and implementation (Ashton 2005:7). This is similar to Model 1, but there is now more open communication between science and government and science and society. Here science can play a fundamental role in supporting governance by mediating the relationship between government and society, through sound research and knowledge creation, assuming there is agreement by both society and government on the findings of science. Conversely, the independent use of science by government and society can result in conflicting opinions and result in disputes over ecosystem management (Gooch 2005b:11). Science is typically 'trusted' more in a mature democracy, often justifiably so because of the strong science base of a mature democracy.

Here the dependence by government upon science in fully understanding the needs and demands of society, within a complex hydrological system, may result in accusations of slow implementation and service delivery by society. The syndrome is one where government needs to understand and comprehend fully all consequences and implications before taking action. As outlined by Mbedzi et al. (2005:7) "*as we find our place along that government-science interface the pressure from society mounts, with some calling for more delivery.*" Such accusations are surfacing in South

Africa with regard to the implementation of the National Water Act (1998). Now, seven years since the Act was promulgated, many fundamental concepts and approaches still need to be developed, e.g. the national classification system, implementation of catchment management agencies (CMAs), etc., resulting in cascading effects that retard implementation and service delivery, e.g. reserve determinations and water allocations.

This may be compared to the situation in Australia, which has already successfully adopted and implemented certain South African integrated water resource management concepts (Doolan 2005a), recognising the need to implement them within an adaptive management framework.

It must be stressed that the two Models (2) and (3) may not necessarily be end-points on the scale of democracy, but rather be stages within a cyclical or spiral evolution of governance.

Rate of Engagement

The rate at which engagement between government and science increases is often a symptom of the culture of the specific government department(s), whether they be local, provincial or national. This is clearly evident in South Africa, where certain government departments have embraced the role that science can play in supporting their function, while others have remained sceptical of the intent of science and prefer to base their engagement with society on their own internal specialist knowledge. Possible reasons for this sceptical approach by some government agencies towards science may be:

- Fear of opposition against policies or programmes;
- Fear of delays resulting in the development and implementation of policies;
- Fear of drawing undue attention to internal institutional weaknesses, particularly as these may relate to technical capacity;
- Uncertainty within government as to what to do with the knowledge/information provided by science; and
- Uncertainty about the relevance of the information provided by science.

In South Africa, a weaker engagement between government and science may be reasonable in a government department that has retained or maintained a high level of internal specialist knowledge, but can prove problematic in a department which underwent radical transformation since 1994, with a loss of core specialist knowledge and understanding. Where the transformation of government departments has resulted in a loss of their technical capacity, it is important that these departments now either re-develop their own internal technical capacity, or establish close working partnerships with other governmental and non-governmental organisations, including science and technology organisations, which can provide this knowledge service (IUCN-ROSA 2001 in: Ashton 2005:5). Here capacity is referred to in the strategic sense of government's ability to "*anticipate and influence change; make informed and intelligent policy decisions; attract, absorb, and manage resources; and evaluate current activities to guide future action*" (Honadle 1981 in: Imperial 2004:16).

With the excessive levels of bureaucracy and administration introduced into government departments in a fledgling democracy, often introduced to deal with corruption or mistrust of senior management, government officials often end up overwhelmed with paper work, becoming project managers rather than technical specialists. To maintain a diverse array of specialist knowledge within a government department is therefore difficult, particularly with the high turnover rate of government officials often experienced in fledgling democracies. Discussions with South African government officials indicate that the average employment period of technical staff in certain provincial government departments is around eight months, while the figure given for a national government department is three years. Because of this turbulent nature of government departments within fledgling democracies, the establishment of partnerships between government and science is seen as being a particularly relevant way to ensure continued sustainable development.

Due to the nature and complexity of the issues facing government and society, it is now no longer just good enough for government simply to have technical expertise, but they must now also have knowledge of the dynamic interactions between the socio-economic system and the ecological system. "*Major challenges in ecosystem and water management include successfully integrating knowledge from the natural and social sciences*" (Gooch 2005b:11). According to Schreiner (2005), the *"difficulty experienced in transforming the Department of Water Affairs and Forestry has been one of transforming from an engineering/technical department, to one which understands and responds to social processes."*

The degree of success in managing South Africa's water resources in a sustainable manner into the future, depends on the establishment and implementation of a good governance system that involves all three partners in the Trialogue Model. The National Water Act (1998) has created the enabling environment for this; future success now depends on the rate and strength of engagement of government with society and science.

The level of engagement between government and society, and between government and science over time, as depicted in Figure 16.8, is never seen to peak at the same level. The engagement between government and society is believed to be much more dynamic over the life of a democratic country, than between government and science. This contention is supported by Doolan (2005b), who sees the three roles of governance (government-society-science) in Australia as being important, but not as equal partners in the governance process.

Does Government Influence Science, or Science Government?

The role that science plays in the governance relationship is also seen to shift between that of fundamental support (reactive) to strategic direction (proactive), which again may be dependent upon the maturity of the democracy. This is due to the nature of the role that science plays in governance in supporting the government-society relationship.

In a country with a weak science base, or where the science community has not been brought into the fold of governance (Model 2, Figure 16.8), science typically has to respond reactively to the needs of government and society in ensuring that policy and programmes are supported technically and maintained in the short- to medium-term. Here science has very little influence over government or society in terms of where policy should be directed in the long-term.

In a mature, or maturing democracy, (Model 3, Figure 16.9), science is believed to have more of a say in terms of the strategic direction of government policies and societies needs. In such an environment, science needs to focus on long-term cutting-edge research, or research that will influence the way in which ecosystems are governed 10 to 20 years from now. According to Doolan (2005b), government departments often have very little leeway in terms of altering the course of policy or programmes, other than through minor '*tweaking*'. Or they may be reluctant to change policy and programmes, which have taken years to develop and implement (Imperial 2003), given the often high levels of scientific uncertainty. Fundamental changes to the way in which ecosystems are governed can only be made through future policy.

In such a situation, the focus of science should be on research that will inform future policy and strategy, thereby providing science with a mechanism to influence the governance of ecosystems. Such research should focus on situations where:

- The problem is known, but solutions are not clear to government or society, and

- Future problems still need to be identified. Such problems are typically identified by allowing science to explore uncharted areas of research.

In reality a combination of both types of science (reactive and proactive), have a critical role to play in supporting ecosystem governance.

The Power Play

In the proposed Trialogue Model (Turton et al. 2005), the question of roles and responsibilities and the position of power is raised. According to Doolan (2005b), on-the-ground action typically only occurs when there is a convergence of the three partners, government-society-science. However, this begs the question: to reach this point of convergence with respect to ecosystem governance, who leads and who follows and how does one ensure that a power struggle does not "*create an obstacle to development*" (Rogers and Hall 2003:9), undermine good governance and thereby effective water resource management? A clear understanding of the roles and mandate of each party is therefore crucial to good ecosystem governance. Rogers and Hall (2003:13) put this very succinctly in the statement "*Modern governance can be about how to maintain some steering capacity in a world full of external (and internal) societal independence.*" It is the opinion of the author that this 'steering' should be provided by government, through the creation of an enabling environment for sound economic development, an environment which listens to and responds to society's needs for the 'common good', and which values the guidance and knowledge provided by science in understanding the complex environment in which we function, and the implications of our decisions.

Conclusion

From the review of presented work and the author's interpretation of ecosystem governance, the role of science in the dialogue between government and society is seen to be a credible one, providing a valued third partner to the Trialogue Model. This is particularly true if the intention of the Trialogue is to identify and nurture research and development to support governance, whether it be in a fledgling or mature democracy. The relationships between the three partners in the Trialogue are believed to be dynamic and complex, oscillating in their strength and activity of engagement, based on the political system of the country, the maturity of the democracy, the culture of the government departments, and the socio-economic conditions of society.

References

Ashton PJ, Chonguica E (2003) Issues and trends in the regional harmonization of EIA processes in Southern Africa, with special reference to trans-boundary and cumulative impacts. In: Assessing the need for a regional approach to environmental impact assessment in Southern Africa. Chonguica E, Brett R (eds) Harare, IUCN Regional Office for Southern Africa, pp 26-59

Ashton PJ (2005) The role of good governance in sustainable development: implications for integrated water resource management in Southern Africa. International Symposium on ecosystems governance, 10-13 October 2005, Kwa Maritane, South Africa

Chambers 20th Century Dictionary, New Edition (1983) Kirkpatrick EM (ed) Edinburgh, W&R Chambers Ltd

Department of Water Affairs and Forestry (1999) Water resources protection policy implementation. Resource directed measures for protection of water resources. Integrated Manual. DWAF Report N/28/99, Version 1.0, 29 September 1999

Doolan J (2005a) An institutional perspective on governance. International Symposium on ecosystems governance, 10-13 October 2005, Kwa Maritane, South Africa

Doolan J (2005b) Personal communication 11 October 2005. Department of Sustainability and Environment. Victoria, Australia

Domoto A (2001) International environmental governance: its impact on social and human development. World Summit for sustainable development international eminent persons meeting on inter-linkages strategies for bridging problems and solutions to work towards sustainable development, 3-4 September 2001. Available online at www.unu.edu/millennium/domoto.pdf [Accessed October 2005]

Gooch G (2005a) Presentation based on the paper: From dialogue to trialogue: sustainable governance and civil society. International Symposium on ecosystems governance, 10-13 October 2005, Kwa Maritane, South Africa

Gooch G (2005b) From dialogue to trialogue: sustainable governance and civil society. International Symposium on ecosystems governance, 10-13 October 2005, Kwa Maritane, South Africa

Hattingh J, Leaner J, Maree G, Strydom W, Turton A, Van Wyk E, Claassen M, Moolman M (2005) Towards innovative ways to ensure successful implementation of government tools. Report ENV-P-I 2004-083. Pretoria, CSIR

Imperial MT (2003) Improving ecosystem governance: lessons from watershed management programs in the United States. Summary of a presentation at the Workshop on the ecosystem approach sponsored by the Sweden Ministry of Environment's Commission on the Marine Environment, 22-24 April 2003.

Vaxholm. Available online at http://people.uncw.edu/imperialm/index.htm [Accessed October 2005]
Imperial MT (2004) Intergovernmental challenges of watershed management: strategies for improving watershed governance. In: Fresh water. Velma Grover (Ed) Oxford & IBH Publishing Co, India & India and Science Publishers, Inc, USA. Available online at http://people.uncw.edu/imperialm/-index.htm [Accessed October 2005]
Imperial MT (2005) Using collaboration as a governance strategy. lessons from six watershed management programs. Administration and Society 37(3):281-320
Magadlela D (2005) Presentation on the capacity development of adaptive governance in a fledgling democracy. International Symposium on ecosystems governance, 10-13 October 2005, Kwa Maritane, South Africa
Mbedzi S, Weston D, Bofilatos E (2005) The role of governance interfaces on the establishment of catchment management agencies. International Symposium on ecosystems governance, 10-13 October 2005, Kwa Maritane, South Africa
Mebratu D (1998) Sustainability and sustainable development: historical and conceptual review. Environ. Impact Assess. Rev (18):493-520
Miller WL, Morris L (1999) Fourth generation R&D. New York, John Wiley & Sons Inc
National Water Act (1998). Act 36 of 1998. Government Gazette, no 19182, 26 August 1998. Pretoria
Parr TW, Sier ARJ, Battarbee RW, Mackay A, Burgess J (2003) Detecting environmental change: science and society – perspectives on long-term research and monitoring in the 21st century. The Science of the Total Environment (310):1–8
Robèrt K-H, Daly H, Hawken P, Holmberg J (1997) A compass for sustainable development. International Journal of Sustainable Development and World Ecology (4):79-92
Rogers P, Hall AW (2003) Effective water governance. Global Water Partnership Technical Committee (TEC) Background papers 7. Available online at http://www.gwpforum.org [Accessed October 2005]
Schreiner B (2005) Presentation on the government and society challenge in a fledgling democracy. International Symposium on ecosystems governance, 10-13 October 2005, Kwa Maritane, South Africa
Turton A, Hattingh H, Claassen M, Roux D, Ashton P (2005) Towards a model for ecosystem governance: an integrated water resource management example. International Symposium on ecosystems governance, 10-13 October 2005, Kwa Maritane, South Africa
UNCED (1992) Earth Summit Agenda 21. The United Nations Programme of Action from Rio. United Nations Conference on environment and development (UNCED), Rio de Janeiro, Brazil. Geneva: United Nations Department of Public Information

The Trialogue Revisited: Quo Vadis Governance?

Anthony R Turton and J Hattingh

Abstract

Governance as a concept has been interrogated in the various chapters of this book. A definition of governance was suggested along with a series of hypotheses that together formed the Trialogue Model of Governance. This new definition is revisited and the different hypotheses are evaluated against the empirical background provided by the respective case studies presented in this book.

Keywords: actor-cluster, governance, Integrated Water Resource Management (IWRM), interface, Trialogue, leadership

Introduction

This book attempts to open up the black box of governance. The chosen strategy for this was the development of a new definition of governance and the presentation of a conceptual model based on three actor-clusters (government, science and society), in which three sets of interfaces occur in a dynamic process called a Trialogue. To support the development of conceptual rigour, a set of hypotheses were proposed and a series of case studies have been presented with the intention of interrogating those hypotheses. All that now remains is to revisit those hypotheses and resultant chapters in order to expand the generic conclusions.

Governance Re-defined

Chapter 1 shows the existence of various definitions for governance and specifically water governance on a global, regional and national level. The subsequent chapters confirm that governance as a concept is complex and it is evident that different perspectives, disciplines and agendas influence the definition and understanding of what we generically call governance.

However, there is a need for re-formulating a definition of governance, with the intention of establishing a shared understanding of the term and to form a common platform for further research and development. The new definition of governance provided in Chapter 1 is *the process of informed decision-making that enables trade-offs between competing users of a given resource so as to balance protection with beneficial use in such a way as to mitigate conflict, enhance equity, ensure sustainability and hold officials accountable.*

From this new definition, it is evident that *governance as process* involves a number of distinct elements, including decision-making about potentially contestable outcomes. This definition also implies *governance as product* and is the perceived quality of those outcomes, specifically with respect to the legitimacy of trade-offs and the level to which these are contested or accepted by society. This is clearly a dynamic process and as such a range of possible results can be expected, but the significant thing is that public perception mediates the outcomes by introducing a value judgement that determines "good" from "bad". Democracy matters in terms of good governance. Good governance can be seen as a subset of democracy – however that might be defined – when applied to ecosystems and water resource management.

The Hypotheses Revisited

Each chapter focussed on a selected case study and presented a specific set of issues, facts and processes, all of which were used to interrogate the Trialogue Model of Governance. Significantly, each chapter came to a specific conclusion that was applicable within a given set of circumstances. We can now revisit the nine hypotheses proposed in Chapter 1 and attempt to aggregate those conclusions. This will be done by presenting each hypothesis, and then discussing the conclusions embedded in the relevant case studies with a view to detecting agreement or disagreement of any form. Stated differently, agreement will be construed as confirming the hypotheses as being valid, whereas disagreement will be interpreted as refuting the hypotheses as being invalid.

Hypothesis 1:

Governance is a Process, with Effective Ecosystem (or IWRM) Management being the Product of "Good" Governance

Agreement on this hypothesis can be detected in the various case studies. The most eloquent case is made by Ashton in Chapter 5 where he unpacks the IWRM paradigm and shows the convergence of five key principles, which when combined through the process of governance, produce effective outcomes as a product. Conca (Chapter 6) deals with this theme from a different dimension, where he shows that effectiveness is not necessarily an outcome in the negotiations around transnational aquatic ecosystems. In this case the bottleneck arises from state sovereignty that hampers the governance as process aspect, producing a sub-optimal outcome. Gooch shows in Chapter 7 that there are indeed nested hierarchies of processes in existence, with good governance outcomes being dependent among other things on the way these processes are mediated by interceding variables. International law can potentially help in governance as process according to Stephan (Chapter 8). This is mostly supported by the North American case presented by Campana et al., in Chapter 9. Corporate cultures play a role in this mediating process according to Nyambe et al., (Chapter 10). Roux et al., show in Chapter 13 that learning processes are critical. Strydom et al (Chapter 14) discusses the role of communication, while perceptions of resource degradation act as potentially powerful triggers as shown by Doolan in Chapter 15. The degree of evolution along a pathway towards democracy is analysed by Godfrey in Chapter 16. There is consequently a great deal of agreement around this aspect to the extent that the authors believe the hypothesis to be validated.

Hypothesis 2:

The Three Clusters of Actors in the Context of Governance as Process – Generically Defined as Government, Society and Science

It is clear that the clustering of actors is a conceptual tool only. The real world is infinitely more complicated, and an analysis of that becomes largely meaningless unless some process is used to order logic and filter evidence. Consequently, the existence of three neat actor-clusters involves a trade-off between real world complexity – and hence the absence of a meaningful outcome based on clear trends and irrefutable evidence – and contrived conformity. The Trialogue Model is a contrived one, being a theoretical approximation of core processes found in an exceedingly complex real world.

A high level of agreement exists around the fact that government is an important actor. Yet we know it is not monolithic. Here the well-known agency-structure debate becomes relevant (Wendt 1987). Decisions are made by people, but these are contextualised within bureaucratic settings involving rules, policies and procedures. This structures those decisions in some way, but does not make them entirely uniform. The very essence of democracy is based on the notion of checks and balances, because it recognises the variability of decision-making and the potential for differing outcomes. Thus the notion of an actor-cluster is valid because it best describes how the key actors are structured and suggests that the way they function is based on recognition of this fact. The government actor-cluster is therefore very important, because it makes rules, applies them to society at large, and adjudicates deviation from those rules.

Similarly, a high degree of agreement exists around the notion of a Society Actor-Cluster, but to a lesser degree than the Government Actor-Cluster. Some argue that civil society should, or indeed does play a bigger role, in governance as a process. This is debatable in some circles, particularly within fledgling democracies, and beyond the scope of this book. What is important however, is to note that society is structured in some way, and that this structure is organic in a sense, evolving over time in response to a given set of social dynamics, in much the same way as an ecosystem evolves, centred on structure, function, hierarchies, relationships and feedback loops. This is where governance becomes relevant, because *governance as process* facilitates that aspect and *governance as product* legitimises the final outcome.

There is less agreement around the notion of a Science Actor-Cluster. Most notably Falkenmark, Chapter 4, suggests the redefinition of that cluster into a Biophysical Process cluster. Our argument is that decisions are made based on perceptions of reality. That reality is complex, specifically when it comes to ecosystem management where anthropogenic impact has resulted in cumulative effects, often being concentrated in aquatic ecosystems. The way to understand that complex reality is through a scientific process, which need not necessarily be a Eurocentric view that is based on a Bacon or Descartian philosophy of science (Bacon 1620; Descartes 1637) and reductionism as typified by Newtonian physics (Newton 1687). Traditional knowledge is important and should feed into the decision-making process. We thus defend the existence of a Science Actor-Cluster by recognising that it need not necessarily be an exclusively natural science-based organisation, and might just as easily be called a "Knowledge Actor-Cluster" instead. In fact we argue that the only way to effectively get to grips with this real-world complexity is to unleash the synergy of transdisciplinarity between the reductionist style of science (as embodied in Newtonian physics (Newton 1687)) and the integrative style of science (as found in contemporary Chaos Theory (Gleick 1987; Stewart 1989)). This can only be achieved in a cluster and in effect involves a series of sub-processes of their own. We conclude by stating that in our view, the only way to inform decision-making processes about complex issues like ecosystem management is by having redundancy in the scientific field. This will enable the best possible decision to be made, by recognising the unintended consequences of specific actions (Tenner 1996), and quantifying the impacts, proposing mitigating actions, alternative management options and identifying possible trade-offs.

Hypothesis 3:

Government, Society and Science Represent Three Different Communities of Practice that are Complementary only to the Degree that they Interface with One Another

A high level of agreement exists around this topic. Clearly the Trialogue Model is a simplification of a complex reality, but as it stands it enables the differences of each Actor-Cluster to be recognised. Government processes are not uniform within the Actor-Cluster, and differ dramatically from Society processes, which in turn are clearly not the same as Science processes. Each represents a specific community of practice, informed by a complex set of norms, values, procedures and processes. Gooch makes an interesting observation in Chapter 7 that there is a Trialogue of Epistemology that defines these different communities of practice. Significantly, that community of practice evolves, resulting in different communities of practice over time. Doolan recognises this aspect in Chapter 15, which is also alluded to by Nyambe et al., in Chapter 10 and Godfrey in Chapter 16. Roux et al., eloquently articulate the complexities of this aspect in Chapter 13. The central role of communication thus becomes relevant, as indicated by Strydom et al., in Chapter 14. Conca shows that clearly different communities of practice exist in Chapter 6, noting that in a transnational context, bridging actors are potentially significant – something we have hitherto not considered as being relevant. With such a high level of agreement, we argue that this specific hypothesis has been validated.

Hypothesis 4:

Governance as Process Involves Decision-Making, but Given the Inherent Complexity of the Real World (Specifically with Respect to Natural Resource Management), Decisions Tend to be Made Against a Background of Different Sets of Norms and Values, which Vary Within Different Stages of Economic Development and Political Evolution

A degree of agreement exists around the fact that norms and values are important. Clearly decisions are mediated by a complex process of value judgement, suggesting that perceptions become reality because they drive decisions that in turn have a series of outcomes, some good and some unintended or unforeseen (Tenner 1996). One way that norms and values become relevant to governance is by being codified into law and bureaucratic procedure. Ashton recognises this eloquently in Chapter 5 where he explains complex processes of balancing demands with capabilities. Conca raises the question of cumulative impact of a selective use of norms and values in the evolution of management instruments for use in transboundary systems in Chapter 6. Gooch shows in Chapter 7 that what he calls the Trialogue of Epistemology filters out different truths, each based on specific sets of norms and values, either implicit or explicit. Stephan focuses specifically on the evolution of key norms and values as they are applicable to transboundary groundwater governance in Chapter 8. This is given more depth and specificity by Campana et al. in Chapter 9 where he analyses the North American case of transboundary groundwater management. At a completely different level of scale, Nyambe et al., show in Chapter 10 that organisational culture is both dynamic and powerful in its own right, and serves to unite norms and values in a specific way. We therefore conclude that the hypothesis is valid given this strong endorsement from various case studies.

Hypothesis 5:

In the Context of Governance as Product, Good Governance Occurs when the Interfaces Between the Three Clusters of Actors – What Can be Called a Trialogue – are Effective, as this Allows for Appropriate Feedback Loops and Exchange of Information with which to Inform the Decision-Making Process – This is Also Reflected in Sound Ecosystem Governance

The whole issue of interfaces has been richly interrogated in all of the case studies and a high degree of consensus exists around the importance of this. Falkenmark (Chapter 4) notes interfaces where she sketches the analogy of water being the bloodstream of the biosphere. How much more deeply interfaced can one become than between different organs and their life-sustaining blood supply? In this regard the high level of interaction, caused by the flow of water through ecosystems and across political economies, is driving governance processes. Ashton shows in Chapter 5 that interfaces exist between a number of systems, most notably the economic, social and ecological systems, which he shows is mediated by the political system that enables the governance system to be brought to bear. This suggests a complex chain of causality where cause-effect linkages are difficult, if not impossible, to disentangle. Gooch shows in Chapter 7 that there are nested hierarchies of interfaces, suggesting a more complex array than initially envisaged by the authors. Law can become a facilitator of these interfaces as shown by Stephan in Chapter 8 and Campana et al. in Chapter 9. There is consequently sufficient agreement to validate this hypothesis although it is clearly more nuanced than originally anticipated by the authors.

Hypothesis 6:

Governance is Dynamic in Nature and is Enmeshed in the Social, Economic, Biophysical and Political Landscape in which it Occurs

Governance can be thought of as being a product of complex social ecology that evolves over time in response to a series of fundamental drivers, while being channelled and constrained by cultures that are enmeshed in norms and values. Seen in this light, governance can be thought of as a form of social-ecological interaction. Just as ecosystems adapt over time, so do governance systems evolve. Just as ecosystems reflect interactions between organisms that are complex, governance systems reflect relationships between entities, some of which are social and some of which are biophysical. Just as ecosystems reach a point of stable equilibrium but have a propensity towards instability (Holling 1973), so too do governance systems reflect similar characteristics, the most notable being a form of dynamic equilibrium that embodies resilience to shocks – or what has earlier been called social adaptive capacity (Ohlsson 1999; Ohlsson and Turton 1999; Turton 1999, 2001; 2002; Turton and Ohlsson 1999). The so-called Trialogue of Epistemology suggested by Gooch in Chapter 7 shows how governance is embedded in different landscapes by processes that are extremely subtle but durable. Nyambe et al., present the notion of organisational change as it applies to ecosystem governance in Chapter 10. Roux et al., highlighted the interfaces that produce learning in Chapter 13, and suggest that good ecosystem governance requires adaptive people with a culture of empathy for other knowledge systems. This raises the issue of social enmeshment and biophysical constraints as determinants of governance structures, processes and principles. Strydom et al., (Chapter 14) present a study of post-Apartheid South African ecosystem governance processes and capture the contemporary political culture of public engagement. Doolan shows in Chapter 15 that Australia has a somewhat different set of drivers. In Chapter 16 Godfrey shows that clever management systems can exploit this situation to manage potentially conflicting outcomes by understanding the dynamics occurring with different pathways of engagement. There is consequently sufficient agreement to validate this hypothesis.

Hypothesis 7:

Good Governance is More Likely to Occur Where There is a Prevailing Political Culture of Democracy

It is here that issues start to become a bit tricky to interpret. What is democracy – a process or a product? Is there one thing called democracy? Hall maps out some of these complexities in Chapter 2 by showing that no clear definition of governance exists because of this fact. Falkenmark shows in Chapter 4 that there is a clear distinction between *what* to govern and *how* to govern. This is clearly dependent on core values of which perceptions of democracy are but one sub-set. Ashton (Chapter 5) shows that core principles include aspects such as openness, participation and accountability, all of which are characteristics of democracy. Conca unpacks the transnational dimension of ecosystem governance in Chapter 6 and shows that sovereignty matters by acting as an obstruction to agreements across international borders. Gooch (Chapter 7) illustrates that his Trialogue of Epistemology acts as an interceding variable, mediating between perceptions of democracy on the one hand, and governance processes and structures on the other. There is consequently sufficient agreement to guardedly validate this hypothesis, but the real world is far more nuanced than this relatively simplistic statement suggests and the range of case studies did not cover all types of political systems, including socialist systems that claim to be democratic.

Hypothesis 8:

Given the Dynamic Nature of Governance, the Evolution of What can be Described as Good Governance Occurs Over Time, with Potentially Different Trajectories or Pathways being Possible in What can Generically be Described as "Mature" Democracies and "Fledgling" Democracies

Time as an element of governance was not considered as an independent variable in the original hypotheses. Doolan makes this point explicit in Chapter 15, showing how important it is from both an analytical and practical perspective. Time, as a unique dimension, is implicit in Chapter 8 where Stephan shows evolutionary tendencies in the articulation of universal norms into customary international water law. A similar trend is implied in Chapter 9 where Campana et al. indicate that evolution of governance is a process that is dynamic and systematic, thus suggesting time as an independent variable. Ashton shows the universality of specific core values of governance that are consistent with notions of democracy in Chapter 5. This suggests that time is involved and that value judgements of "good" or "weak" are best made around the degree of application of these core values. Conca laments the resilience of some potentially non-useful aspects of transboundary governance in Chapter 6, suggesting that the maturity of a given "democracy" is largely irrelevant. Gooch indicates that different systems, spatial levels and geographic contexts are all relevant in Chapter 7. Nyambe et al., make a case in Chapter 10 that the harmonisation of goals and responsibilities of the three actor-clusters in the Trialogue Model of Governance underpins the concept of "good" governance, which they conclude is a pedestal of democracy. Doolan notes in Chapter 15 that while the Trialogue is correct at a high level, time is needed to establish the necessary feedback loops, suggesting that democracy is not universal or monolithic. Godfrey analyses this explicitly in Chapter 16, where she shows different forms of democracy could potentially have different relationships within the various interfaces of the Trialogue Model. Reading these critically it becomes apparent that there is some agreement, but also significant disagreement around this specific hypothesis, to the extent that it has not been fully validated by the evidence provided in the case studies.

Hypothesis 9:

Governance is Not the Same as Government, Nor is It Solely the Mandate of Government Authorities

This is certainly a subtle aspect that is deeply hidden within the various case studies. The South African Constitution makes reference to this fact, and it is possibly in this context that the team conceptualising the original Trialogue Hypotheses – all of them South African scientists deeply engaged in implementation processes – became fixated on this issue. Hall (Chapter 2) claims that governance is a process that has no universal definition.

To address this obvious shortfall in the literature, the definition by the authors of Chapter 1 has been offered (*governance is the process of informed decision-making that enables trade-offs between competing users of a given resource to balance protection with beneficial use in such a way to mitigate conflict, enhance equity, ensure sustainability and hold officials accountable*).

Falkenmark shows that the interconnectedness of landscapes and water drives the process of governance, suggesting that a range of actors is needed. From this it is evident that government is but one actor if governance is understood to be a process. The modern mantra of Integrated Water Resource Management (IWRM) drives the need for governance as a process according to Ashton in Chapter 5. This suggests again that government is an important actor in the process of governance. The transnational environment poses a different arena notes Conca in Chapter 6, in which government as actor engages in governance as process. Gooch shows in Chapter 7 that civil society is a key actor in the governance process, again implying that government as actor is different from governance as process. Customary international water law can facilitate the process of governance by providing rules that have been sufficiently codified to the extent that they impact on the behaviour of government according to Stephan in Chapter 8. The governance of complex transboundary groundwater aquifers is a process in which governments engage according to Campana et al., in Chapter 9. Nyambe et al., suggest in Chapter 10 that organisational cultures drive governance structures, which in turn make them resilient to an extent when engaging with government as actor. Roux et al., show in Chapter 13 that learning and adaptation is an inherent characteristic of governance as process, although they say little about how this differs from government. Strydom et al., argue in Chapter 14 that communicating complex facts to "a decision-making elite", usually sitting in government as actor, is a vital function of governance as process. Godfrey shows in Chapter 16 that government as actor can engage in different governance pathways or processes in order to achieve a possible outcome.

The agreement within this set of conclusions seems to be centred on the fact that governance is more about process, whereas government is more about being an actor capable of authoritative decision-making. This seems to be the difference. Governance as process supports the making of complex decisions in a way that facilitates trade-offs between competing users in a potentially contestable way. Good governance therefore facilitates that decision-making process by legitimising the outcome, which is normally an authoritative decision by government as actor that reduces the conflict potential to the minimum that can be managed.

Towards a New Research Agenda

The participants in the International Symposium on Ecosystem Governance proposed, during a facilitated process, 10 research themes on ecosystem governance, all of which are embedded in the various chapters of this book and the Special Edition of the journal *Water Policy* (in press). These were prioritised and the results are presented in Figure 17.1 as research themes. The size of the bubble indicates the size of priority attributed to the specific research theme by the participants of the facilitated process, and supported by the conclusions of each chapter in this book.

The research theme with the highest priority was "defining governance as process". Evidently, more research is required on specific aspects of governance in general, with the definition of governance offered in Chapter 1 being a potential point of departure. Better understanding is needed of governance as a concept, specifically explaining the link between governance and democracy, and the linkage of socio-economic change to ecosystem reform. The economies of scale should be explored, with emphasis on the implications of severity, intensity, age and human capacity. Methods, practices and strategies applicable across all scales need to be documented. At a slightly lower priority, "Groundwater governance" was a research theme with various sub-components that still requires investigation. "Science communication, learning and knowledge sharing" and "Integration of traditional and formal systems" were identified as the third highest priority research themes. Institutional arrangements, values, traditional knowledge and inclusive participation of traditional systems in formal systems are included in this theme.

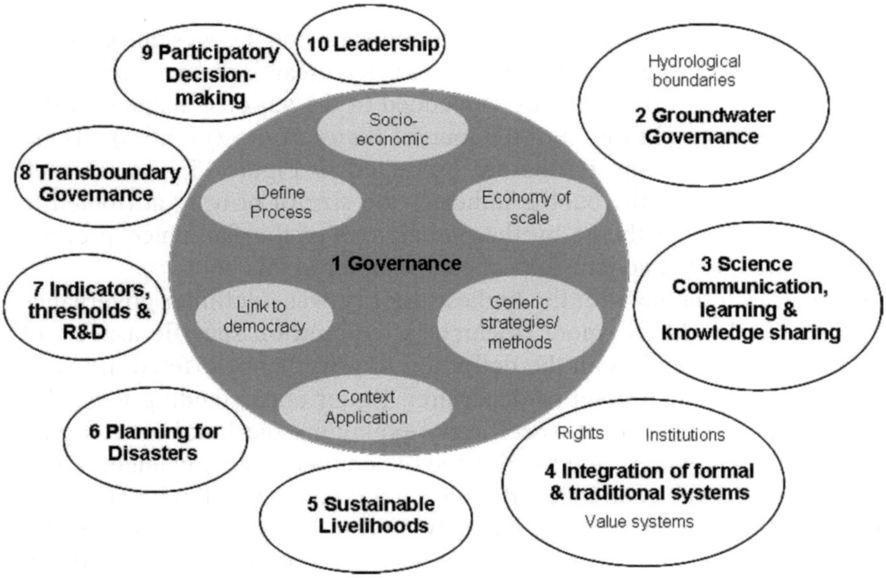

Fig. 17.1. *The prioritised 10 research themes as developed during the International Symposium on Ecosystem Governance, KwaMaritane, October 2005*

"Sustainable livelihoods" and linking governance, economics and business came out as the fourth highest research priority. The role of partnerships should be investigated as a component of this theme. Strongly linked to the sustainable livelihoods theme, is "Planning for disasters". There is still a need for the development of "Indicators and thresholds", as was highlighted by Research Theme 7. The work by Doolan (Chapter 15) and Strydom et al., (Chapter 14) explicitly supports this theme. An investigation of the roles of institutions and implementation of conventions and law in the "Transboundary governance" research theme are just some aspects to be researched further. Campana et al., (Chapter 9) and Stephan (Chapter 8) highlight these requirements.

"Participative decision-making" was identified as Research Theme 9, with a strong focus on involvement, empowerment, representivity, cost effectiveness, trust and power relations. This theme was supported by most of the Chapters.

"Leadership" is important to ensure good governance and the concept should be unpacked to promote a better understanding of the concept specifically as it applies to fledgling democracies emerging from periods of political transition.

These combined research themes were converted to a high-level research agenda and published as a special edition of *Water Policy*, entitled "Case studies of Government-Society-Science as a Trialogue: Towards a Governance Research Agenda" (Hattingh et al. in prep).

Conclusion

This chapter shows that a high degree of agreement exists in the different case studies to validate the hypotheses that were presented in Chapter 1 of this book. This is encouraging, because it suggests that the Trialogue Model of Governance is a potentially useful analytic tool. It also provides sufficient evidence that the definition of governance that was presented in both Chapter 1 and this chapter is a potentially useful one, because it takes us forward in a collective way. This being the case, a broad research agenda has been presented to guide possible future collaborative efforts. This is not offered as an exhaustive agenda, but rather as a point of departure in an ongoing process of engagement, informed by a rich discussion held at Kwa Maritane on 10-14 October 2006 and subsequently reinforced by the contributions to this book.

References

Bacon F (1620) The novum organon. Republished in: Kitchen GW (ed) (1855) Oxford, Oxford University Press

Descartes R (1637) Discourse on method. In: Anscombe E, Geach P (eds) (1954) Descartes: philosophical writings. Edinburgh, Nelson & Sons

Gleick J (1987) Chaos: making of a new science. New York, Viking

Hattingh J, Maree GA, Ashton PJ, Leaner JJ, Strydom WF, Turton AR (in prep) Case studies of government-society-science as a trialogue: towards a governance research agenda. Special Edition of Water Policy. World Water Council. In prep

Holling CS (1973) Resilience and stability of ecological systems. Annual Review of Ecology and Systematics 4:1-23

Newton I (1687) Philosophiae naturalis principia mathematica. London, Publisher unknown

Ohlsson L (1999) Environment, scarcity and conflict: a study of Malthusian concerns. Department of Peace and Development Research, University of Göteborg

Ohlsson L, Turton AR (1999) The turning of a screw. Paper presented in: Plenary session, 9th Stockholm Water Symposium, Urban stability through integrated water-related management, hosted 9-12 August by the Stockholm International Water Institute (SIWI), Sweden. Also available as MEWREW Occasional Paper 19 online at http://www.soas.ac.uk/Geography/WaterIssues/OcsionalPapers/home.html and from http://www.up.ac.za/academic/libarts/-polsci/awiru

Stewart I (1989) Does God play dice? the mathematics of chaos. Harmondsworth, Middlesex, Penguin Books Ltd

Tenner E (1996) Why things bite back: technology and the revenge of unintended consequences. New York, Knopf

Turton AR (1999) Water scarcity and social adaptive capacity: towards an understanding of the social dynamics of managing water scarcity in developing countries. MEWREW Occasional Paper 9. SOAS Water Issues Study Group. Also in: Proceedings of the Conference sustainability, risk and nature: the political ecology of water in advanced societies, held at Oxford University, 15-17 April 1999. Available online at http://www.soas.ac.uk/Geography/-WaterIssues/OccasionalPapers/home.html

Turton AR (2001) Water and social stability: the Southern African dilemma. In: Rotblat J (ed) Confronting the challenges of the 21st century. Proceedings of the 49th Pugwash Conference, Science and world affairs. London, World Scientific, pp 363-362

Turton AR (2002) 'Water demand management' (WDM), 'natural resource reconstruction' and 'adaptive capacity': establishing the linkage between variables. Final report. WARFSA Research Project PJ02/99. Pretoria, Harare, AWIRU, WARFSA. Available online at http://www.up.ac.za/academic/libarts/polsci/awiru

Turton AR, Ohlsson L (1999) Water scarcity and social adaptive capacity: towards an understanding of the social dynamics of managing water scarcity in developing countries. Paper presented at Workshop No 4, Water and social stability of the 9th Stockholm Water Symposium urban stability through integrated water-related management, 9-12 August, hosted by the Stockholm International Water Institute (SIWI) in Sweden. Also available as MEWREW Occasional Paper 18 online at http://www.soas.ac.uk/Geography/WaterIssues/OccasionalPapers/home.html and from http://www.up.ac.za/academic/libarts/polsci/-awiru

Wendt A (1987) The agent-structure problem. In: International Organization 41(3:335-370)

Index

F

G

H

I

K

MIX
Papier aus verantwortungsvollen Quellen
Paper from responsible sources
FSC® C105338

If you have any concerns about our products,
you can contact us on
ProductSafety@springernature.com

In case Publisher is established outside the EU,
the EU authorized representative is:
Springer Nature Customer Service Center GmbH
Europaplatz 3, 69115 Heidelberg, Germany

Printed by Libri Plureos GmbH
in Hamburg, Germany